Knowledge Acquisition for Knowledge-Based Systems

Knowledge-Based Systems

One of the most successful and engaging initiatives in Artificial Intelligence has been the development of knowledge-based systems (or, expert systems) encoding human expertise in the computer and making it more widely available. Knowledge-based system developments are at the leading edge of the move from information processing to knowledge processing in Fifth Generation Computing.

The Knowledge-Based Systems Book Series publishes the work of leading international scientists addressing themselves to the spectrum of problems associated with the development of knowledge-based systems. The series will be an important source for researchers and advanced students working on knowledge-based systems as well as introducing those embarking on expert systems development to the state-of-the-art.

Volume 1 has been compiled from the following issues of International Journal of Man–Machine Studies:

Volume 26	Number 1	January 1987
Volume 26	Number 2	February 1987
Volume 26	Number 4	April 1987
Volume 27	Number 2	August 1987
Volume 27	Number 3	September 1987
Volume 27	Number 4	October 1987

Knowledge Acquisition for Knowledge-Based Systems

Knowledge-Based Systems
Volume 1

edited by

B. R. Gaines
Department of Computer Science,
University of Calgary,
Calgary, Alberta T2N 1N4
Canada

and

J. H. Boose
Boeing Computer Services,
P.O. Box 24346,
Seattle, Washington 98124–0346,
U.S.A.

1988

ACADEMIC PRESS

Harcourt Brace Jovanovich, Publishers
London San Diego New York Berkeley Boston
Sydney Tokyo Toronto

ACADEMIC PRESS LIMITED
24/28 Oval Road
LONDON NW1 7DX

United States Edition published by
ACADEMIC PRESS INC.
San Diego, CA 92101

British Library Cataloguing in Publication Data

Gaines, B. R. (Brian R)
Knowledge acquisition tools for expert
systems.—(Knowledge-based systems;
v.1).
1. Artifical intelligence. Applications
of computer systems
I. Title II. Boose, John H. III. Series
006.3

ISBN 0-12-273251-0

Typeset by The Universities Press (Belfast) Ltd.
Printed in Great Britain at the University Press, Cambridge

Contributors

N. J. BELKIN *The School of Communication, Information, and Library Studies, Rutgers University, New Brunswick, NJ 08903, U.S.A.*
J. H. BOOSE *Boeing Computer Services, P.O. Box 24346, Seattle, Washington 98124–0346, U.S.A.*
H. M. BROOKS *The School of Communicative Information, and Library Studies, Rutgers University, New Brunswick, NJ 08903, U.S.A.*
B. G. BUCHANAN *Department of Computer Science, Stanford University, Stanford, CA 94305, U.S.A.*
W. BUNTINE *New South Wales Institute of Technology and Macquarie University, Computing Science, N.S.W. I.T., P. O. Box 123, Broadway, 2007 Australia*
T. BYLANDER *Laboratory for Artificial Intelligence Research, Department of Computer and Information Science, The Ohio State University, Columbus, Ohio 43210, U.S.A.*
J. CENDROWSKA *c/o The Faculty of Mathematics, The Open University, Walton Hall, Milton Keynes, MK7 6AA, U.K.*
B. CHANDRASEKARAN *Laboratory for Artificial Intelligence Research, Department for Computer and Information Science, The Ohio State University, Columbus, Ohio 43210, U.S.A.*
W. J. CLANCEY *Department of Computer Science, Stanford University, Stanford, CA 94305, U.S.A.*
J. G. CLEARY *Department of Computer Science, University of Calgary, 2500 University Drive, Alberta T2N 1N4, Canada*
D. A. CLEAVES *Fire Economics and Management Research, USDA Forest Science, Pacific Southwest Forest and Range Experiment Station, 4955 Canyon Crest Drive, Riverside, CA 92507, U.S.A.*
P. J. DANIELS *Admiralty Research Establishment, Ministry of Defence, Queens Road, Teddington, Middx. TW11 0L9, U.K.*
J. P. DELGRANDE *School of Computing Science, Simon Fraser University, Burnaby, B.C., Canada V5A 1S6*
B. R. GAINES *Department of Computer Science, University of Calgary, Calgary, Alberta T2N 1N4, Canada*
S. GARBER *3M, St Paul, Minnesota, U.S.A.*
E. HOLLNAGEL *Computer Resources International, Copenhagen, Denmark*
P. E. JOHNSON *Department of Management Sciences, University of Minnesota, Minneapolis, Minnesota, U.S.A.*
M. KLEYN *Schlumberger-Doll Research, Old Quarry Road, Ridgefield, CT 06877–4108, U.S.A.*
Y. KODRATOFF *Inference and Learning Group, Laboratoire de Recherche en Informatique, A. 410 du CNRS, Université de Paris-Sud, Bât. 490, 91404 Orsay, France*
J. KORNELL *General Research Corporation, P.O. Box 6770, Santa Barbara, CA 93111–6770, U.S.A.*
M. LAFRANCE *Department of Psychology, Boston College, Chestnut Hill, MA 02167, U.S.A.*
D. C. LITTMAN *Cognition and Programming Project, Department of Computer Science, Yale University, New Haven, CT 06520, U.S.A.*
M. MANAGO *Inference and Learning Group, Laboratoire de Recherche en Informatique, A. 410 du CNRS, Université de Paris-Sud, Bât. 490, 91405 Orsay, France*
A. A. MITCHELL *Faculty of Management Studies, University of Toronto, Toronto, Ontario M56 IV4, Canada*
M. J. PAZZANI *UCLA Artificial Intelligence Laboratory, 3531 Boelter Hall, Los Angeles, CA 90024 and The Aerospace Corporation, P.O. Box 92957, Los Angeles, CA 90009, U.S.A.*

E. J. PETTIT *Merit Technology, Inc., 5068 W. Plano Parkway, Plano, TX 75075–5009, U.S.A.*
M. J. PETTIT *Software Consulting Services, 2733 Riviera Drive, Garland, TX 75040*
J. R. QUINLAN *Artificial Intelligence Laboratory, Massachusetts Institute of Technology, 545 Technology Square, Cambridge, MA 02139, U.S.A.*
R. SMITH *Schlumberger-Doll Research, Old Quarry Road, Ridgefield, CT 06877–4108, U.S.A.*
D. C. WILKINS *Department of Computer Science, Stanford University, Stanford, CA 94305, U.S.A.*
H. WINSTON *Schlumberger-Doll Research, Old Quarry Road, Ridgefield, CT 06877–4108, U.S.A.*
E. WISNIEWSKI *Schlumberger-Doll Research, Old Quarry Road, Ridgefield, CT 06877–4108, U.S.A.*
D. D. WOODS *Westinghouse Research & Development Center, Pittsburgh, PA 15235, U.S.A.*
I. ZUALKERNAN *Department of Computer Science, University of Minnesota, Minneapolis, Minnesota, U.S.A.*

Preface

The initial success of expert system developments and the development of a number of reasonably domain-independent software support systems for the encoding and application of knowledge have opened up the possibility of widespread usage of expert systems. In particular, Fifth Generation Computing System development programs worldwide assume this will happen and are targeted on knowledge processing rather than information processing. However, what Feigenbaum has termed *knowledge engineering,* the reduction of a large body of knowledge to a precise set of facts and rules, has already become a major bottleneck impeding the application of expert systems in new domains. We need to understand more about the nature of expertise in itself and to be able to apply this knowledge to the elicitation of expertise in specific domains.

The problems of knowledge engineering have been stated clearly:

> "Knowledge acquisition is a bottleneck in the construction of expert systems. The knowledge engineer's job is to act as a go-between to help an expert build a system. Since the knowledge engineer has far less knowledge of the domain than the expert, however, communication problems impede the process of transferring expertise into a program. The vocabulary initially used by the expert to talk about the domain with a novice is often inadequate for problem-solving; thus the knowledge engineer and expert must work together to extend and refine it. One of the most difficult aspects of the knowledge engineer's task is helping the expert to structure the domain knowledge, to identify and formalize the domain concepts." (Hayes-Roth, Waterman & Lenat 1983)

The knowledge acquisition bottleneck has become the major impediment to the development and application of effective knowledge-based systems. Many research groups around the world have been working on knowledge acquisition methodologies, techniques and tools to overcome this problem. In 1985, members of a number of these groups realized that there had been rapid progress in knowledge acquisition research and application. However there was subtantial duplication of effort and limited communication between researchers, and therefore it would be valuable for a workshop to be held that would encourage the sharing of results and experience.

The American Association for Artificial Intelligence agreed to sponsor such a workshop. John Boose of Boeing Advanced Technology Centre and Brian Gaines of the Knowledge Science Institute at the University of Calgary agreed to organize it. Other researchers agreed to contribute effort to the organization and refereeing of papers, resulting in a program and local arrangements committee of: Jeffrey Bradshaw, Boeing Advanced Technology Centre, William Clancey, Stanford University, Catherine Kitto, Boeing Advanced Technology Centre, Janusz Kowalik, Boeing Advanced Technology Centre, John McDermott, Carnegie-Mellon University, Ryszard Michalski, University of Illinois, Art Nagai, Boeing Advanced Technology Centre, Gavriel Salvendy, Purdue University, and Mildred Shaw, University of Calgary.

The response to the call for papers for the Workshop on Knowledge Acquisition for Knowledge-Based Systems (KAW) was overwhelming. The intention was to hold a discussion-intensive meeting of some 35 highly involved researchers. In practice over 120 papers were submitted and some 500 applications to attend were received from some 30 countries. Apart from increasing the refereeing and organizational problems beyond all expected bounds, this response indicated the magnitude and impact of the knowledge acquisition bottleneck and the worldwide interest.

These submissions resulted in 60 people attending the first *Knowledge Acquisition for Knowledge-Based Systems Workshop* (*KAW*) from November 3–7, 1986, at the Banff Centre in Banff, Canada. Each of the 120 papers submitted was refereed by five to seven referees and 42 papers were finally selected. Much of the rejected material was of high-quality and it would have been possible to base a major conference on the material and requests to attend. However, it was decided that the priority at that stage should remain that of establishing in-depth communication between research groups.

It was also clear that it was important to disseminate the workshop material as widely as possible, and arrangements were made to publish revised versions of the papers in the *International Journal of Man–Machine Studies* after the Workshop. These papers have now been gathered together as the first two volumes of the *Knowledge-Based Systems* series.

The table below shows the format of the first KAW. It was very effective in establishing a network linking the community of knowledge acquisition researchers

Knowledge Acquisition for Knowledge-Based Systems *AAAI Workshop, Banff, November 1986*	
Structure	Residential workshop Accomodation, meals and sessions together Attendance limited to 60 (originally 35) 120 papers submitted, 43 accepted Several hundred requests to attend
Overview/ Summary Papers	Gaines—Overview of Knowledge Acquisition Clancey—Cognition and Expertise McDermott—Interactive Interviewing Tools I Boose—Interactive Interviewing Tools II Salvendy—Analysis of Knowledge Structures Michalski—Learning
Mini-Conference	Cognition & Expertise 6, Learning 8 Analysis of Knowledge Structures 7 Interactive Interviewing Tools 16
Workshops on Major Issues	Cognition & Expertise Interactive Interviewing Tools Learning Knowledge Representation
Panels on other Issues	Knowledge Acquisition Methodology/Training Reasoning with Uncertainty
Papers and Books	Preprint volume of all papers to attendees Four special issues of IJMMS in 1987 Two books, Academic Press 1988

worldwide. It resulted in two further KAWs in 1987, a second one at Banff again sponsored by the American Association for Artificial Intelligence (AAAI), and the first European KAW (EKAW) in London and Reading, England, sponsored by the Institute of Electrical Engineers. Papers from these workshops have again been published in the *International Journal of Man–Machine Studies* and constitute the third volume of the *Knowledge-Based Systems* series.

In 1988, the third AAAI-KAW was held at Banff in November with a theme of integration of methodologies, and the second EKAW was held at Bonn, Germany, in June with sponsorship from the Gesselschaft für Mathematik and the German Chapter of the ACM. A specialist workshop on the Integration of Knowledge Acquisition and Performance tools was held at the AAAI Annual Conference in St Paul in August. Sessions and tutorials on knowledge acquisition have become prevalent at a wide variety of conferences concerned with knowledge-based systems worldwide.

These two volumes based on the first AAAI-KAW at Banff contain a wide range of material representing foundational work in knowledge acquisition problems, methodologies, techniques and tools from the major research groups worldwide. All those contributing hope that access to this material will enable other researchers and practitioners to expedite their own developments through the shared knowledge and experience documented here.

Knowledge acquisition research is still in its early stages and there are many fundamental problems to be solved, new perspectives to be generated, tools to be developed, refined and disseminated, and so on—the work seems endless. Like many modern technologies, knowledge acquisition requires a large-scale cooperative international effort. It is virtually impossible for one research and dvelopment group to have world-class expertise in all the issues, technologies and experience necessary to develop integrated knowledge acquisition tools for a wide range of knowledge-based systems.

We wish to thank the many people who have been involved in organizing these workshops and the organizations that have given them sponsorship and publicity. We have a fundamental debt to those who put in place the computer communication networks worldwide, such as UseNet, that have made the world a global village and enable networks such as ours to operate effectively. We are particularly grateful to the AAAI for its role in sponsoring the North American Workshops and for providing such effective means of disseminating information to the massive community of those now involved in knowledge-based systems research.

We sometimes wonder how we have become so involved in the bureaucracy of organizing workshops and networks when our personal priorities are hacking new knowledge acquisition tools. However, the stimulation of discussions with colleagues at the workshops and across the networks is vital to the direction of our own research. We hope the books will make this stimulation widely available and bring a new generation of researchers into the knowledge acquisition network.

We have attempted to structure the material by dividing it between two books. However, we must make it very clear that the division between the books and into sections in this volume is our own. It is somewhat arbitrary in places, and was not discussed with the contributors. There are many cross-connections between papers in different sections. There is fundamental material in this volume and tool-oriented material in the other. The reader will find it worthwhile to browse through both volumes to get a feel for the many different perspectives present and interactions possible.

Brian Gaines and John Boose

Contents

Learning Under Uncertainty

Knowledge Acquisition for Knowledge-Based Systems

B. R. GAINES
Department of Computer Science, University of Calgary, Calgary, Alberta T2N 1N4, Canada

J. H. BOOSE
Boeing Computer Services, P.O. Box 24346, Seattle, Washington 98124–0346, U.S.A.

1. Introduction

This volume contains the papers concerned with foundations, knowledge engineering and inductive methodologies from the AAAI Knowledge Acquisition for Knowledge-Based Systems Workshop in November 1986, in Banff, Canada. We have grouped them under subject headings although there is much overlap between themes. There were keynote addresses, panels and group discussions at the workshop that addressed major themes but did not result in published papers. This paper attempts to capture the essential issues raised in these other presentations.

2. Plenary Papers

Plenary talks were given on the first day by members of the program committee. The speakers were asked to summarize papers in their area and give an overview of their views on the area. Topics covered relevant to this volume were:

2.1 AN OVERVIEW OF KNOWLEDGE-ACQUISITION AND TRANSFER, BRIAN GAINES, UNIVERSITY OF CALGARY

An overview of knowledge acquisiton research was presented. The world was described in terms of distinctions, and models of the world were described in terms of epistemological, construction, and action hierarchies. Knowledge support systems were seen as anticipatory systems that observed events and produced actions. The full paper is reproduced in this volume.

2.2 COGNITION AND EXPERTISE, WILLIAM CLANCEY, STANFORD UNIVERSITY

How do knowledge bases relate to what people know? What is knowledge engineering? How do its goals and methods relate to traditional science and engineering? How does a knowledge base—today and in principle—relate to what experts know? Given this, what knowledge acquisition methods will be most effective?

Knowledge engineering is a methodology for acquiring, representing and using computational, qualitative models of systems. Knowledge acquisition is an investigative, experimental process involving interviews, protocol analysis, and reformulation of written materials in order to design computational, qualitative models of systems.

Knowledge engineering methodology:

- Start with concrete cases, knowledge expressed in some representation.
- Search for patterns, specifically unexplained regularities.
- List patterns.
- Ask about origins, generative principles—laws, processes—that could generate all instances of patterns.
- What generative principles could generate the patterns themselves?

Examples from NEOMYCIN:

- Knowledge bases representations are becoming more fine-grained than experts can state.
- There is no reason to believe that there are principled, but inaccessible type and causal networks encoded in the brain.
- Behaviour may be regular without the expert's awareness and does not necessarily reflect instantiation of a single procedure or principle encoded in the brain.
- Background and efficiency constraints shape and bias expert behaviour, but representations of these are post-hoc and never 'complete'.

Implications for knowledge acquisition:

- It is unclear how well manipulation of representations can approximate human reasoning.
- Be aware that an expert's justifications actually combine considered representations with preconceptions, 'authoritative rationalizations', and models derived by observing his own behaviour.
- Learning research should focus on representational breakdown (failure-driven).

All knowledge bases contain qualitative models (primarily non-numeric representations of processes). Human reasoning involves the use of representations, but reasoning behaviour is not generated directly from representations. Therefore, the knowledge acquisition bottleneck is not a problem of accessing and translating what is already known, but the familiar scientific and engineering problem of formalizing models for the first time.

Conclusions:

- Representation: Behaviour patterns are not necessarily generated from predefined, fixed criteria. Representation arises in explaining the breakdown of previous 'conceptions'.
- Expertise: Experts know how to solve problems and they know partial models of how they solve problems, but these are inherently distinct sources and forms of knowledge.
- Knowledge engineering: Knowledge base construction requires scientists, not mere scribes or programmers. Knowledge engineering proceeds most successfully when there is sufficient recurrence of problem situations and solutions to allow formalization of fixed, readily agreed upon, and apparently objective concepts and relations.

Summary of papers presented in this area:

Problems addressed in papers:
- Expert bias.
- Knowledge acquisition interview difficulty.
- Joint person-machine problem-solving.
- Formalizing initial model.

Approaches taken:
- Study correlation of belief with data and experience.
- Study knowledge engineering process.
- Study relation of system functionality to cognitive demand.
- Improve heuristics by explanation-based learning.
- Develop systematic notation.
- Apply logic representation to structure facts.

2.3 APPROACHES TO THE ANALYSIS OF KNOWLEDGE STRUCTURES, GAVRIEL SALVENDY, PURDUE UNIVERSITY

Thinking models were compared in terms of mental operators, strategy, declarative representation, content of representation, structure of procedural knowledge, and content of structures. Models included were problem solving, decision theory, network models, and stimulus-response.

2.4 LEARNING, RYSZARD MICHALSKI, UNIVERSITY OF ILLINOIS AT URBANA

Learning is constructing or modifying representations of what is being experienced (with the intention of being able to use them in the future). Dimensions of representations include validity (truthfulness, accuracy), effectiveness (usefulness), and abstraction level and type (explanatory power). These affect the quality of learning. Learning is building, modifying, and improving descriptions. Descriptions can be declarative descriptions, control systems, algorithms, simulation models, and theories. Different learning strategies were compared and contrasted. Relationships between similarity-based learning, explanation-based learning, constructive inductive learning, and apprentice systems were shown.

3. Working Group Summaries

Attendees participated in several small working groups that attempted to define the aims, objectives, problems, state-of-the-art, and future directions in their areas. The following summaries were presented at the workshop.

3.1 COGNITION AND EXPERTISE, WILLIAM CLANCEY, SPOKESPERSON

Aim:
- What aspects of human cognition are relevant to expertise?

Objectives:
- To determine how the study of cognition and expertise can contribute to the development of knowledge systems for support/collaborative-solving, replacement of people, and theory formation (new types of problem-solving).
- To determine how the development of knowledge systems can contribute to cognitive studies.

Problems and issues:

- People are imperfect and different.
- Modelling an aggregate versus an individual can be confusing.
- What aids/methods will help persons reveal/articulate their experience?
- Why do people fail and how can we best meet their needs?
- How can you interact with people in cooperative problem-solving to keep them engaged and responsible?
- How can consensus be reached when experience is distributed?
- How can tools, methodologies, and knowledge systems be evaluated?
- How can knowledge be modelled context-dependent in a dynamic environment?
- How can you decide what system, if any, is appropriate for a given person or situation?
- What is the space of performance niches?
- What type of discourse is a 'consultation'?
- How can known alternatives by synthesized/brought together in order to decide which is the best? How do we know we have seen all the possible alternatives?
- The 'real' discussion in the working group centred on: Can there be experts without novices? Could you be an expert if you could not adapt? How are skills and knowledge related? What is intuition?

State-of-the-art:

- The evolving distinction between consultation, expert, and knowledge systems is settling out.
- Many studies in expertise have been performed with feedback to knowledge acquisition tools and methods.
- Many candidate knowledge acquisition methodologies are ready for testing and distribution.
- We are realizing the essential need for abstractions, separate from the implemented model, and that languages and tools can blind us to alternative ways of viewing problem-solving.

Aspects of problem-solving to be identified and exploited in knowledge acquisition:

- Identify recurrence in social interaction and case history (background).
- Immediate context drives nature of expertise (data, competing activities).
- Goals, system functionality, and interface (person-machine interactions).
- Cognitive resources (memory, attention).
- Cognitive biases (associational and intuitional).
- Representational breakdown (failure-driven formalization)—watch experts in situations where they fail or have trouble.

Future directions:

- Might proliferate to many fields; growth just starting.
- Engineering, business, agriculture, education.
- Develop systems to aid experts in problem conceptualization, then as independent problem-solvers.
- Strong impact on natural language processing.
- Should respect history of innovation diffusion.

- Should emphasize study of computer-human interface and cooperative problem-solving.

Slogans (competing):

- "We want to know where you are coming from."
- "If you do not know what you are doing, you are probably doing it wrong."
- "If you know what you are doing, you are not learning anything."
- "The intelligence in an intelligent system lies in the tool builder or user, but not the programme itself."
- "No expertise without cognition."
- "Cognition and expertise—the link between theory and practice."
- "Leveraging intuition with a cognitive mirror."

3.2 LEARNING, RYSZARD MICHALSKI, SPOKESPERSON

Aims:

- Intelligent information system.
- Human learning; modelling sequences of learning tasks.
- Learning by analogy for defining primitives of the domain.
- Incremental knowledge construction in an imperfect world.
- Autonomous learning system; universal data compression.
- Explanation-based learning; problems of access to the constructed knowledge; the trade-off between storing and redefining the knowledge.
- Learning multiple concepts; learning prototypes; what is similarity?
- What is self-organizing activity?; changing representations to facilitate learning; constructive induction.
- Knowledge debugging and refinement: the 'end-game'; the inductive apprentice system.
- Interaction and combination of learning strategies; cognitive economy.

Objectives/issues:

What is learning? Can there be learning without improving performance (yes)? Can there be learning with a decrease in performance? Can there be learning without intention of storing the organized knowledge for future use? Can there be learning without the possibility of recalling or retrieving the representation?

Problems:

- Combining explanation-based learning and similarity-based learning systems.
- Dealing with noise and inconsistency.
- Evaluating representations.
- What can people learn easily?
- What can machines learn easily?
- Global and local credit (blame) assignment for knowledge bases; building an expert system for diagnosing a knowledge base.
- Trade-offs between different learning strategies.

State-of-the-art:

- Inductive learning—programs have been built.
- Analogy—programs have also been built, but less has been achieved.
- Achievements—inductive learning programs applied to practical problems; discovery systems.

- Explanation-based learning—examples more for guidance (need to be correct).

Future directions:

- Inductive learning systems are ready to be applied.
- Expert system shells with learning capabilities.
- Significant discovery (estimates from '20–100 years' to '5–10 years')
- Understanding fielding of problems (five years).
- Problem; not only to discover, but to explain.

Slogan:

- 'AI = Machine Learning. There is no future without learning'.

Panel Discussions

Panel discussion were held in two areas:

4.1 KNOWLEDGE ACQUISITION METHODOLOGY AND TRAINING

Panel:

- Tom Bylander, Michael Freiling, and Marianne LaFrance

Goal of work:

- Effective development of knowledge engineering skills.

Objectives of work:

- Develop awareness of the nature of expertise.
- Develop knowledge of knowledge acquisition sources.
- Develop awareness of experts' problems in transferring expertise.
- Develop skills in the application of specific methodologies.

Problems in achieving objectives:

- Lack of models of expertise.
- Piece meal nature of techniques and tools.
- Short training courses required.
- Inadequate interdisciplinary backgrounds of students.

State-of-the-art

- Structure methodologies for knowledge representation and acquisition.
- Interactive knowledge acquisition systems.
- Grid technique (LaFrance).

Future Developments

- Integration of techniques.

4.2 REASONING WITH UNCERTAINTY: IMPLICATIONS FOR KNOWLEDGE ACQUISITION

Panel:

- Brian Gaines, Ryszard Michalski, and Ross Quinlan.

Goal of Work:

- Effective acquisition and inference with uncertain knowledge.

Objectives of Work:

- Develop solid foundations for reasoning with uncertainty including representation, deductive reasoning and inductive knowledge acquisition.

- Relate reasoning to domain in such a way as to give maximum power without distortion.
- Implement representation and reasoning in expert system shells with parameters that enable them to be tailored to specific domains, knowledge islands, and so on.

Problems in achieving objectives:

- Lack of formal foundations for reasoning with uncertainty.
- Lack of comprehensive empirical studies evaluating different approaches in a wide range of domains.

State-of-the-art

- Probability theory, Bayesian inference.
- Fuzzy logic.
- Dempster–Schaeffer evidence.
- Constraint propagation.

Cognitive Foundations of Knowledge Acquisition

An overview of knowledge-acquisition and transfer

BRIAN R. GAINES
Department of Computer Science, University of Calgary, Calgary, Alberta, Canada T2N 1N4

A distributed anticipatory system formulation of knowledge acquisition and transfer processes is presented which provides scientific foundations for knowledge engineering. The formulation gives an operational model of the notion of expertise and the role it plays in our society. It suggests that the basic cognitive system that should be considered is a social organization, rather than an individual. Computational models of inductive inference already developed can be applied directly to the social model. One practical consequence of the model is a hierarchy of knowledge transfer methodologies which defines the areas of application of the knowledge-engineering techniques already in use. This analysis clarifies some of the problems of expertise transfer noted in the literature, in particular, what forms of knowledge are accessible through what methodologies. The model is being used as a framework within which to extend and develop a family of knowledge-support systems to expedite the development of expert-system applications.

Introduction

Growing recognition of the significance of knowledge-based computing systems has focused attention on processes of knowledge acquisition and transfer. Commercial application of expert systems is being impeded by the knowledge-engineering bottleneck and has led to the development of rapid prototyping tools. These have proved practically useful but the range of applications of existing tools is limited, and it is not clear what limitations are inherent in the prototyping tools, what arise from the shells and their knowledge representation and inferencing procedures, and what arise from our lack of understanding of the underlying processes of knowledge acquisition and transfer.

Knowledge acquisition and transfer processes in human society are complex and poorly understood. Many disciplines are involved in their study but none provides a comprehensive framework let alone overall theory. Neurology, psychology, linguistics, education, sociology, anthropology, philosophy and systems theory all have significant contributions to make. However, integrating them, disambiguating jargon, and combining different objectives and perspectives are major intellectual problems. The task is attracting increasing effort but this is unlikely to have significant short-term impact on knowledge-based system development.

Knowledge acquisition and transfer are now important aspects of advanced computer technology, and neither the pragmatic development of rapid prototyping techniques nor the dependence on partial theories from other disciplines can be regarded as satisfactory foundations. We need to be eclectic in acquiring techniques from all possible sources, but we also need to complement our pragmatism with effort to develop scientific foundations for our activities. We cannot be satisfied until

KNOWLEDGE-BASED SYSTEMS Vol. 1
ISBN 0-12-273251-0

we have a precise understanding of the processes underlying expertise, its operation, acquisition and transfer. This is not a simple requirement since it entails understanding the nature of knowledge, its dynamics and application. The foundations of computer technology in the physical sciences are no longer adequate and need extension into the humanities. The philosophy, psychology and sociology of knowledge processes are highly significant to future computing, and we have to make theories operational and obtain precise answers to questions that have long been regarded as certainly controversial and possibly intractable.

This paper sketches a framework for encompassing the variety of processes and phenomena of knowledge acquisition and transfer that provides a convenient classification, an indication of overall theoretical foundations, and a guide to research methodologies.

Phenomena of expertise-acquisition and -transfer

The problem of expertise elicitation from a skilled person is well known in the literature of psychology (Nisbett & Wilson, 1977; Broadbent, Fitzgerald & Broadbent, 1986). Hawkins (1983) has analysed the nature of expertise and emphasizes its severe limitations and dependence on critical assumptions which are often implicit. Dixon (1981) has surveyed studies showing that much human activity is not accessible to awareness. Collins (1985) has studied knowledge-transfer processes among scientists and suggests that some knowledge may not be accessible through the expert, not only because he cannot express it, but also because he may not be aware of its significance to his activity. Bainbridge (1979, 1986) has reviewed the difficulties of verbal debriefing and notes that there is no necessary correlation between verbal reports and mental behavior, and that many psychologists feel strongly that verbal data are useless. Clinical psychologists see the problem as one of cognitive defences that impede internal communication, and have developed techniques of verbal interaction to identify underlying cognitive processes (Freud, 1914; Kelly, 1955; Rogers, 1967). These can be used to bypass cognitive defences, including those resulting from automization of skilled behavior. Hayes-Roth, Waterman & Lenat (1983) have given guidelines about knowledge transfer from experts to expert systems, and Welbank (1983) has surveyed the psychological problems of doing this.

The main problems identified in accessing an expert's knowledge are:

"Expertise" may be fortuitous. Results obtained may be dependent on features of the situation which the expert is not controlling;

Expertise may not be available to awareness. An expert may not be able to transmit the expertise by critiquing the performance of others because he is not able to evaluate it;

Expertise may not be expressible in language. An expert may not be able to transmit the expertise explicitly because he is unable to express it;

Expertise may not be understandable when expressed in language. An apprentice may not be able to understand the language in which the expertise is expressed;

Expertise may not be applicable even when expressed in language. An apprentice may not be able to convert verbal comprehension of the basis of a skill into skilled performance;

"Expertise" expressed may be irrelevant. Much of what is learnt, particularly under random reinforcement schedules, is superstitious behavior that neither contributes nor detracts from performance;

Expertise expressed may incomplete. There will usually be implicit situational dependencies that make explicit expertise inadequate for performance;

"Expertise" expressed may incorrect. Experts may make explicit statements which do not correspond to their actual behavior and lead to incorrect performance.

In the development of knowledge engineering methodologies and rapid prototyping techniques for expert systems the emphasis has been on interviewing experts, and hence on linguistic transmission of expertise. A rich variety of alternative methods for expertise transfer exists in human society:

Expertise may be transmitted by managing the learning environment. A trainer may be able to establish effective conditions for an apprentice to acquire expertise without necessarily understanding the skill or himself being expert in it;

Expertise may be transmitted by evaluation. A trainer may be able to induce expertise by indicating correct and incorrect behavior without necessarily understanding the skill in detail or himself being expert in its performance;

Expertise may be transmitted by example. An expert may be able to transmit a skill by showing his own performance without necessarily understanding the basis of his expertise.

There are many sources of expertise that do not involve others:

Expertise may be acquired by trial and error learning. This is the basic inductive knowledge acquisition process that is always in operation although heavily overlaid by the social transfer processes already discussed;

Expertise may be acquired by analogical reasoning. The transfer of models and skills from one situation to another is an important source of knowledge;

Expertise may be acquired by the application of general laws and principles to new situations. The use of physical laws and systemic principles to generate specific expertise is the basis of scientific and engineering expertise.

These many aspects of the problems of knowledge acquisition and transfer are further confounded by the combinatorial possibilities resulting from the recursive nature of knowledge processes. We can acquire knowledge about knowledge acquisition. We can mimic the behavior of an expert coach managing a learning environment. We can give evaluative feedback on analogical reasoning or the application of principles and laws. We can express the principles behind the effective management of learning in specific domains. The knowledge processes of human society are rich and complex, and, in practice, many of these possibilities will be instantiated as parallel, interrelated activities in any knowledge acquisition and transfer situation.

Socio-systemic foundations of expertise

Figure 1 shows some of the relations between a knowledge engineer, or knowledge transfer system, an expert in a problem domain, and the physical, social, knowledge and problem environments.

An "expert role" has been shown rather than an expert person to emphasize that expertise is focused on specific task domains and is not an indication of global capability. Within that role, various significant psychological processes have been noted: perception, action and language as interfaces with the world; memory, awareness, intentions, emotions and attention as associated internal phenomena. Some of these processes go beyond the basic information-processing models of current cognitive science (Norman, 1980), but they are essential to knowledge acquisition and transfer.

The specific problem environment for the expert is shown as intersecting three general environments that are common to most problems: physical, social and knowledge. The knowledge engineer is part of the social environment but outside the problem environment. He interacts with the expert and also with his environments. For example, if the expert is performing a task in the physical environment then the knowledge engineer will attempt to understand the basis of that task, observe the expert performing it, and may manage the task environment to create scenarios in which to observe specific features of the skill.

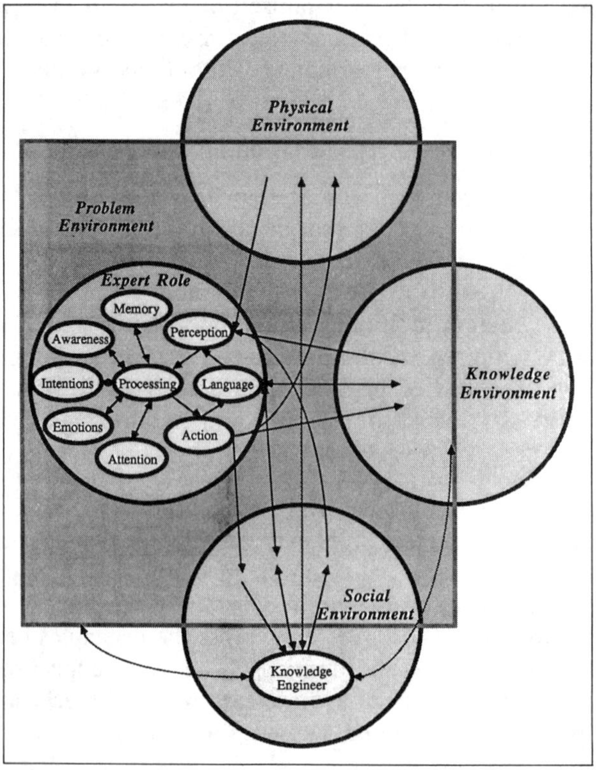

FIG. 1. Relations between knowledge engineer, expert and environments.

The three environments correspond to Popper's (1968) basic delineation of three *worlds*:

World 1, the physical environment, is the world of "things", of physical objects and their dynamics;
World 2, the social environment, is the world of "subjective experience", of people and the dynamics of their mental experience;
World 3, the knowledge environment, is the world of "statements in themselves", of theories and the dynamics of their logical development.

All the phenomena of expertise acquisition and transfer discussed in the previous section can be represented and analysed in terms of this diagram. The expert role develops through interaction with the three environments shown: direct experience of the dynamics of the physical, social and knowledge environments; learning by example, evaluation and linguistic communication from other experts in the social environment; managed experience and transmission; and so on. Figure 1 provides a general framework for knowledge acquisition and transfer, including knowledge engineering for expert systems.

However, the diagram only gives expression to the phenomena, making some significant distinctions, but giving no underlying model for the processes involved. The remainder of the paper develops such an overall model of knowledge acquisition and transfer.

Humanity as a distributed anticipatory system

It is difficult to know where to begin in analysing human knowledge processes. The theoretical foundations to be developed are based on the concept that humanity can be characterized as a *distributed anticipatory system*—a coordinated collection of autonomous systems whose primary dynamic is the anticipation of the future. The motivation for this concept links back to the processes underlying the survival of the species. To survive, a living system must have access to the necessary resources and be capable of coping with threats to its survival. Systems evolve to maximize their access to resources and minimize their vulnerability to threats.

If the universe were static a simple model of resource availability and prey/predator relations would determine the dynamics of living systems. Until the advent of humanity, this planet was a static universe over long periods of evolutionary time. The beginnings of the humanity were set in this static universe but its activities soon began to change that universe so that uncertainty about the future began to dominate our survival processes. Changes in the earth that would have taken millions of years began to occur over millenia, and now they occur within our lifetimes. Humanity developed resources far beyond their natural availability, extinguished most predators other than itself, and changed the ecology of the earth. Much of this planet is now a human construct and the distinction between natural and artificial has become meaningless.

Figure 2 shows the system dynamics underlying humanity. The shaded areas show the response to uncertainty. The keys to survival in an uncertain world are three-fold: first to maximize territorial dispersion so that some part of the system is outside the range of a threat; second to maximize cultural diversity so that some part

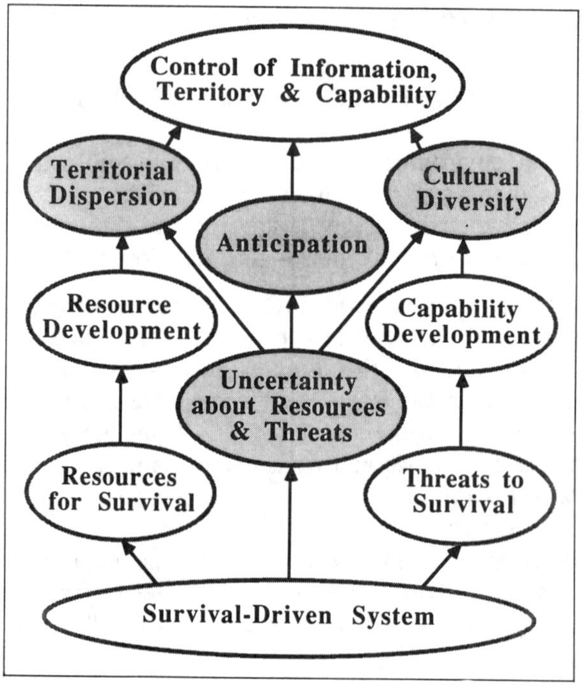

FIG. 2. Dynamics of a survival-driven system.

of the system has the capability to survive a threat; third to improve anticipation of the future, passively to predict threats in advance, and actively to rebuild the universe so that they do not occur.

Figure 2 encompasses the key technological phenomena of our time. Information technology increases control of information, giving computing when implemented in silicon and genetic engineering when implemented in DNA. The space program aims to achieve territorial dispersion and access to new resources. Genetic engineering aims to achieve cultural diversity and improvement of existing resources. Information technology is being used to model our anticipatory processes, leading to their greater understanding through cognitive science and their enhancement through artificial intelligence.

Characterizing humanity as a distributed anticipatory system follows from the logic of Fig. 2. The survival-driven system is the species rather than the individual, and the main knowledge-acquisition system is the species. Humanity has achieved territorial dispersion and cultural diversity through its structure as distributed collection of autonomous entities. Language has evolved as a means of coordinating the activities of these autonomous entities. It is essential to many of the processes of knowledge acquisition and transfer. Interaction with the social and knowledge environments is at least as important as interaction with the physical world. The evolutionary pressures have been very strong in selecting genes giving the capability for the species to act as a single distributed individual, combining autonomy and cohesion through enhanced linguistic communication.

Knowledge acquisition as a modeling process

The characterization of humanity as a distributed anticipatory system, modeling the world so as to increase its probability of survival, places the systematic foundations of knowledge processes in theories of modeling. Klir (1976, 1985) has provided the most general account of modeling, and proposed an epistemological hierarchy accounting for the main components of any modeling systems and their interrelations. Gaines (1977) gave a mathematical formulation of the general problem of modeling as a relation between order relations at different levels of this hierarchy. The hierarchy has proved a valuable conceptual tool in analysing a wide variety of modeling systems both in terms of their ontological presuppositions and their epistemological processes.

Underlying the modeling hierarchy is the notion of a distinction as the primitive concept underlying the representation of knowledge. It is a sufficient primitive to give foundations for systems theory including the notion of a system itself (Gaines, 1980). In its psychological form, as a personal construct (Kelly, 1955), the notion has been used to derive very effective techniques for knowledge transfer from experts to expert systems (Boose, 1985). Its foundational role in knowledge acquisition is evident in the hierarchical representation of distinctions in a modeling system shown in Fig. 3. The levels of the hierarchy itself are the results of distinctions that we make so that no additional primitives are introduced—in Klir's (1976) terminology:

The source system is distinguished as those distinctions that the particular modeling system makes—it is a distinction about distinctions defining the construct system of an individual;

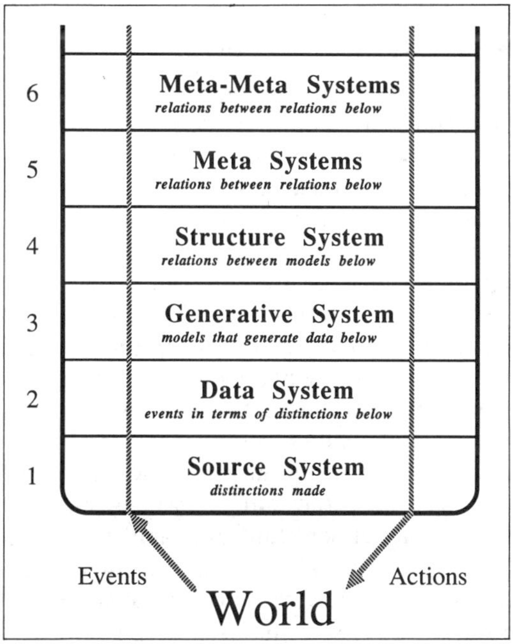

FIG. 3. Epistemological hierarchy of a system modeling a world.

The data system is distinguished as those distinctions that have been made about a particular event—a distinction about distinctions defining an event;

The generative system is distinguished as a set of distinctions that also defines an event—these are model-generated rather than event-generated—it is the match between the model-generated and event-generated distinctions that determines the degree of approximation of the model to the world—this is a distinction about distinctions among distinctions that defines goodness of fit;

The structure system is distinguished as a set of distinctions that compare models—it is the order relation of simplicity/complexity on models that determines the preference for the simplest model that is an adequate approximation to the world—this is a distinction about distinctions that defines our preference for simple models;

The meta system is distinguished as a set of distinctions that specify the basis of these comparisons;

The meta–meta system, and higher levels, are distinguished as sets of distinctions that specify further relations among the distinctions on the level below.

Note that the upper levels of modeling are totally dependent on the system of distinctions used to express experience through the source system.

Distinctions are not just static partitions of experience. They may be operations: actions in psychological terms; processes in computational terms. Whether a system finds a distinction in the world, imposes it passively as a view of the world, or imposes it actively as a change in the world, is irrelevant to the basic modeling theory. It makes no difference to the theory whether distinctions are instantiated through sensation or action. In knowledge engineering we have to incorporate both the expert's prediction and control processes.

Knowledge acquisition in the hierarchy

The hierarchy of Fig. 3 accounts for knowledge acquisition as the modeling of events enabling adequate prediction and action. A modeling schema results from distinctions about distinctions at each level in the hierarchy. In prediction the key distinction is to what degree a level accounts for the information flowing through it and hence this distinction may be termed one of *surprise* (Gaines, 1977), in the sense used by the economist Shackle (1955). Surprise goes in opposition to the degree of membership of a predicted event to an actual event and the expected surprise is a form of entropy. Surprise at the lowest level of the hierarchy corresponds to distinctions being inadequate to capture events; surprise at the next level to inadequate variety to experience events; at the next level to inadequate approximation to predict events; at the next level to inadequate simplicity to explain events; at the next level to inadequate comprehensiveness to account for events.

The formal theory of modeling is one in which models are selected at each level down the hierarchy to minimize the rate at which surprise is passing up the hierarchy. The criteria for model selection independent of the data are generally thought of as being ones of simplicity/complexity: of two models which fit the data equally well choose the simplest. However, notions of simplicity/complexity are not well-defined nor intrinsic to the class of models. The simplicity/complexity ordering is arbitrary and in its most general form is just one of *preference*. Hence the general modeling schema is one in which surprise flows up the hierarchy and preference flows down. In situations that are mathematically well defined, such as determining the structure of a stochastic automaton from its behavior, such a model schema gives the correct results (Gaines, 1977). Conversely, the success of the schema in stabilizing with regard to a given world defines the characteristics of that world.

Thus the basic modeling schema for learning from experience is one in which surprise flows up the hierarchy and preferences flow down. In primitive organisms only the lower levels of the hierarchy are developed, surprise is generated from experience and preference is genetically encoded. In higher organisms the modeling process generalizes both surprise and preference to cope with novel environments. Humanity has developed the upper levels of the hierarchy and detached surprise from experience and preference from its genetic roots. Surprise can flow up from a level without flowing into it from below because the processes at that level have generated novelty. Preference can be generated at a high level detached from both experience and genetic origins and flow down to affect the relations of the organism to the world.

There is neurological and behavioral evidence of the existence within the brain of the two channels of communication shown in Fig. 3 (Tucker & Williamson, 1984). The arousal system passes surprise up to the cortex from the limbic region as unexpected events occur. The activation system passes preferences down from the cortex to the motor regions.

Cognitive-science interpretation of the hierarchy

The loop in Fig. 1 from events through distinctions up through the modeling hierarchy and then down again to predictions and actions corresponds to the epistemological model of man as an anticipatory system developed by Kelly (1955). Thus, the systemic hierarchy of Fig. 3 has an analog in psychological terminology as shown in Fig. 4:

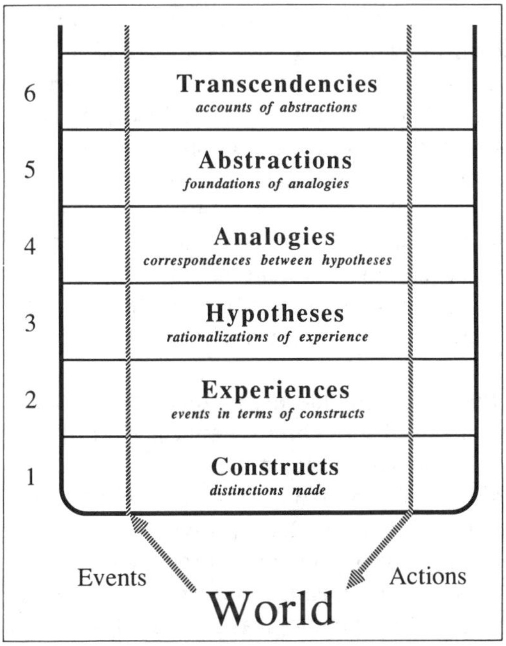

FIG. 4. Construction hierarchy of a person modeling a world.

The source level is one of constructs, distinctions made in interacting with the world;
The data level is one of experiences, events which happen to us, and we make happen, in terms of the distinctions already made;
The generative level is one of hypotheses which are rationalizations of experience;
The structure level is one of analogies which are correspondences between these rationalizations;
The meta level is one of abstractions which are foundations of analogies;
The meta–meta level is one of transcendencies which are accounts of abstractions.

Expertise, roles, groups and society

The anticipatory processes of the modeling hierarchy may be extended to the operation of society by viewing groups as cross-sections comprising multiple individuals. This concept may be given deeper significance by considering the inductive inference process underlying knowledge acquisition. Whereas deductive logical inference is well understood and well founded, the inductive inference that underlies knowledge acquisition is not. Deduction guarantees to take us from valid data to valid interfaces, but the inferences are thereby part of the data—no new knowledge is generated. Induction takes us from valid data to models of that data go beyond it—by predicting data we have not yet observed, and by giving explanations of the data in terms of concepts that are unobservable. Induction generates new knowledge but, as Hume (1739) pointed out over 200 years ago, the process is not deductively valid and it is a circular argument to claim that it is inductively valid.

Philosophers have continued to debate Hume's arguments and search for justification of the inductive process. Goodman (1973) proposed that we accept the circularity but note that it involves a dynamic equilibrium between data and inference rules as shown in Fig. 5: "A rule is amended if it yields an inference we are unwilling to accept; an inference is rejected if it violates a rule we are unwilling to amend". Rawls (1971) in his theory of justice terms this a reflective equilibrium. Stich & Nisbett (1984) noted flaws in Goodman's argument and repaired them by proposing that the equilibrium is social not individual: "a rule of inference is justified if it captures the reflective practice not of the person using it but of the appropriate experts in our society". This gives an operational definition of the role of experts within an anticipatory system as referential anticipatory sub-systems coordinating the overall knowledge-acquisition process. It also leads to an concept of the role of expert systems in society, as making the referential process more overt and widely available.

The extension of the modeling hierarchy to social processes is straightforward since Fig. 3 presents a general modeling schema and applies as much to groups of people, companies and societies as it does to the roles of a person. The epistemological hierarchy of a person is a cross-section of the epistemological hierarchy of the society generating their life-world. Pask's (1975) concept of P-Individuals as the basic units of psycho-socio-processes allows roles, people, groups, organizations and societies to be treated in a uniform framework. An individual is defined in cognitive terms as a psychological process (Pask, 1980) and more complex

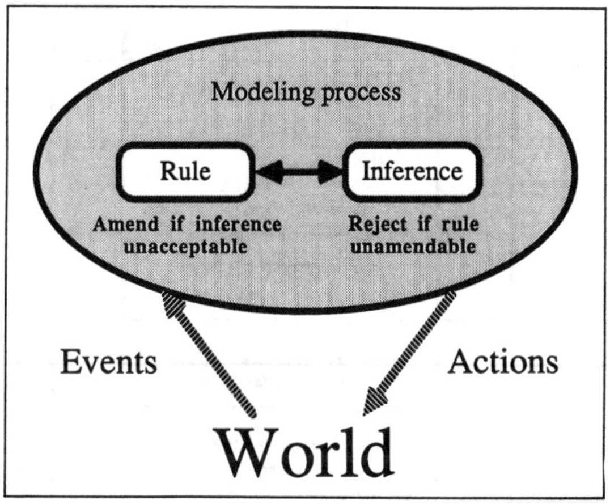

FIG. 5. Reflective equilibrium in inductive inference.

psychological and social structures may be defined similarly by taking into account the possibilities of timesharing, process switching and distributed processing with psychological processors. For example, one person may assume many psychological roles (process switching), whereas a group of people working together may act as a single goal-seeking entity and hence behave as one process (distributed processing).

Representation of expert skills within the hierarchy

In the analysis of expertise, the skills to achieve goals in the world are the crucial capabilities of the modeling system, and the usual hierarchical models of skills (Powers, 1973; Rasmussen, 1983) are naturally subsumed within the modeling formulation. Figure 6 shows the basis for action at different levels in the modeling hierarchy:

At level one, the activation of a construct may be linked directly to a primitive act, another construct. This corresponds to reflex actions and stimulus-reponse connections. In system-theoretic terms this level might be implemented by conditional probability calculations giving confirmation-theoretic inductive inference;

At level two, constellations of experience may be linked to complete action sequences through rules derived from similar experience. In system-theoretic terms this level might be implemented by fuzzy production rules giving generalization-based indictive inference. These constellations may be regarded as prototypical schema, or frame-like structures in computational terms;

At level three, a generative model of experience, may be used to compute an optimal action sequence. In system-theoretic terms this level might be implemented by a state-based modeling scheme giving model-based inductive inference.

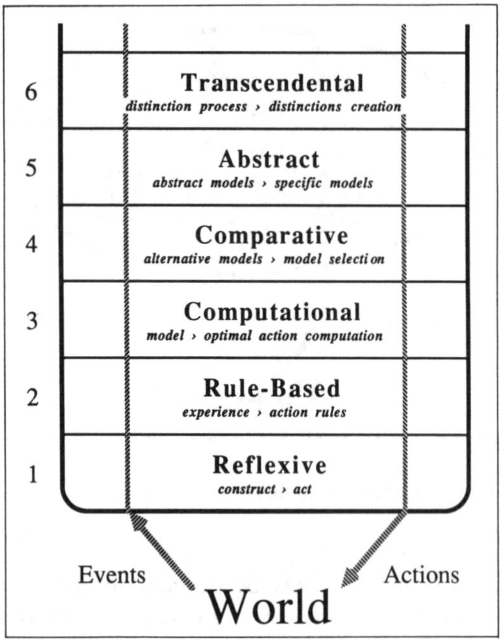

FIG. 6. Action hierarchy of system modeling a world.

At level four, a variety of alternative models may be compared as a basis for selecting one appropriate to the required goals. In system-theoretic terms this level might be implemented by a category-theoretic functional analysis scheme giving analogical inductive inference.

At level five, generalized abstract models may be used as templets from which to instantiate one appropriate to the required goals. In system-theoretic terms this level might be implemented by a category-theoretic factoring scheme abstracting the mathematical form of an analogy and giving abstractional inductive inference.

At level six, the entire process described may be transected through a recognition that it is based on distinctions being made at various level, and an attempt to rationalize these distinctions and create new ones. In system-theoretic terms this level might be implemented by a distinction-based analysis scheme giving what might be termed transcendental inductive inference.

It is an interesting comment on the state of the art in computer science that it has proceeds "middle–outward" in its representation of the knowledge involved at different levels of the hierarchy. Information technology has been primarily concerned with level three activities, and is only now beginning through expert systems to emulate level two activities.

Language and culture in knowledge acquisition

The creation of new knowledge takes place through the surprise/preference flows within the hierarchy and it is these processes that determine the rate of scientific discovery, technological invention and product innovation. The capability for an entire society to act as a distributed knowledge acquisition system depends on communication processes to coordinate activity at a given level of the hierarchy across different people. This process whereby each person does not have to undertake all aspects of the inductive process but can share the results of such processing by others supports what is generally termed the culture of a society (Vanderburg, 1985). People use language for much of this communication but they also have in common with other animals the capability to communicate cultural information without the use of language. Mimicry is an important mechanism for knowledge acquisition as is reinforcement through reward and punishment.

The human development of language enables coordination to take place in a rich and subtle fashion that greatly enhances, but does not replace, the more basic mechanisms in common with other species. It is particularly important at the upper levels of the hierarchy where direct expression is difficult. From an individual point of view, language is a way of bypassing the normal modeling procedures and interacting directly with the system at any level. In particular it can directly affect the preference system. Much skilled activity is not directly accessible through language, but even when language cannot mediate the direct transmission of knowledge it may be used to achieve the same effect by the indirect support of other mechanisms. For example, one can describe a good learning environment, or a behavior in sufficient detail for mimicry.

Figure 7 shows the cultural support for knowledge acquisition at different levels in the modeling hierarchy:

The reflexive knowledge at level one has no verbal component and comes directly from experience, often that of mimicking the behavior of others. This level has been termed informal to correspond to Hall's (1959) definition of cultural transmission of behavior of this type.

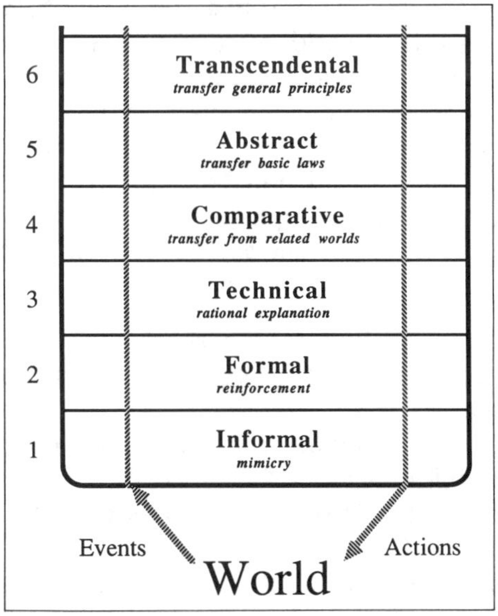

FIG. 7. Cultural transmission hierarchy of people modeling a world.

The rule-based knowledge at level two is usually transmitted by reinforcement of behavior, verbal rules, or is induced from the patterns of knowledge at level 1. This level has been termed formal to correspond to Hall's definition of cultural transmission of behavior of this type.

The computational knowledge at level three is usually transmitted by technical explanation, or is induced from the patterns of knowledge at level two. This level has been termed technical to correspond to Hall's definition of cultural transmission of behavior of this type.

The comparative knowledge at level four is usually transmitted by simile and metaphorical analysis, or is induced from the patterns of knowledge at level three. Hall does give a name to this level but the term comparative captures his own activity of highlighting the features of one culture by contrasting it with others.

The abstract knowledge at level five is usually transmitted through mathematical representation, or is induced from the patterns of knowledge at level four.

The transcendental knowledge at level six is usually transmitted by general system-theoretic analysis, or is induced from the patterns of knowledge at level five. Many mystical and consciousness-raising techniques may be seen as attempts to communicate knowledge at this level when formal analysis is impossible. It involves moving outside the framework established at the lower levels. Pope (1984) has given examples of this process in a wide range of cultures.

Knowledge-transfer processes

The action and cultural transmission hierarchies of Figs 6 and 7 may be combined to give an overall framework for knowledge representation, processing, acquisition and transmission as shown in Fig. 8. This gives detailed foundations for the processes of knowledge engineering discussed in the first part of this paper and shown in Fig. 1. The primary learning process is one of direct acquisition based on attempts to solve tasks through interaction with a world, as shown in the lower part of Fig. 8. This is applicable to all three worlds: physical, social and knowledge.

This primary process is the only one open to humanity as a species and is distributed across partially autonomous individuals coordinated by cultural communication processes. From an individual perspective it is these communication processes that dominate knowledge acquisition and transfer.

The right hand side of Fig. 8 shows the cultural communication processes supporting knowledge transfer:

At level one mimicry by watching experts is the main transfer mechanism;

At level two reinforcement by working under expert supervision is the main transfer mechanism;

At level three rational explanation from interviewing experts and books is the main transfer mechanism;

At level four use of analogous models to derive knowledge from similar situations is the main transfer mechanism;

At level five use of basic laws to derive specific models is the main transfer mechanism.

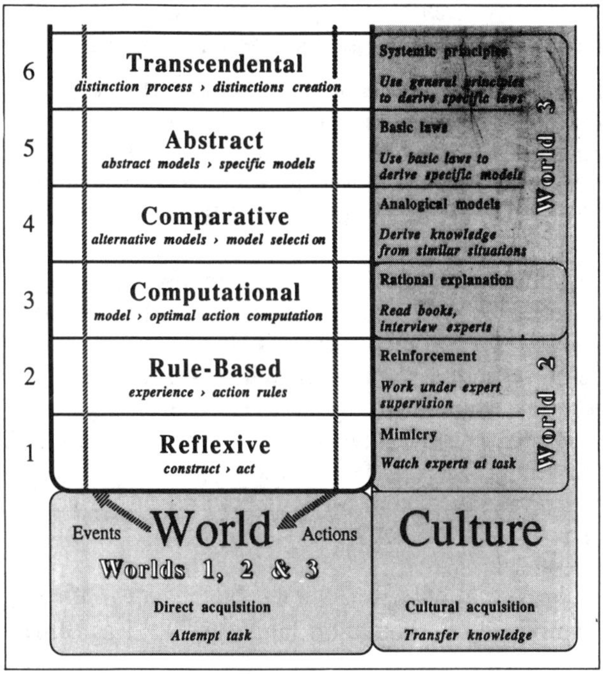

FIG. 8. Knowledge transfer processes in three worlds.

At level six use of system principles to derive specific laws is the main transfer mechanism.

Levels 1 and 2 of the cultural transmission process, mimicry and reinforcement, clearly belong to world 2, the social environment of subjective experience. Levels 4 through 6, analogies, laws and principles, clearly belong to world 3, the knowledge environment of statements in themselves. Level 3, of rational explanation, bridges both worlds.

Knowledge transfer methodologies

One practical consequence of the analysis of knowledge acquisition and transfer given above is a hierarchy of knowledge transfer methodologies based on Fig. 8 which defines the areas of application of the knowledge engineering techniques already in use. Figure 9 shows this hierarchy in relation to the basic architecture of an expert system:

Knowledge base: the central feature is a *knowledge base* of facts and inference rules;

Expert-system shell: the lower oval shows the operational system for applying the knowledge base to knowledge support.

The inference system derives the consequences of facts about a particular situation.

The planning system determines how to use the inference system to satisfy the specified objectives.

The explanation system answers questions about the basis of the inferences made.

Knowledge-acquisition system: the upper oval shows the knowledge transfer processes that may be used to establish the knowledge base.

The knowledge generation system implements knowledge acquisition by the raw induction of models from experience without cultural support as exemplified by ATOM (Gaines, 1977) and AM (Davis & Lenat, 1982).

The expertise modeling system implements knowledge acquisition by mimicking an expert's behavior as exemplified by INDUCE (Michalski & Chilausky, 1980).

The performance reinforcement system implements knowledge acquisition from performance feedback as exemplified by the PERCEPTRON (Rosenblatt, 1958) and STeLLA (Gaines & Andreae, 1966).

The knowledge-elicitation system implements knowledge acquisition by interviewing experts as exemplified by ETS (Boose, 1985), KITTEN (Gaines & Shaw, 1986), MORE (Kahn, Nowlan & McDermott, 1985) and SALT (Marcus, McDermott & Wang, 1985).

The knowledge-structuring system implements knowledge acquisition by analogy as exemplified by TEIRESIAS (Davis & Lenat, 1982) and CYC (Lenat, Prakash & Shepherd, 1986).

The basic laws system implements knowledge acquisition by building physical models as exemplified by simulation languages such as SIMULA (Nygaard & Dahl, 1981).

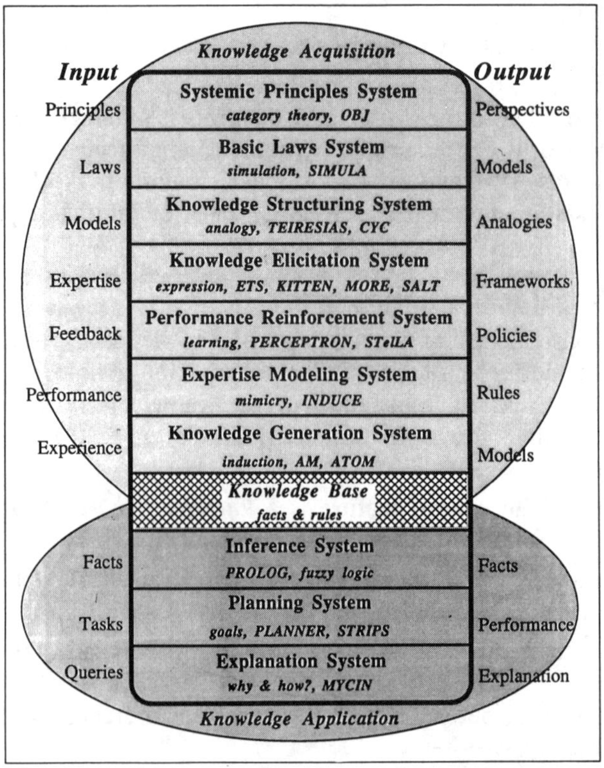

FIG. 9. Knowledge acquisition hierarchy for expert systems.

The systemic principles system implements knowledge acquisition by derivation from abstract principles as exemplified by the category-theoretic language OBJ (Goguen & Meseguer, 1983).

Most current expert-system shells offer a limited knowledge base represented in terms of production rules and frames, have weak planning capabilities limited to deriving consequences and testing assertions, and give simple explanations in terms of the facts and rules used. However, rapid improvements are taking place in knowledge representation, inferencing and explanation capabilities of expert systems (Hayes-Roth, 1985; Michalski & Winston, 1986; Pearl, 1986; Swartout, 1985).

Most current knowledge acquisition is done by manual knowledge elicitation. However, rapid improvements are taking place in the knowledge automatic knowledge-acquisition capabilities of expert systems (Boose, 1985, 1986; Bradshaw & Boose, 1986; Gaines & Shaw, 1986; Kahn *et al.*, 1985; Marcus *et al.*, 1985; Shaw & Gaines, in press; van de Brug, Bachant & McDermott, 1985).

The knowledge base and shell have been incorporated with the acquisition methodologies in Fig. 9 to show the natural relationships between all three. Shell structure and knowledge-acquisition research and development have so far been treated as separate enterprises. Integrating the two is essential to future knowledge-based system applications so that the combined system provides maximal and coherent support to the expert and knowledge engineer.

Conclusions

The distributed anticipatory system formulation of knowledge acquisition and transfer processes outlined in this paper provides scientific foundations for knowledge engineering:

It provides an operational model of the notion of expertise and the role it plays in our society, and hence provides formal foundations for expert systems;

It provides a detailed structure for the cognitive processes involved in knowledge acquisition and transfer;

There are links between the model and neurological processes in the brain;

It suggests that the basic cognitive system that should be considered is a social organization, rather than an individual;

Computational models of inductive inference already developed can be applied directly to the social model;

Cultural knowledge acquisition and transmission processes are explicitly identified and structured within the model;

Linguistic communication coordinates the distributed autonomous units forming the anticipatory system;

It suggests that groups of experts, rather than individuals, should be used in developing expert systems.

It provides a framework within which all the methodologies for knowledge acquisition and transfer currently under developed can be incorporated, analysed and compared.

There are several implications for future developments in Figs 8 and 9. Whereas past knowledge acquisition and transfer systems have been developed using techniques at one level of the hierarchy, future systems will increasingly follow the example of human apprenticeship and operate at multiple levels. The combination of teaching by example, performance feedback and explicit statement is ubiquitous in human knowledge transmission and will become increasingly so in expertise transfer systems. The combination of rule-based, model-based, analogical and law-based inference is ubiquitous in human knowledge processing and will become increasingly so in expert systems. A companion paper describes the development of an integrated knowledge acquisition and transfer system, KITTEN, based on these concepts (Shaw & Gaines, in press).

The range and variety of expertise transfer systems being developed is very important. They are complementary rather than competitive. We are entering a phase of system integration when the best features of many systems will be melded to provide a new generation of knowledge acquisition and transfer systems.

Financial assistance for this work has been made available by the National Sciences and Engineering Research Council of Canada. Many of the developments described in this paper have resulted from joint research with my colleague Mildred Shaw.

References

BAINBRIDGE, L. (1979). Verbal reports as evidence of the process operator's knowledge. *International Journal of Man–Machine Studies,* **11,** 411–436.

BAINBRIDGE, L. (1986). Asking questions and accessing knowledge. *Future Computing Systems,* **1,** 143–149.

BOOSE, J. H. (1985). A knowledge acquisition program for expert systems based on personal construct psychology. *International Journal of Man–Machine Studies* **20,** 21–43.

BOOSE, J. H. (1986). Rapid acquisition and combination of knowledge from multiple experts in the same domain. *Future Computing Systems* **1,** 191–216.

BRADSHAW, J. M. & BOOSE, J. H. (1986). NeoETS. *Proceedings of the North American Personal Construct Network Second Biennial Conference,* University of Calgary, Department of Computer Science, pp. 27–41.

BROADBENT, D. E., FITZGERALD, P. & BROADBENT, M. H. P. (1986). Implicit and explicit knowledge in the control of complex systems. *British Journal of Psychology,* **77,** 33–50.

VAN DE BRUG, A, BACHANT, J. & MCDERMOTT, J. (1985). Doing R1 with style. *Proceedings of the Second Conference on Artificial Intelligence Applications,* IEEE 85CH2215-2, pp. 244–249.

COLLINS, H. M. (1985). *Changing Order: Replication and Induction in Scientific Practice*: London: Sage.

DAVIS, R. & LENAT, D. B. (1982). *Knowledge-based Systems in Artificial Intelligence.* New York: McGraw–Hill.

DIXON, N. (1981). *Preconscious Processing.* Chichester: Wiley.

FREUD, S. (1914). *Psychopathology of Everyday Life.* London: Benn.

GAINES, B. R. (1977). System identification, approximation and complexity. *International Journal of General Systems,* **3,** 145–174.

GAINES, B. R. (1980). General systems research: quo vadis?. In GAINES, B. R. Ed., *General Systems 1979,* Vol. 24, pp. 1–9. Kentucky: Society for General Systems Research.

GAINES, B. R. & ANDREAE, J. H. (1966). A learning machine in the context of the general control problem. *Proceedings of the 3rd Congress of the International Federation for Automatic Control.*

GAINES, B. R. & SHAW, M. L. G. (1986). Knowledge engineering techniques. *Proceedings of AUTOFACT'86.* pp. 8–79—8–96. Dearborn, Michigan: Society of Manufacturing Engineers.

GOGUEN, J. A. & MESEGUER, J. (1983). Programming with parametrized abstract objects in OBJ. In *Theory and Practice of Programming Technology.* Amsterdam: North-Holland.

Goodman, N. (1973). *Fact, Fiction and Forecast.* Indianapolis: Bobbs–Merrill.
Hall, E. T. (1959). *The Silent Language.* New York: Doubleday.
Hawkins, D. (1983). An analysis of expert thinking. *International Journal of Man–Machine Studies,* **18,** 1–47.
Hayes-Roth, B. (1985). A blackboard architecture for control. *Artificial Intelligence,* **26,** 251–321.
Hayes–Roth, F., Waterman, D. A. & Lenat, D. B., Eds. (1983). *Building Expert Systems.* Reading, Massachusetts: Addison–Wesley.
Hume, D. (1739). *A Treatise of Human Nature.* London: John Noon.
Kahn, G., Nowlan, S. & McDermott, J. (1985). MORE: an intelligent knowledge acquisition tool. *Proceedings of the Ninth International Joint Conference on Artificial Intelligence,* pp. 581–584.
Kelly, G. A. (1955). *The Psychology of Personal Constructs.* New York: Norton.
Klir, G. J. (1976). Identification of generative structures in empirical data. *International Journal of General Systems,* **3,** 89–104.
Klir, G. J. (1985). *Architecture of Systems Problem Solving.* New York: Plenum Press.
Lenat, D., Prakash, M. & Shepherd, M. (1986). CYC: using common sense knowledge to overcome brittleness and knowledge acquisition bottlenecks. *AI Magazine* **6,** 65–85.
Marcus, S., McDermott, J. & Wang, T. (1985). Knowledge acquisition for constructive systems. *Proceedings of the Ninth International Joint Conference on Artificial Intelligence,* pp. 637–639.
Michalski, R. S. & Chilausky, R. L. (1980). Knowledge acquisition by encoding expert rules versus computer induction from examples—a case study involving soyabean pathology. *International Journal of Man–Machine Studies,* **12,** 63–87.
Michalski, R. S. & Winston, P. H. (1985). Variable precision logic. *Artificial Intelligence,* **29,** 121–146.
Nisbett, R. E. & Wilson, T. D. (1977). Telling more than we can know: verbal reports on mental processes. *Psychological Review* **84,** 231–259.
Norman, D. (1980). Twelve issues for cognitive science. *Cognitive Science,* **4,** 1–32.
Nygaard, K. & Dahl, O.-J. (1980). The development of the SIMULA languages.
In Wexelblat, R. L., Ed. (1981). *History of Programming Languages,* pp. 439–480. New York: Academic Press.
Pask, G. (1975). *Conversation, Cognition and Learning.* Amsterdam: Elsevier.
Pask, G. (1980). Developments in conversation theory—Part I. *International Journal of Man–Machine Studies,* **13,** 357–411 (November).
Pearl, J. (1985). Fusion, propagation and structuring in belief networks. *Artificial Intelligence,* **29,** 241–288.
Pope, S. (1984). Conceptual synthesis: beating at the ivory gate?. In Smith, W. Ed., *Systems Methodologies and Isomorphies,* pp. 31–40. California: Intersystems.
Popper, K. R. (1968). Eipstemology without a knowing subject. In Van Rootselaar, B. Ed., *Logic, Methodology and Philosophy of Science III,* pp. 333–373. Amsterdam, Holland: North-Holland Publishing Co..
Powers, W. T. (1973). *Behavior: the Control of Perception.* New York: Aldine.
Rasmussen, J. (1983). Skills, rules and knowledge; signals, signs and symbols, and other distinctions in human performance models. *IEEE Transactions on Systems, Man & Cybernetics,* **SMC-13,** 257–266.
Rawls, J. (1971). *A Theory of Justice.* Cambridge, Massachusetts: Harvard University Press.
Rogers, C. R. (1967). *On Becoming a Person: a Therapist's View of Psychotherapy.* London: Constable.
Rosenblatt, F. (1958). The Perceptron: a probabilistic model for information storage and organization in the brain. *Psychological Review,* **65,** 386–407.
Shackle, G. L. S. (1955). *Uncertainty in Economics.* Cambridge University Press.
Shaw, M. L. G. & Gaines, B. R. (1987). KITTEN: knowledge initiation and transfer tools for experts and novices, *International Journal of Man–Machine Studies.* In press.
Shortliffe, E. H. (1976). *Computer-based Medical Consultations: MYCIN.* New York: Elsevier.

STICH, S. P. & NISBETT, R. E. (1984). Expertise, justification and the psychology of inductive reasoning. In HASKELL, T. L. Ed., *The Authority of Experts,* pp. 226–241. Bloomington, Indiana: Indiana University Press.

SWARTOUT, W. R. (1985). Knowledge needed for expert system explanation. *Proceedings of the 1985 National Computer Conference,* Vol. 54, pp. 93–98.

TUCKER, D. M. & WILLIAMSON, P. A. (1984). Asymmetric neural control systems in human self-regulation. *Psychological Review,* **91,** 185–215.

VANDERBURG, W. H. (1985). *The Growth of Minds and Cultures: a Unified Theory of the Structure of Human Experience.* Toronto: University Press.

WELBANK, M. (1983). *A Review of Knowledge Acquisition Techniques for Expert Systems.* BTRL, Ipswich: Martlesham Consultancy Services.

Cognitive biases and corrective techniques: proposals for improving elicitation procedures for knowledge-based systems

DAVID A. CLEAVES

Fire Economics and Management Research, USDA Forest Service, Pacific Southwest Forest and Range Experiment Station, 4955 Canyon Crest Drive, Riverside, CA 92507, U.S.A.

Expert system output is only as good as the expert judgments on which the system is based. One component of expertise may be defined as the ability to distinguish casual from random occurrence. Judgments of expert and novice alike have been shown to reflect systematic biases in comparison with normative statistical logic. These biases may be important where accuracy, consistency, and coherence are important attributes of the required judgment. Several biases and their cognitive origins are discussed in the context of building knowledge-based systems for wildland fire control. Preliminary guidelines are offered for recognizing and correcting biases during the knowledge-elicitation process.

Introduction

The elicitation of knowledge from experts is a time-consuming process and is usually conducted in the absence of a systematic conceptual design. Few guidelines are available to help the knowledge engineer map out his/her course and pursue it efficiently. Improvements that are stumbled on in knowledge elicitation practice may escape recognition or may be difficult to transfer from one system-building experience to another.

One goal of knowledge elicitation should be to insure that the process is extracting the experts' best judgment. The programming phase of prototype development takes the quality of knowledge as given. This assumption should be constantly reexamined in light of two questions. What does "best" mean? Can we elicit rules and judgments that fully reflect what the expert knows?

The first question rests on the understanding that expertise is multi-dimensional and is accessible through a rigorous assessment of the particular dimensions that are most important for a domain. Defining expertise is an important prerequisite not only in selecting experts but also in choosing approaches for knowledge elicitation. The proliferation of knowledge-based systems has created a need for flexibility in these definitions. The best expert for a particular task probably exhibits an optimum mix of performance in different dimensions. Different task domains require different mixes with emphasis on certain dimensions. Basically, a knowledge-based system should exhibit calibration, consistency, credibility, speed, understandability, and compatibility with the values of the individual or organization that uses it.

Maintaining the quality of expertise represented in any of these dimensions depends on fully capturing the best judgments and the important knowledge responsible for them. Often overlooked is how to detect, avoid, and correct errors

KNOWLEDGE-BASED SYSTEMS Vol. 1
ISBN 0-12-273251-0

in judgment, by the expert. Errors can occur not only in making an expert judgment, but also in explaining the underlying reasoning process. If left undetected, these errors can be preserved and perhaps exaggerated in the representation of decision rules and uncertainties in the prototype.

Calibration, or the degree of agreement between a judged value (or relationship) and a real value, is one measure of expert performance. Lack of calibration can be caused by lack of knowledge or skill or from contextual and cognitive factors which prevent the expert from manifesting extant knowledge or skill on a particular problem. Training and experience may be the difference between the skill levels of the novice and expert. However, both may be subject to cognitive limitations. Error that results from these limitations will be called expert bias. Research comparing expert and novice performance has begun to identify knowledge differences, but expert bias has been relatively unexplored in expert systems development work.

Expert bias is a research topic in cognitive science and behavioral decision theory in which the focus has been on subjective assessments of uncertain quantities and on perceptions of risk. Judgment research, as it is called, has produced many findings that could be useful in building knowledge elicitation systems or guidelines for manual interviewing techniques. The purpose of this paper is to review some of these findings and relate them to the knowledge elicitation process. The following questions will be addressed:

(1) Are experts biased in their judgments?
(2) Why do these biases occur?
(3) Where would the knowledge engineer look for expert bias in the judgment process being elicited?
(4) What might be done to correct or adjust for expert biases?

The research on judgmental bias is somewhat controversial within the field of psychology. Critics of the line of inquiry say that most biases have been discovered in experiments using unmotivated subjects and that implications drawn from such work have tended to overgeneralize. Not enough attention is given to individual differences. Bias is defined in terms of deviation from the normative logic of probabilistic reasoning, but fails to address the variety of other cognitive tasks to which statistical reasoning is poorly suited. Nevertheless, the evidence of bias in the form of different responses to the way questions are framed or information is presented has been found in many replicated studies. The concept of expert bias is recognized by the public and decision makers. Associated questions of expert trust may be extremely important in future knowledge-based applications.

Why be concerned about expert biases? We need criteria for comparing and selecting expertise. Some experts with great knowledge may produce judgments so biased that better answers could be obtained from less "expert" subjects. The building of systems from multiple experts could use levels of calibration or measures on other dimensions of expertise with which to build portfolios of expertise to minimize the risk of bias. Expert bias may also be important in validating and evaluating expert systems. Better understanding of expert bias may enhance the ability to provide knowledge elicitation procedures and systems that improve, rather than merely mimic judgment processes. We may someday be able to test and correct for biases within the expert system itself, perhaps producing "purer" estimates on a consistent basis than the subject experts on which they are based.

Our perspective on these problems stem from our struggles in eliciting subjective assessments from experts in wildland fire control. Earlier efforts to model fire control processes involved decision analysis and real-time simulation, or fire gaming exercises. More recently, we have been developing three knowledge-based systems. The first seeks to estimate first- and second-order uncertainty of wildfire escapes on our controlled burn operations. The second system is designed to use multiple expertise to diagnose fire prevention problems in specific locations and to prescribe educational, enforcement, or other activities to address the problem The third system is being designed to estimate how fast our crews and machinery can build fireline under various conditions of terrain complexity, fatigue, and other factors.

ARE EXPERTS BIASED?

Bias can occur in any of the generic cognitive tasks which comprise problem solving. Fischoff (1984) describes four types of judgment tasks: (1) identifying active elements; (2) characterizing interrelationships; (3) assessing (estimating or predicting) parameter values; and (4) evaluating quality. Each of these tasks require different cognitive skills and are subject to different types of bias. In the expert-system parlance, these tasks translate into: (1) hypothesis and solution generation; (2) decision rule articulation; (3) assessing levels of uncertainty and confidence; and (4) hypothesis testing.

We might expect the expert to be better than the novice in all cognitive tasks. The *de facto* classification of expert is sometimes assigned to individuals solely on the basis of degrees and professional longevity, not necessarily on performance. However, we often do not know the conditions under which bona fide experts in actual decision settings provide information and reasoning or how to measure its reliability. Furthermore, the process of knowledge elicitation forces cognitive tasks into formats that may be unfamiliar to experts and novices alike. A prime example is probabilistic reasoning, assessing probabilities as well as using probabilities or other uncertainty expressions in making inferences. The essence of this component of expertise is to distinguish between causal and random occurrence. However it does not follow that the longer a person is exposed to chance processes, the better he/she will become at recognizing them. Research has shown basic differences between clinical and probabilistic reasoning in diagnosis, even though probabilistic reasoning is often expected of many experts. Research (Winkler & Murphy, 1968; Kahneman, Slovic & Tversky, 1982) has shown that experts in some fields do not produce any better estimates or forecasts in probabilistic terms than do relative novices. In addition, the expert is much less likely to say: "I don't know," which can lead to disastrous consequences in subsequent decisions.

Although the concept of bias is based on some notion of a true or complete value, set of solutions, probability or relationship, the true value is either not known or perhaps revealed only after the expert judgment has been made. This often leads the knowledge engineer to accept prematurely those values produced by the subject expert or to validate them only against those of another "better" expert. This develops the tendency to embed limitations of human performance in expert systems without any knowledge of their influence on system performance.

Expert bias that is defined for a particular dimension of expertise can be understood by the nature of the cognitive process. Popular topics of judgment research have been the degree and nature of calibration in the assessment of uncertain quantities. Additional effort has been directed at understanding human abilities in generating alternative hypotheses and problem solutions. Much could be borrowed from these reseach areas and developed into operational concepts for building valid knowledge-based systems.

Biases can occur in several ways (Table 1). Bias can be introduced in the method or the response format in which the judgment task is presented to the expert. For example, eliciting judgments in the form of decision rules or probabilities may create bias if the expert does not normally articulate his thoughts about the problem in those terms. Conceptual bias results either from motivational processes, such as wishful thinking or having other vested interests in the judgments' consequences, or from cognitive processes which are the gears of the expert's intuition. Motivational bias is very situational and can be difficult to detect and interpret without a strong familiarity with the organizational environment. At least some cognitive biases, on the other hand, tend to be systematic and can be recognized. The following discussion will focus on cognitive rather than task-based and motivational biases. Particular emphasis will be placed on the tasks of assessing uncertain quantities (probabilities and consequences). This is an important group of tasks because ultimately many quantities that enter into expert judgment are treated as uncertain quantities.

TABLE 1
Biases in judgement

Task
Conceptual
Motivational
Cognitive

WHY DO BIASES OCCUR?

Biases result from the human's use of cognitive heuristics (Tversky & Kahneman 1974, 1981) that allow shortcuts through complex tasks. These heuristics should not be confused with the specific "if . . . then" decision rules that are extracted during the knowledge-elicitation process. Cognitive heuristics are metalevel modes of judgment which may occur outside the awareness of the individual but, nevertheless, influence reasoning and judgment. Described below are several major cognitive heuristics which have been observed in experimental and real-life judgments.

Anchoring

Individuals tend to search for a first approximation or natural starting point and then to adjust judgments of subsequent outcomes or probabilities using the initial estimate as a base. Available information which would support extreme adjustments is seldom utilized (Tversky & Kahneman, 1974).

Availability

Events, quantities, or relationships that are more easily brought to mind are more apt to be judged as likely. Unlike purely probabilistic reasoning, frequency is not the only determinant of how psychologically available an event may be. Familiarity, saliency and imaginability strongly enhance the retrievability of an event, even though the event itself may be rare. Overreliance on availability may lead to systematic disagreement with more objectively determined estimates of likelihood. Very recent experiences and observations or those that are extremely vivid or memorable may have inordinate weight in expert responses, preventing the expert from fully relying on the depth of long years of experience. Associated biases include the preference of concrete over abstract information, extreme sensitivity to present conditions, and illusory correlation among various components and factors in a problem (Tversky & Kahneman, 1973).

Representativeness

Likelihoods of events and risks are judged by their degree of similarity to familiar or stereotypic events rather than statistical frequency. The human tends to extrapolate occurrences from a small sample of events to what he/she perceives to be a similar, larger class of events. Even those experts with deep knowledge of the subject area tend to predict outcomes that are most representative of the evidence at hand with relative insensitivity to its reliability or to what is generally known about prior probabilities (Kahneman & Tversky, 1973). Representativeness can result in stereotyping of outcomes, e.g. imagining that a set of conditions will result in certain consequences because it exhibits "classic" characteristics. Bias results when objective information is forsaken for a more narrowly defined set of these characteristics. The more detailed a scenario is, the more likely it seems, even though the combination of characteristics in an actual scenario is less likely to actually occur.

Internal Coherence

Information or judgments consistent with an individual's previous beliefs are favored over those that are less consistent or suggest the need for additional testing. For example, the plausibility of a scenario or an estimate of variability becomes more heavily weighted in its perceived likelihood than its probability of occurrence. This effect is especially pernicious when the expert has expressed the same judgment in the past even without validation of his/her prediction. Experts may give short shrift to conflicting evidence or scenarios outside their range of previous experience and approval.

Consistency

Consistency of information sources can lead to increases in the confidence in judgment, but not to increased predictive accuracy or other performance. A small body of consistent data has much greater effect than a larger body of less consistent data, especially where the task (e.g. planning) is nonrepetitive and there is little feedback on the outcomes (Hogarth, 1980). People may actually collect further information that is biased toward initial preferences in order to increase the consistency.

WHERE IN THE JUDGMENT PROCESS DO BIASES OCCUR?

Cognitive heuristics do not inevitably lead to poor judgments. Indeed, these are natural processes with which the expert develops his/her clinical abilities. However, excessive reliance on certain heuristics can result in systematic biases (Hogarth & Makridakis, 1981, Kahneman *et al.*, 1982). Different cognitive tasks are prone to different types of biases (Table 2). These biases have been identified in experimental and operational contexts.

TABLE 2
Cognitive biases in expert judgment tasks

Hypothesis and solution generation	Decision rule articulation	Uncertainty assessment	Hypothesis evaluation
	Attribution errors	Overconfidence	Selection bias
Option generation deficiency	Illusory correlation	Hindsight	Information
	Order effects	Conservatism	
	Information bias		Order effect
Selection bias	"Law of small numbers"	Rare, conjunctive event bias	Hindsight
		Gambler's fallacy	Overestimation of completeness
		Regression bias	
		Base rate ignorance	

HYPOTHESIS AND SOLUTION GENERATION

Option Generation Deficiency
Omitting important hypotheses or elements of scenarios. Our fire experts are most vulnerable to scenarios that were totally unforseen during their planning process. Group simulation techniques have generated a richer variety of scenarios, but the problem is still acute where single experts are unaided by divergent perspectives on a problem.

Selection
Limiting selection to only options that the individual has experienced or expects to occur. Because of frequent personnel transfers, our experts have gained knowledge from experience in local areas as well as other geographical settings. Nevertheless, many base their judgments almost entirely on local experience and tend to have great difficulty translating their experience from other settings.

DECISION RULE ARTICULATION

Attribution errors
Attributing too much responsibility to one condition or factor; attributing success to skill and failure to chance, or vice versa. Human error and the lack of proper firefighting resources are the two factors most quickly cited by our experts as

influencing firefighting success. Without prompting, they often fail to consider the influence of a variety of other environmental conditions such as wind, air temperature, and the arrangement of vegetative fuels.

Illusory correlation
Belief that two or more variables co-vary when in fact they do not or that patterns exist in what is actually a random process. Changes in fire occurrence patterns are especially difficult to diagnose. There is a tendency for the fire-prevention person to see causal patterns that are not there, simply because fire incidence may increase at the same time certain human activities do.

Order effects
Assigning undue importance to the first items in a sequential presentation of conditions (primacy), or to the last items (recency). Because of the dynamics of wildland fire, firemen learn to treat the most recent information as the most valuable, whether or not it is more reliable. In the knowledge-elicitation context, they respond strongly to new pieces of information that may be, on further questioning, unimportant.

Information bias
Overweighting conditions described by information that is concrete, absolute, and consistent with initial inferences while underweighting conditions with abstract, relative, and conflicting evidence.

Law of small numbers
Mistakenly expecting small samples to be highly representative of a population, leading to undue confidence in early results, premature truncation of the search process, and little appreciation for variability in the sample of conditions given.

ASSESSMENT OF UNCERTAINTY AND CONFIDENCE

Overconfidence
Estimating much higher discrete probabilities (levels of certainty) or much tighter disributions (and intervals) than the actual. Assessed probabilites of fire escape and distributions of fireline production rates and escape levels are almost always higher and tighter than are shown in data. Typically, our human assessors are keying their judgments on similarity to recent or vividly remembered catastrophes.

Hindsight
Overestimating the probability of events just because they have occurred in the past. The mistaken belief that an event was inevitable leads to undervaluing new facts and research. Experts who have a long history in one narrowly circumscribed area often weave their entire assessment around the range in events they have personally experienced.

Conservatism
Failure to revise an opinion on receipt of new information. Once a judgment is made, some of our experts are extremely stubborn about even considering new

information. In their defense is the fact that survival has depended on decisiveness and avoidance of constant revision. It is difficult, however, for the knowledge engineer to determine where and when the initial judgment is made. Important information that is delivered late in the assessment task is often given short shrift or completely ignored.

Rare and conjunctive events
Overestimation of the probability of infrequent single events or the joint probability of independent events.

Gambler's fallacy
Overestimation of the probability of one event or value when an unexpected number of other events or values have occurred. Strings of escaping fires or "hot streaks" (successes) are not often predicted although they may be just as probable as strings of no escapes.

Regression bias
Using extreme values of a variable to predict opposite extreme values for the next observation (failing to allow for regression to the mean). This can be especially active in sequential judgments. Another manifestation is that high performance is often followed by lower performance even in the presence of reward, merely because of the random element, but individuals often mistakenly assume it is cause-and-effect.

Base rate effects
Relying exclusively on specific, descriptive information and neglecting prior probabilities (base rates) of the general class of events or values.

HYPOTHESIS TESTING

Note: The process of testing hypotheses is subject to many of the biases described above under other cognitive tasks. Most important are probably:

overestimation of completeness;
order effects;
selection;
hindsight;
information bias.

MONITORING AND CORRECTIVE PROCEDURES

Corrective procedures have been designed to combat biases in individual and group judgment (Table 3). Our research is aimed at bringing valid procedures into the realm of expert-systems knowledge elicitation. Mechanical procedures manipulate the task or adjust the judgments after elicitation. Behavioral procedures use interviewing and group interaction techniques to encourage full employment of the assessor(s)' knowledge. Most techniques have been developed to address particular biases, so the knowledge engineer should have some *a priori* notion of what types of biases may be more active at various stages in the judgment task that is being

TABLE 3
Corrective procedures for decision rule and uncertainty assessment

Mechanical		
Task	Assessment	Behavioral
Visual props Response mode	Scoring rules Consensus weighting*	Focusing Decomposition Training Logic challenges Consensus interaction*

* Applies to group assessment.

modeled. Not all procedures have been subjected to rigorous testing in knowledge-elicitation practice, so caution and exposure to a wide variety of techniques is urged. The literature on these procedures is extensive, although the references cited herein provide a good introduction.

Described below are some classes of corrective techniques which have been applied to the tasks of articulating decision rules or uncertainty levels. This listing is not an exhaustive comparison of techniques, but rather a taxonomy to guide the interested knowledge engineer. This list does not address techniques designed to improve hypothesis generation and selection.

MECHANICAL TECHNIQUES

Visual props
Letting the individual select, create, and compare visual patterns rather than verbal expressions to represent his judgment. An example is the probability wheel (Spetzler & Stael Von Holstein, 1975) which allows the respondent to work with the concept of probability without having to consciously wrestle with the numerical intricacies.

Response mode changes
Varying the format of the requested judgment, hypothesis, or uncertainty level can identify inconsistencies and find more natural means of expression. For example, rare event assessments are less biased when they are elicited as odds (1 in 1000) than when they are elicited in probabilities or decimals (0·001). People can more easily express larger probabilities and quantities in an indirect mode by comparing or ranking values. Indirect mode has been shown to be less biased than the direct expression of each assessment, say in physical units. Asking for the likelihood of a given value usually produces better calibrated answers than asking for the value for a given likelihood level.

Scoring rules
Using some function of the difference between actual and assessed values either to motivate individuals or to recalibrate assessments *post facto*.

Consensus weighting
Combining judgments from several individuals by creating weighted means of their responses on the same judgment task. Weighting bases may include differences in technical ability, self-ratings of confidence, prior calibration, or other factors.

BEHAVIORAL TECHNIQUES

Focusing
Structuring both the task and the interviewing environment so that specific biases are identified and corrected as they become symptomatic. One focusing technique is providing "reminder" information on base rates and requiring individuals to respond as to how they assimilated this information into their assessments.

Decomposition
Breaking down the quantity or relationship into subevents or important factors, each of which may be more probable and easier to assess than the original composite. These subcomponents can be weighed individually for their contribution to the overall strength of relationship or likelihood. Assessment of rare conjunctive uncertainties have been shown to be amenable to this approach.

Training
Individuals can sometimes remove bias if they understand and can manipulate the representation format (certainty factor, production rule, frames, etc., themselves). Training in probability assessment has been successful in improving calibration, especially when feedback on performance was frequent, immediate, and specific to the task.

Logic challenges
Exhorting individuals to justify reasons (and to offer opposing explanations or assessments) for their own answers. Biases are sometimes revealed when an individual must list as many reasons for as those against a (higher) best guess. Answers often change in response to this upfront checking.

Consensus interaction
Presenting, in a group setting, each individual with feedback from the other members about his/her judgments. When the level of controversy is controlled by the knowledge engineer, this can encourage reconsideration and better judgments. The degree of face-to-face interaction distinguishes the various techniques in this class. The two best examples are Delphi and nominal group techniques. There is no best technique, but a commonly recommended approach consists of nominal group application for initial estimates, structured feedback to the group, and silent voting on two or three estimates. This estimate-talk-estimate sequence works better than other sequences for most tasks.

Conclusions

Expert judgment is human judgment and as such can be improved. Effective knowledge elicitation should incorporate strategies for ensuring that the best judgments are passed on to the knowledge representation stage. Unless this is done, expert systems will perpetuate and perhaps exacerbate some of the limitations of human judgment which have been recognized in research on cognition. Recognizing potential flaws in judgment requires the knowledge engineer to be familiar with the domain and to develop explicit performance and selection criteria on the various dimensions of expertise. Some dimensions may warrant more monitoring and debiasing than others.

Research on judgment has tested concepts and techniques for correcting judgmental bias. Some of this work has used experts as subjects. Surprisingly little has been done to apply these approaches in knowledge-elicitation practice, much less to integrate them into a framework for interviewing or designing knowledge-elicitation tools. The cognitive heuristics that have been identified in judgment research also appear to be the very mechanisms through which expert performance is distinguished. Paradoxically, the same processes may be responsible for expert performance and commonsense bias. Little research has been done to determine the conditions under which this paradox occurs. Using a model of judgment to guide the knowledge-elicitation process and during the codification of knowledge structures could help to identify where biases may lurk.

During subsequent rounds of prototyping and elicitation, these biases could be sought out with questions and corrective protocols. The utility of expert systems could be enhanced through the development of enhanced or "leveraged" intuition. Systems developed on experts for experts could incorporate interactive procedures for checking and correcting biases and inconsistencies in judgmental inputs. Continued use of such systems in advisory roles or as training devices could actually improve human expert judgment in some domains.

We are testing various approaches for identifying and correcting biases in our knowledge-elicitation efforts. Our goal is to develop a set of general guidelines for knowledge engineers to use in manual or automated elicitation. Preliminary observations of our own knowledge-elicitation experiences have pointed out the need to separate the knowledge-elicitation and representation phases and to delay prototyping until the possible biases and knowledge structures can be explored fully. Premature prototyping captures biases that may be difficult for the expert or some truth-maintenance systems to detect.

Open-ended interviewing should be mixed with more highly structured styles as the process converges on specific judgments. For example, overuse of anchoring can be combated by asking the expert for extreme judgments before most of his/her likely ones. Overreliance on representativeness appears to be partially corrected by providing the expert with base-rate information and actively seeking verbal consideration of it. Many scenarios should be presented, including those with inconsistent information. Care should be taken not to let classic or well-publicized incidents or scenarios become prominent. Confidence factor assessments should not be accepted without close scrutiny. They can act as guides for more indepth questioning, but high confidence levels can hide as many problems as low ones turn up, depending on the expert's ability to introspect.

References

Fischoff, B. (1983). Judgmental aspects of risk analysis. Eugene, Oregon: Decision Research Inc.

FISCHOFF, B. (1984). Eliciting information from experts. In *Research Needs for Human Factors* Washington, D.C: National Academy Press, pp. 33–48.
HOGARTH, R. M. (1980). *Judgment and Choice: Strategies for Decision*. New York: Wiley
HOGARTH, R. M. & MAKRIDAKIS, S. (1981)). Forecasting and planning—an evaluation. *Management Science* **27,** 115–138
KAHNEMAN & TVERSKY, A. (1973). On the psychology of prediction. *Psychological Review*, **80,** 751–757
KAHNEMAN, D., SLOVIC, P. & TVERSKY, A. (1982). *Judgment Under Uncertainty: Heuristics and Biases* Cambridge: Cambridge University Press
SPETZLER, C. S. & STAEL VON HOLSTEIN, C. S. (1975). Probability encoding in decision analysis. *Management Science* **22,** 340–358
TVERSKY, A. & KAHNEMAN, D. (1973). Availability: a heuristic for judging frequency and probability. *Cognitive Psychology*, **5,** 207–232
TVERSKY, A. & KAHNEMAN, D. (1974). Judgments under uncertainty: heuristics and biases. *Science* **185,** 1124–1131
TVERSKY, A. & KAHNEMAN, D. (1981). The framing of decisions and the psychology of choice. *Science*, **211,** 453–458
WINKLER, R. L. & MURPHY, A. H. (1968). "Good" probability assessors. *Journal of Applied Meteorology*, **7,** 751–757

Formal thought and narrative thought in knowledge acquisition

JIM KORNELL
General Research Corporation, PO Box 6770, Santa Barbara, CA 93111-6770, U.S.A.

There are two different kinds of thought of interest to knowledge engineers. One is *formal thought,* as exemplified by logic and mathematics; the evaluation criterion for formal thought is *truth.* The second is *narrative thought,* as exemplified by metaphors, analogies, and *gestalts*; the evaluation criterion for narrative thought is *verisimilitude.* To build a knowledge system which operates in the realm of narrative thought, one must build a model of expert knowledge of the domain, and such a model must contain not only the facts and heuristics used by the expert, but also the *patterns of reasoning* and most importantly the *kinds of reasoning* used by the expert. Three representative knowledge acquisition tools—RuleMaster, MOLE, and Aquinas—are briefly reviewed. It is suggested that the tools discussed, and the members of the classes each represents, fail to effectively support identifying and acquiring the patterns of reasoning and especially the kinds of reasoning used by experts in narrative domains.

1. Introduction

Two modes of thinking can be differentiated as the *formal* and the *narrative.* Formal thinking has to do with constructing publicly scrutable chains of reasoning, using explicit premises and methods of combination. The power of formal thinking is in allowing us to be as confident in the outcome of a chain of reasoning as we are in the premises. Narrative thinking has to do with implicit assumptions and plausible (or habitual) methods of combination. The power of narrative thinking is in allowing us to act reasonably in the real world. The purpose of knowledge-based systems is to model narrative thinking, but the distinction between the formal and the narrative modes, and the implications of the distinction, are not well reflected in our discourse or in our tools.

The next section of this essay delineates the differences between the formal and the narrative modes of thinking. The following section discusses some of the implications for knowledge acquisition (and knowledge representation): *patterns of reasoning* and *kinds of reasoning* are characterized, and it is asserted that our failure to sufficiently take them into account impairs our ability to model expert problem solving. Section 4 discusses some representative knowledge-acquisition tools to provide a rough assessment of where we now stand on automated support for dealing with the problems which are the subject of this essay. Finally, a brief conclusion summarizes the arguments of the preceding sections.

Articulating the distinction between modes of thought helps us better understand some of the difficulties of knowledge acquisition, and can provide direction in the

KNOWLEDGE-BASED SYSTEMS Vol. 1
ISBN 0-12-273251-0

development of our knowledge-acquisition tools. Gregory Bateson called evolution the process of responding to the differences that make a difference; the distinction between the formal and the narrative modes of thought is such a difference.

2. Two modes of thought

Knowledge-based systems are appropriate for problems which are algorithmically solvable, but for which heuristic solutions are better because the algorithms are excessively cumbersome, difficult, or computationally expensive. DENDRAL (Lindsay, Buchanan, Feigenbaum & Lederberg, 1980) and R1 (McDermott, 1982) are examples of systems built to deal with this kind of problem. Also, knowledge systems can be used for problems for which algorithms are not known, and solutions require some creativity. AM (Lenat, 1982) and BACON (Langley, Zytkow, Simon & Bradshaw, 1986) are examples of systems built for this kind of problem. The difficulty of these problems results from *epistemic boundedness* (Fodor, 1983) and *sphexishness* (Hofstadter, 1985) respectively.† This essay will focus on issues of knowledge acquisition for domains needing creative (i.e. non-algorithmic) problem solving. In such domains the structuring of the patterns of reasoning and the kinds of reasoning is at the heart of the decision process; since the topic of this essay is the acquisition of patterns and kinds of reasoning, it is appropriate to focus on this class of problems.

The cognitive processes involved in non-algorithmic problem solving in art and in science are basically the same (Koestler, 1964; Guilford, 1959). They rely on *narrative* rather than *formal* thought processes. The narrative and the formal are different modes of thought.‡ In formal thought, axioms and legal transformations allow derivation of (systemic) truth. Structure comes from the composition of basic elements which are atomic. Formal thought deals with closed worlds, in which overall models emerge from elements. The predicate calculus is an exemplar. In narrative thought, the goal is less clearly defined: verisimilitude, plausibility, useful hypotheses. The elements of such models do not have to be self-sufficient: the whole taken together can be lifelike even if none of the elements is particularly likely. The overall model determines the elements. A narrative *requires* a gestalt, which the formal does not (Wertheimer, 1959).

The distinction between the formal and the narrative is *not* the same as the distinction between operational and descriptive modes of thinking (and behavior) (Johnson & Gruber, 1986). Operational models represent what experts *do* in decision processes, and descriptive models represent how experts *describe* decision processes: operational models are effective procedures and descriptive models are scrutable procedures. The distinction between narrative and formal modes of thought is a distinction between different *ways of thinking*. Either the operational or the descriptive can be based on the narrative mode of thought. [Given the difficulty

† Epistemic boundedness is a fancy way of saying that there are thoughts which are just too big for our brains. Sphexishness (a word coined by Douglas Hofstadter in honor of the apparently rational but severely learning-impaired sphex wasp) has to do with the limits of our ability to creatively reassess and modify behavior patterns which may inhibit, or be superfluous to, adaption.

‡ The inspiration for this essay came from a book by Jerome Bruner, *Actual Minds, Possible Worlds* (Bruner, 1986), in which this difference is very strongly argued. I disagree with Bruner's formulation, but credit him with prompting this essay.

of learning and consistently employing formal methods of thought, one could speculate that operational models are likely to be very heavily biased away from the formal—the *mental models* work of Philip Johnson-Laird certainly suggests this (Johnson-Laird, 1983)].

An example from Robert Cummins (Cummins, 1983) makes the difference between narrative and formal thought more vivid. In context, Cummins was making a point about explanatory force as distinct from (formal) truth, but the example serves just as well to illustrate the difference between narrative and formal truths.

> Conrast · · · the theory that molecules are held together by "hooks-and-eyes" with the theory that molecules are held together by some unknown force. By the mid-nineteenth century the former theory was strongly disconfirmed, while the latter theory turned out to be true (because molecules *are* held together by a force unknown at the time). Still, the "hooks-and-eyes" theory is at least capable of explaining such facts as the occurrence of H_2O but not H_4O, and the occurrence of H_2—e.g., on the hypothesis that each hydrogen atom has one free hook, and each oxygen atom has two free eyes. The "unknown force" theory, however, has no explanatory force whatever, being equivalent to the theory that the molecules that do occur do occur, and the ones that don't don't. So explanatory force is quite independent of confirmation and truth.

The important point here is that the totality of a (narrative) model can have (narrative) truth, *even if basic elements of the model are known to be untrue.* Clearly, this is not so for formal models.

At the most basic level, the concerns of the two modes of thought are different. Formal thought is about how to know truth. Narrative thought is about how to construct meaning. Formal thought seeks closed, well-defined systems; narrative thought, open, dynamic systems. The operations of formal thought are "syntactic": conjunction and disjunction, deduction and induction, strict implication, instantiation and idealization, all operating on the formal aspects of the subject. The operations of narrative thought are "semantic": representativeness, plausibility, cultural appropriateness, all depending on the context and the specific content of the subject. All operate on the meaning of the subject.

Consider Thomas Kuhn's assertion that science moves through paradigm shifts, and that the scope of the shifts distinguishes evolutionary and revolutionary science (Kuhn, 1970). The task of evolutionary science is to instantiate the current explanatory framework for the subject class of phenomena—to fill in the details of the paradigm. Revolutionary science is seen when the degree of adaption of a paradigm to new information or new metaphors is such that the field is perceived as transformed—a new vision is present, in which the old concepts will be seen in new light, and in which new concepts will assert their importance. But the difference in type between evolutionary and revolutionary science highlights their identical ways of employing narrative and formal modes of thought. In both cases it is the narrative—either the current or the revolutionary paradigm—which sets the stage for and provides the context of interpretation of the data generated through use of the formal. Evolutionary science is concerned with validating or disproving a paradigm which emerges from, and whose meaning is endowed by, narrative thinking. Revolutionary science comes through the fundamental transformation of a narrative. Neither evolutionary nor revolutionary science calls for a fundamental change in formal methods. The narrative and the formal modes of thought are of different types, and the way we use both modes show our implicit understanding of the difference.

It seems to me that the formal is a tool of the narrative. The formal is too narrow an approach for day-to-day problems, and in most cases for creative and imaginative thinking. However, when the conditions can be carefully set for its employment, it offers a kind of power that the narrative cannot. Very long chains of reasoning are usually suspect, but when formal, truth-preserving methods are employed one can travel very far indeed with confidence that one's conclusions are no less valid than one's premises. Even in areas where formal methods are at the heart of the enterprise, it is narrative thought that sets the agenda and the goals, and that assesses the meaning of the results. And, in the areas where knowledge systems are most desirable—characterized both by epistemic boundedness and by sphexishness—expert problem-solving relies heavily on narrative thought.

3. What is in it for knowledge acquisition

To build knowledge systems in domains where reasoning relies predominantly on narrative modes of thought, a model of the expert's problem-solving behavior in the domain must be built. The subject of knowledge acquisition is the elicitation and articulation of such models.

Models include facts, heuristics, relation-structures, and transformations.† Transformations for formal systems include legal changes of state, substitutions, and derivations. Transformations for narrative systems include legal changes of state; in addition, *root metaphors* (e.g. "cognition is clockwork", "cognition is computation") and some ways of distinguishing productive from unproductive analogies are likely to be included. In narrative thought, transformations are instantiated through *patterns of reasoning* and *kinds of reasoning*. A pattern of reasoning corresponds to meta-level rules, strategies of inference, scripts and plans, etc. Kinds of reasoning are frame-of-reference specific ways of ordering and relating information.

The distinction between a pattern of reasoning and a kind of reasoning is important. A particular domain might require that all time-critical possibilities be checked before more likely but less urgent possibilities. This is a pattern of reasoning; it provides a strategy and rationale for arranging and processing rules, frame structures, or procedures. At a higher level, *group-and-differentiate* or *propose-and-revise* characterize classes of patterns of reasoning. A pattern embodies assumptions about contexts and goals, and provides an organization or a sequence of steps to follow in seeking a goal. Scripts (Shank & Abelson, 1977) and schemata (Rumelhart, 1980) were developed expressly to represent patterns of reasoning.

Another domain might be rich in possibilities but sparse in examples and without detailed theory; such a domain might require reasoning by analogy. This is a kind of reasoning; it suggests a strategy and rationale for choosing a class of inference operators and a set of well-formedness criteria. A number of authors (Lakoff & Johnson, 1980; Kahneman, Slovic & Tversky, 1982; Clement, 1986; Cleaves, 1986) have discussed different kinds of reasoning. A kind of reasoning emerges from

† This is a characterization, not a definition. For a definition, see (Craik, 1943) or (Williams, Hollan & Stevens 1983); obviously, my discussion is also influenced by Johnson-Laird and by Wertheimer.

assumptions about the nature of the content and evaluation criteria of a particular "chunk" of reasoning, as these interact with an expert's "mental shortcuts" (or, less kindly, biases)† and provides the means for processing a particular chunk. Patterns have to do with decomposing and arranging chunks of reasoning; kinds with handling each chunk. I hope the distinction between patterns and kinds of reasoning is clear; each will be discussed in the examination of current knowledge-acquisition tools.

Typical knowledge-acquisition practice seems to focus on facts and heuristics. When a knowledge engineer happens to be using a frame-based tool, structure-relations are incorporated. Particularly careful knowledge engineers will consciously elicit patterns of reasoning. In knowledge acquisition, kinds of reasoning are ignored; for the most part, the kind of reasoning employed results solely from the choice of formalism or tool for knowledge representation.

This raises problems of two sorts. First, this hole in our methodology prevents us from doing some interesting and otherwise accessible problems. Second, systems are built in which the content of the rules and the uncertainty management operations are forced to simulate kinds of reasoning which would be much better if represented explicitly.

The kinds of problems which are interesting and which seem accessible from the current state of our technology include:

domains typified by frequent use of analogy;
domains where problems are solved through recourse to one or more root metaphors;
domains typified by correlational (rather than causal) associations;
problems of recognizing and distinguishing gestalts;
problems of reconciling conflicting gestalts (e.g., "least damage" problems involving whole systems).

As for forcing the content of the rules to compensate for the formal inference procedures, often in narrative thought people reason with *prototypes* (Smith & Medin, 1981)—reasoning by judging to what extent perceived circumstances match an ideal (or a class member held to be typical‡). To simulate prototype reasoning, formally matching systems have to use the content of their rules (or frames or nets) to compensate for rigid inference procedures. Shaping a rule's content to fit a pre-selected inference scheme distorts the domain expert's model to just the extent that the a priori inference scheme differs from the reasoning processes actually used by the expert. Using UNLESS clauses and asking for "strength of association" rather than a numeric certainty value indicate an acknowledgement that human experts find many of our representation requirements foreign to their ways of thinking. That should be signal enough that we should reconsider how we are framing their reasoning.

† Examples of "shortcuts" include anchoring, the availability bias, the representativeness heuristic, overvaluing evidence based on internal coherence, ignoring evidence to preserve consistency, order effects, attribution errors, illusory correlations, and all sorts of other habits domain experts (and we) evidence in our reasoning.

‡ Most prototype theorists believe that reasoning with prototypes does *not* require matching on necessary and sufficient conditions. See also Kahneman *et al.* (1982) for a discussion of the representativeness heuristic.

To summarize: an early belief in knowledge-based systems work was that large amounts of knowledge could be coupled with simple inference schemes to model expert problem solving. While our terminology has become more refined (e.g. "structured selection using heuristic classification"), we are still using simple inference procedures. We have become much more sophisticated in our understanding of the complexities of applying knowledge effectively, but with very few exceptions (Michalaski, 1986; Bylander & Chandrasekaran, 1986) there seems to be little acknowledgement that people *necessarily* reason with varied and often delicate combinations of knowledge and inference. In this and the previous section I have been arguing that application of knowledge relies on coordination of facts and heuristics with patterns of reasoning and kinds of reasoning. Patterns of reasoning are discussed in our literature, but little automated assistance exists for identifying and eliciting them. Kinds of reasoning are rarely discussed in our literature, and no automated assistance exists for identifying and eliciting them.

4. How do current knowledge acquisition tools stack up?

How do current knowledge acquisition tools do in regard to the issues associated with acquiring and representing patterns and kinds of reasoning? We will examine three systems: RuleMaster, MOLE, and Aquinas. These were chosen as representative of three different attitudes toward knowledge acquisition, and the discussion of each will include reference to other tools of similar philosophy. Each will be assessed in three ways:

(1) support for identifying, acquiring, and representing patterns of reasoning;
(2) support for identifying, acquiring, and representing kinds of reasoning; and,
(3) support for integrating kinds of reasoning with patterns of reasoning.

RuleMaster (Michie, Muggleton, Reise & Zubrick, 1984) is a knowledge system application generator. RuleMaster appears to be better suited to classification than to construction problems, although it is claimed to be useful for both. Its focus is on flexibility, on ease of incorporation in a variety of environments, and on its capability to be employed on both diagnostic and procedural (e.g. planning) problems. Knowledge is represented as rules, and as procedures in the Radial language. RuleMaster operates using deductive inference, embedded in a control language (Radial) in which the outputs of rule sets direct the processes of reasoning. RuleMaster also supports inductive generation of general rules from specific examples as part of its knowledge acquisition facility. Tools similar to RuleMaster include TIMM (Kornell, 1984) (inductive generalization and use of the domain expert's vocabulary), S.1 (Teknowledge, 1986) (an explicit control language), and ART (Williams, 1984) and EMYCIN (Van Melle, 1979) (a multi-level production-rule formalism).

In common with each of the named similar tools, RuleMaster does not provide facilities for choosing or mixing reasoning processes. It provides a language for representing reasoning patterns, but provides no assistance to the knowledge engineer in identifying and acquiring those patterns. RuleMaster makes no distinctions regarding kinds of reasoning, and there is no facility for composing kinds with patterns of reasoning.

MOLE (Eshelman & McDermott, 1986) is a knowledge acquisition and application shell that supports (predominantly diagnostic) heuristic classification problem solving. Its focus is on minimizing the demands made on the human expert by providing both static and dynamic analysis of the knowledge base to allow "filling-out" of underspecified domain characterizations and to enhance efficiency in interactive refinement of knowledge bases. It does this by embodying an explicit domain model. From the perspective of this essay, it is representative of a class of knowledge acquisition tools which narrow their focus to very particular kinds of problems, with the intent of using specificity as a lever to gain power. Besides MOLE, one could include MOLE's predecessor MORE (Kahn, Nowlan & McDermott, 1985), SALT (Marcus, 1986), KNACK (Klinker, Bentolila, Genetet, Grimes & McDermott 1986), TKAW (Kahn, Breaux, Joseph & DeKlerk, 1986), and OPAL (Musen, Fagan, Combs & Shortliffe, 1986).

MOLE represents knowledge as a network of marked nodes and links, instantiated as production rules. It bases its inference procedure on explicit heuristic assumptions about the world. It makes an epistemological distinction between *covering* and *circumstantial* knowledge. Covering knowledge is that which "needs to be explained or covered by some hypothesis" (i.e. in a causal relation), circumstantial knowledge is that which is "merely correlated with a hypothesis". During knowledge acquisition, when associations between evidence and hypotheses are indicated, MOLE expressly attempts to understand the type of the association; it is able to use this information in its reasoning and, during dynamic analysis, during interviews with the domain expert. So, while MOLE lacks facilities for choosing or mixing reasoning strategies, it has the advantage of making its epistemological biases explicit. MOLE does not support identifying and acquiring reasoning patterns; however, this is because of the definite stand MOLE takes on how reasoning is to proceed. MOLE's authors might claim to have attempted to explicitly characterize the *generic* reasoning patterns in diagnostic, heuristic classification domains (or at least that subset where the covering and circumstantial distinction seems to hold). MOLE does not provide a facility for identifying and acquiring the kind(s) of reasoning within a domain, nor, clearly, is the integration of kinds and patterns of reasoning an issue.

Aquinas (Boose & Bradshaw, 1986) is the successor to ETS (Boose, 1986). ETS was developed specifically to be a knowledge-acquisition tool, and Aquinas is an extension of ETS meant both to compensate for some perceived weaknesses in ETS and to integrate capabilities seen as useful by the knowledge systems research community. The goal of Aquinas is to provide a powerful testbed for rapidly prototyping portions of many kinds of complex knowledge-based systems. Aquinas can be seen as representative of both *repertory grid* (Kelly, 1955) and general-purpose knowledge-acquisition tools. PLANET (Shaw & Gaines, 1986) and the work of Garg-Janardan and Salvendy (Garg-Janardan & Salvendy, 1986) are examples of the use of repertory grids; INFORM (Moore & Agogino, 1986), BLIP (Morik, 1986), KRITON (Diederich, Ruhmann & May, 1986), HERACLES (Clancey, 1986*a*) and ODYSSEUS (Wilkins, 1986), and NEXPERT (Rappaport, 1986) are examples of general-purpose tools.

Aquinas is at the opposite end of the spectrum from MOLE, in trying to be a truly general knowledge-acquisition tool for structured selection domains. A number of ways of eliciting and classifying associations are employed, and the entailments of an expert's expressed beliefs are explored. A strong effort is made to illuminate inconsistency and to elicit the expert's internal, operational model (as contrasted with the expert's conscious, articulate model). Boundary conditions are explored, attempts are made to unify similar traits, and consistency and completeness are sought.

Aquinas basically follows Clancey's heuristic classification model in its reasoning (Clancey, 1986*b*). Briefly, one abstracts specific circumstances into trait classes, which are mapped to abstract solution classes, which are then refined into specific solutions. The strength of Aquinas is in the variety of ways that experts are allowed (or cajoled) into viewing their problem-solving knowledge, which holds out the hope that an effective way of expressing (and representing) the problem-solving knowledge can be found.

Aquinas does not provide facilities for choosing particular reasoning processes. Patterns of reasoning can be represented through the arrangement of the knowledge networks (structures made up of individual rule sets), and a major advance of Aquinas over ETS is the explicit support for eliciting these networks. (The contrast though between these networks and the procedural language used by RuleMaster is instructive. In Radial, one writes procedures for applying knowledge, and the patterns are explicit as procedures. In Aquinas, the networks only allow certain patterns of reasoning to be actuated; reasoning is implicitly "channeled". Among application tools, S.1 is like RuleMaster in supplying a procedural language, and KEE is like Aquinas in supporting networks). There is no support in Aquinas for identifying and acquiring kinds of reasoning, nor for melding kinds with patterns of reasoning.

5. Conclusion

I have argued that there are two different kinds of thought processes of interest to knowledge engineers. Formal thought is characterized by closed worlds, carefully defined laws of operation, and by a concern for consistency and truth. Narrative thought is characterized by open worlds, interacting gestalts, and by evaluation by verisimilitude. I have further argued that there are different *kinds* of reasoning associated with each, and that the capabilities to identify and acquire the *patterns of reasoning* and particularly the *kinds of reasoning* used within a domain have been neglected by the knowledge-acquisition community.

Three representative knowledge-acquisition tools were assessed. While all can in some way represent patterns of reasoning, none provided automated support for identifying and acquiring patterns of reasoning. Neither the systems discussed nor any of the referenced systems even acknowledges the existence of different kinds of reasoning, and so, none provides any help to the knowledge engineer in identifying and acquiring such knowledge.

Dennis Cooper, Larry Eshelman, Elizabeth Liles, and Dana Nance each provided useful discussion or criticism (or both) of the ideas presented in this essay, for which I am grateful. However, they should (at least in the present case) be held blameless.

References

Boose, J. (1986). ETS: a PCP-based program for building knowledge-based systems. *WESTEX-86 Proceedings, IEEE Computer Society Press,* Washington, D.C., 1986.

BOOSE, J. & BRADSHAW, J. (1986). Expertise transfer and complex problems using AQUINAS as a knowledge acquisition workbench for expert systems. *AAAI Knowledge Acquisition for Knowledge-based Systems Workshop,* 1986.

BRUNER, J. (1986). *Actual Minds, Possible Worlds.* Cambridge MA: Harvard.

BYLANDER, T. & CHANDRASEKARAN, B. (1986). Generic tasks in knowledge-based reasoning: the 'right' level of abstraction for knowledge acquisition. *AAAI Knowledge Acquisition for Knowledge-based Systems Workshop.*

CLEAVES, D. (1986). Cognitive biases and corrective techniques: proposals for improving elicitation procedures for knowledge-based systems. *AAAI Knowledge Acquisition for Knowledge-based Systems Workshop.*

CLANCEY, W. (1986*a*). From GUIDON to NEOMYCIN and HERACLES in twenty short lessons. *AI Magazine,* **7,** (3), Conference 1986.

CLANCEY, W. (1986*b*). Heuristic classification. In KOWALIK, J., Ed., *Knowledge-based Problem Solving.* New York: Prentice–Hall.

CLEMENT, J. (1986). Methods for evaluating the validity of hypothesized analogies. *Proceedings of the 1986 Cognitive Science Society Conference,* Hillsdale New Jersey: Lawrence Erlbaum Associates.

CRAIK, K. (1943). *The Nature of Explanation.* Cambridge, U.K. Cambridge University Press.

CUMMINS, R. (1983). *The Nature of Psychological Explanation,* Cambridge, Massachusetts: MIT.

DIEDERICH, J., RUHMANN, I. & MAY, M. (1986). KRITON: a knowledge acquisition tool for expert systems. *AAAI Knowledge Acquisition for Knowledge-based Systems Workshop* 1986.

ESHELMAN, L. & MCDERMOTT, J. (1986). MOLE: a knowledge acquisition tool that uses its head. *AAAI-86 Proceedings.* Los Altos, California: Morgan Kaufmann.

FODOR, J., (1983). *The Modularity of Mind.* Cambridge, Massachusetts: MIT.

GARG-JANARDAN, C. & SALVENDY, G., (1986). A conceptual framework for knowledge elicitation. *AAAI Knowledge Acquisition for Knowledge-based Systems Workshop,* 1986.

GUILFORD, J. P. (1959). Traits of creativity. In *Creativity and Its Cultivation.* New York: Harper & Row.

HOFSTADTER, D. (1985). *Metamagical Themas.* Basic Books, New York, 1985.

JOHNSON, P. & GRUBER, S. (1986). Specification of expertise: knowledge acquisition for expert systems. *AAAI Knowledge Acquisition for Knowledge-based Systems Workshop,* 1986.

JOHNSON-LAIRD, P. (1983). *Mental Models,* Cambridge, Massachusetts: Harvard.

KAHN, G., BREAUX, E., JOSEPH, R. & DEKLERK, P. (1986). An intelligent mixed-initiative workbench for knowledge acquisition. *AAAI Knowledge Acquisition for Knowledge-based Systems Workshop,* 1986.

KAHN, G., NOWLAN, S. & MCDERMOTT, J. (1985). MORE: an intelligent knowledge acquisition tool. *Proceedings of the International Joint Conference on Artificial Intelligence,* 1985.

KAHNEMAN, D., SLOVIC, P. TVERSKY, A. (1982). *Judgement Under Uncertainty: Heuristics and Biases.* Cambridge U.K: Cambridge University Press.

KELLY, G. (1955). *The Psychology of Personal Constructs.* New York: Norton.

KLINKER, G., BENTOLILA, J., GENETET, S., GRIMES, M. & MCDERMOTT, J. (1986). KNACK: report-driven knowledge acquisition, *AAAI Knowledge Acquisition for Knowledge-based Systems Workshop.*

KOESTLER, A. (1964). *The Act of Creation.* New York: Macmillan.

KORNELL, J. (1984). A VAX tuning expert built using automated knowledge acquisition. *IEEE Conference on Artificial Intelligence Applications.*

KUHN, T. (1970). *The Structure of Scientific Revolutions.* Chicago, Illinois: University of Chicago.

LAKOFF, G. & JOHNSON, M. (1980). *Metaphors We Live By.* Chicago Illinois: University of Chicago.

LANGLEY, P., ZYTKOW, J., SIMON, H. & BRADSHAW, G. (1986). The search for regularity: four aspects of scientific discovery. In [MICHALSKI, R., CARBONELL, J. & MITCHELL, T. Eds, *Machine Learning, Vol. II.* Los Altos, California: Morgan Kaufmann.

LENAT, D. (1982). AM: discovery in mathematics as heuristic search. In DAVIS & LENAT Eds, *Knowledge-based Systems in Artificial Intelligence,* New York: McGraw–Hill.
LENAT, D., PRAKASH, M. & SHEPHERD, M. (1986). CYC: using common sense knowledge to overcome brittleness and knowledge acquisition bottlenecks. *AI Magazine,* Vol. 6 (4), Winter 1985.
LINDSAY, R., BUCHANAN, B., FEIGENBAUM, E. & LEDERBERG, J. (1980). *Applications of Artificial Intelligence for Organic Chemistry: The DENDRAL Project,* New York: McGraw-Hill.
MARCUS, S. (1986). Taking backtracking with a grain of SALT. *AAAI Knowledge Acquisition for Knowledge-based Systems Workshop.*
MCDERMOTT, J. (1982). R1: a rule-based configurer of computer systems. *Artificial Intelligence,* Vol. 19 (1).
MICHALSKI, R. (1986). Two-tiered concept meaning, inferential matching, and conceptual cohesiveness. *Allerton Conference on Analogy and Similarity.*
MICHIE, D., MUGGLETON, S., RIESE, C. & ZUBRICK, S. (1984). RuleMaster: a second-generation knowledge engineering facility. *IEEE Conference on Artificial Intelligence Applications.*
MOORE, E. & AGOGINO, A. (1986). INFORM: an architecture for expert directed knowledge acquisition. *AAAI Knowledge Acquisition for Knowledge-based Systems Workshop.*
MORIK, K. (1986). Acquiring domain models. *AAAI Knowledge Acquisition for Knowledge-based Systems Workshop.*
MUSEN, M., FAGAN, L., COMBS, D. & SHORTLIFFE, E. (1986). Using a domain model to drive an interactive knowledge editing tool. *AAAI Knowledge Acquisition for Knowledge-based Systems Workshop.*
RAPPAPORT, A. (1986). Multiple problem spaces in the knowledge design process. *AAAI Knowledge Acquisition for Knowledge-based Systems Workshop.*
RUMELHART, D. (1980). *Schemata: the building blocks of cognition.* In SPRIO, BRUCE & BREWER Eds, *Theoretical Issues in Reading Comprehension.* Hillsdale, New Jersey: Lawrence Erlbaum Associates.
SHANK, R. & ABELSON, R. (1977). *Scripts Plans Goals and Understanding.* Hillsdale, New Jersey: Lawrence Erlbaum Associates.
SHAW, M. & GAINES, B. (1986). PLANET: a computer-based system for personal learning, analysis, negotiation, and elicitation techniques. In *Cognition and Personal Structure: Computer Access and Analysis.* Praeger Press. In press.
SMITH, E. & MEDIN, D. (1981). *Categories and Concepts.* Cambridge, Massachusetts: Harvard.
Teknowledge Technical Presentations. Palo Alto, California: Teknowledge.
VAN MELLE, W. (1979). A domain-independent production rule system for consultation programs. *Proceedings of the International Joint Conference on Artificial Intelligence.*
WERTHEIMER, M. (1959). *Productive Thinking,* New York: Harper Torchbooks.
WILKINS, D. (1986). Knowledge base debugging using apprenticeship learning techniques. *AAAI Knowledge Acquisition for Knowledge-based Systems Workshop.*
WILLIAMS, C. (1984). ART conceptual overview. Los Angeles, California: Inference Corporation.
WILLIAMS, M., HOLLAN, J. & STEVENS, A. (1983). Human reasoning about a simple physical system. In GENTNER, D. & STEVENS, A. Eds, *Mental Models.* Hillsdale, New Jersey: Lawrence Erlbaum Associates.

Mapping cognitive demands in complex problem-solving worlds

David D. Woods
Westinghouse Research & Development Center, Pittsburgh, PA 15235, U.S.A.

and

Erik Hollnagel
Computer Resources International, Copenhagen, Denmark

Building a cognitive description of a complex world

Tool builders have focused, not improperly, on tool building—how to build better performing machine problem-solvers, where the implicit model is a human expert solving a problem in isolation. A critical task then for the designer working in this paradigm is to collect human knowledge for computerization in the stand alone machine problem-solver.

But tool use involves more. Building systems that are "good" problem-solvers in isolation does not guarantee high performance in actual work contexts where the performance of the joint person–machine system is the relevant criterion. The key to the effective application of computational technology is to conceive, model, design, and evaluate the joint human–machine cognitive system (Hollnagel & Woods, 1983). Like Gestalt principles in perception, a decision system is not merely the sum of its parts, human and machine. The configuration or organization of the human and machine components is a critical determinant of the performance of the system as a whole (e.g. Sorkin & Woods, 1985). The joint cognitive system paradigm (Woods, 1986; Woods, Roth & Bennett, in press) demands a problem-driven, rather than technology-driven, approach where the requirements and bottlenecks in cognitive task performance drive the development of tools to support the human problem-solver.

In this paper we describe an approach to understand the cognitive activities performed by joint human–machine cognitive systems. The real impediment to effective knowledge acquisition is the lack of an adequate language to describe cognitive activities in particular domains—what are the cognitive implications of some application's task demands and of the aids and interfaces available to the people in the system; how do people behave/perform in the cognitive situations defined by these demands and tools. Because this independent cognitive description has been missing, an uneasy mixture of other types of description of a complex situation has been substituted—descriptions in terms of the application itself, of the implementation technology of the interfaces/aids, of the user's physical activities or user psychometrics.

We describe one approach to provide an independent cognitive description of complex situations that can be used to understand the sources of both good and poor performance, i.e. the cognitive problems to be solved or challenges to be met.

KNOWLEDGE-BASED SYSTEMS Vol. 1
ISBN 0-12-273251-0

The results from this analysis help to define the kind of solutions (tools) that are needed for successful performance, and the results can be deployed in many possible ways—exploration training worlds, new information, representation aids, advisory systems, or machine problem-solvers. The method can address existing joint cognitive systems in order to identify deficiencies that cognitive system redesign can correct and prospective joint cognitive systems as a design tool during the allocation of cognitive tasks and the development of an effective joint architecture (e.g. Pew *et al.*, 1986).

To build a cognitive description of a complex world the first hurdle is to escape from the language of the application and to characterize the kinds of cognitive demands that arise in the course of performing domain tasks (Newell, 1982; Clancey, 1985; Gruber & Cohen, 1986). To carry out this translation we start from the constraints imposed by the design and proper function of the system in question (Lind, 1981) via a kind of goal-directed analysis to map a functional representation of the domain (e.g. Davis, 1983; Rasmussen, 1983, 1986). The resulting functional representation serves as a framework and guide to identify—what kinds of problem-solving situations can arise in this world? what must people know and how they must use that knowledge to solve these problems (what are the difficult cognitive demands)? how do people actually respond in these types of problem-solving situations (where do breakdowns such as fixation effects occur)?

Thus, the approach combines an analysis of the domain to determine psychological, particularly cognitive, demands and a psychological analysis of human, especially problem-solving, performance given those demands. One can think of problem solving situations as composed of three basic elements; the world to be acted on, the agent who acts on the world, and the representation of the world utilized by the problem-solving agent. To build a cognitive description of a world we need to capture the relationships between the cognitive demands of domain tasks, how the available representation of the world effects problem-solver's information processing activities and problem-solving performance, and the characteristics of the problem solver (Rasmussen, 1985; Woods, in press; Woods, in prep.). There is an intimate relationship between these three elements so that cognitive description is at once a psychological description and a domain description. In this paper we focus on mapping the demands imposed by the domain, independent of representation and cognitive agent. However, a complete cognitive description must specify the relationships between all three sets of factors. For example, Woods (1984) contains a cognitive analysis of the effect of a class of representational techniques on human problem-solving in the context of some general cognitive demands, and Sorkin & Woods (1985) contains a cognitive analysis of one kind of multiple cognitive agent system given human information-processing characteristics and a range of cognitive demands. We first confronted the need for a description of cognitive demands in the course of trying to gather and interpret data on human problem-solving performance in a complex dynamic world (power plant emergencies)—investigations that focused on the cognitive agent apex of the above triangle in the context of specific representations of a specific world. We found (Hollnagel, 1981; Pew, Miller & Feehrer, 1981; Woods, Wise & Hanes, 1982) the need for a description of the demands imposed on the human problem-solver that was independent of the language of the application in order to interpret human performance data in terms of cognitive bottlenecks and that was independent of languages of various tool building technologies in order to identify or evaluate potential performance aids. Mapping the demands of domain tasks in cognitive terms is analogous to the Gibsonian position in the study of perception. Understanding how people perceive [solve problems] requires, as a first step, understanding the information the world makes available as it relates to the perceiver's [problem-solver's] goals and opportunities for action. As a new description of the world to be perceived (what is called ecological physics) is needed to study human perception in complex situations, so a new description of the problem-solving situations that can arise is needed to study problem-solving in complex worlds. Without such an understanding, only tool-driven and not principle-driven development of decision support will be possible.

One tactic for building a description of cognitive activities begins with and builds on *an analysis of the cognitive demands* imposed by the requirements of the domain. The particular technique which we will focus on here has been developed and refined over a number of years in the context of thermodynamic systems, e.g. problem-solving during power-plant emergencies (Lind, 1981; Rasmussen, 1983; Woods & Roth, 1986). It has also been applied on a smaller scale to electrical distribution systems, data-communication network management, and logistics maintenance systems. A variant on this technique has been applied to emergency management, CAD/CAM, office systems and library systems (Rasmussen, 1986). Similar techniques have been used to aid designers of aerospace cockpits (Pew *et al.*, 1986) and to map the cognitive demands to build representation aids (Mitchell & Miller, 1986; Mitchell & Saisi, in press). Embrey & Humphreys (1985) and Boel & Daniellou (1984) have developed specific techniques for building goal-directed structures starting from analyses and interviews of domain problem-solvers rather than from knowledge about proper system function (in part, to map errors in people's knowledge about system structure and function).

As the analysis progresses it helps the cognitive technologist interpret answers to questions or observations and decide where to direct further investigation of the domain. The approach is indifferent to particular sources of domain information. Information to carry out the analysis is gathered opportunistically: some questions relate to domain technical knowledge so the knowledge-acquisition problem is finding the right specialist to talk to or to point you to the right documents/analyses; sometimes the path is look empirically at how the problem is solved, e.g. critical incident analysis (Pew *et al.*, 1981) or putting the problem-solver in the situation of interest or simulations of it either formally (experiments) or informally; sometimes the path is to interview people who perform the task or people whose specialties intersect within the task. All of these specific acquisition tactics are potentially useful; the analytical framework helps the cognitive technologist ask meaningful questions and integrate the information acquired from multiple diverse sources. The choice and combination of tactics depends on the resources and constraints of the knowledge-acquisition enterprise and on where information about the domain in question resides.

The result of using this approach is a characterization of the kinds of problems to be solved in the domain and what aspects of the psychology of human performance are relevant in those situations. We will describe the approach primarily in the context of the cognitive activities that arise in managing power-plant emergencies. The approach was developed and has been used in more than a dozen applications within this domain to provide support systems that improve joint cognitive system performance—in exploratory training systems, in information management to aid problem formulation (Woods, Elm & Easter, 1986), in graphic representations to aid situation assessment (Woods, Roth, in press).

Mapping the problem space: goal-directed representation

The paper is addressed to complex worlds, not because of limited applicability, but because complex worlds are the strongest tests of a method's effectiveness. Woods (in press) describes four dimensions that contribute to the complexity of problem solving: dynamism, the number and interactions between parts, uncertainty and risk. A domain's position (or the position of an application within a domain) along these four dimensions begins the description of its cognitive characteristics. The approach has been utilized most extensively in a world that is high on all four of these dimensions—power-plant emergencies, and is therefore well suited for dynamic worlds where large numbers of parts interact, where multiple failures can occur, and where the problem-solver must cope with consequences of disturbances as well as repair causes of disturbances (disturbance management tasks). The technique for goal-directed representation is designed to map the relationship between parts, how the parts work, how evidence testifies to the state of these parts, and how each can change as a function of the state of the domain. This serves as a framework to describe the cognitive situations that arise in the course of carrying out domain tasks.

The mapping of cognitive demands begins with a goal-directed analysis that consists of defining the units of description (goals) and the relationships between those units that are appropriate for domain tasks (cf. Wimsatt, 1972; Warfield, 1973; Lind, 1981; Pearl, Leal & Saleh, 1982; Rasmussen, 1983, for descriptions of this type of analysis). A goal-directed representation is constructed by structuring domain tasks in terms of the goals to be accomplished, the relationships between goals (e.g. causal relations, constraints) and the means to achieve goals (how a process functions, alternative means, pre-conditions, post-conditions).

The basic unit of description is a goal–means or functional interrelationship. A function (the means) provides some commodity (the goal) that is needed or required in order for a higher level function to work properly. A function consists of the set of processes that affect this goal. Processes that make up a function, in turn, need to be supplied with commodities by lower level functions in order to work properly themselves [Fig. 1(a)].†

- *Goal:* What the function should provide; some value of some commodity, e.g. the amount of material within control volume A (a reservoir) greater than Y.

† All of the specific examples used to illustrate the technique come from thermodynamic systems, in particular, emergency operations in nuclear power plants. The technique is general to any large complex system including logistics maintenance systems, power plants, chemical processes, air traffic control, flightdecks, medicine, data communications network management, and electrical distribution centers.

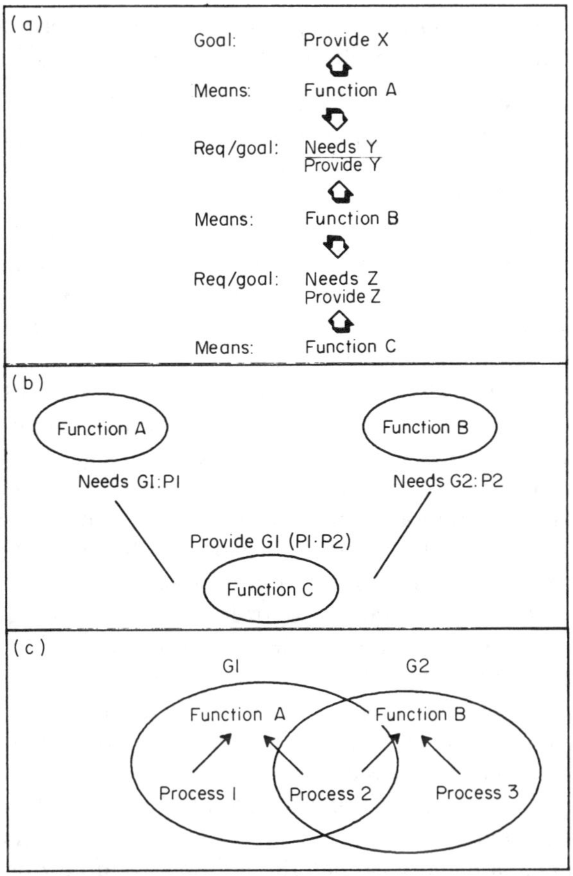

FIG. 1. (a), structure of the goal–means network: the goals to be accomplished, the relationships between goals and the means to achieve goals. (b), different functions may place different requirements on a single goal (G1); or from a different point of view, function C must satisfy multiple criteria on G1. (c), given G1 as goal-of-interest, Process 2 provides G1 and affects G2 (G2 is a side effect of, or a constraint on process 2 if G1 is the object of interest and vice versa if G2 is the object of interest.

- *Function:* The set of *processes* that provide the commodity and knowledge about how they work to achieve this goal (e.g. a balance of material inflow and outflow processes around the reservoir). A process consists of a set of alternative versions and their interrelationships with respect to the goal (e.g. mutually exclusive, simultaneously active) and a model, in terms of functional elements, of how the process effects the goal; the functional elements of a model can themselves be processes (subprocesses) which possess models and alternatives.
- *Requirements:* What is needed for a process to be able to work properly if it is demanded, e.g. a source of material is needed for an inflow process to be capable of functioning (source reservoir level greater than Y). A requirement serves as a goal for a lower-order function, e.g. the set of processes that can provide material to replenish the source reservoir. Requirements can occur at

any level of decomposition of a process—the entire process, alternatives, a model functional element, or alternative versions of a functional element.

To see how this network is built, consider a heat exchanger. Williams, Hollan & Stevens (1983) examined the mental models average people have of how a heat exchanger functions. Here our objective is to explore constraints on the representation an experienced practitioner should have of a particular heat exchanger in a specific system context. Let us begin with the assumption that the designer of a complex thermodynamic system such as a power plant requires that the temperature within some control volume stay below some limit value. He or she then designs a process or processes, in this case a heat exchanger, to achieve this goal. In principle there are many aspects of the thermodynamics of a heat exchanger that the designer could use to control the temperature of interest—the amount of mass, the rate of circulation, the kind of mass (different materials have different heat transfer properties), or the energy of the mass on either the primary or secondary side of the heat exchanger. In practice the designer chooses to use only some of these possibilities to control temperature in a particular case and the remainder are fixed at some value or range of values as a part of the environment of the heat exchange/temperature control process. For example, the primary side mass and circulation are fixed and temperature control is achieved through variations in the secondary side variables in commercial electric power plants. These decisions define what is included in the temperature control process, how this particular process works, and what are the requirements needed for the process to be able to work properly (the primary side mass and circulation variables in this example). The variables that are fixed as features of the environment define other goals which must be achieved. For example, a function is needed to keep the amount of mass on the primary side of the heat exchanger above the minimum value necessary for proper heat transfer; namely a balance of material inflow and outflow processes around the primary side of the heat exchanger. These judgements about what constitutes the "system" under consideration vs exogenous factors are often required when decomposing complex systems (e.g. Iwasaki & Simon, 1986).† Note that the term goal here is used to indicate domain goals, not the current objective of the problem solver.

This analysis can capture the interconnections between goals. Different functions can impose criteria to be achieved on a single commodity [Fig. 1(b)]. This means that breakdowns in the lower order function can result in violations of multiple criteria placed on a single commodity and therefore the disturbance can propagate in multiple directions in the functional topology of the domain. Second, a process may affect more than one goal, i.e. participate in several functions [Fig. 1(c)]. For example, a mass balance process will affect the goal of system material inventory, but because of thermodynamics, it also affects goals on system pressure and temperature. Each goal specifies a different reason to be interested in a process or a context in which to view that process. Goals can also be interconnected because of shared physical equipment, for example, a chemical carried by water when the

† Note that in the case of an existing design this technique is a *post hoc* rationalization of the design and not a psychological description of the design process. The technique can also be used as a design evolves to aid design decision-making.

inventory of both materials is relevant. Thus, the goal-directed analysis can capture the kinds and extent of *inter-goal constraints* or side effects that exist in a domain: given a goal-of-interest, what constraints imposed by related goals govern the operation of a process to achieve the goal-of-interest.

A function is mapped by specifying the set of processes that can affect the goal in question. Each process is modeled qualitatively to represent *how* it works to provide a goal by decomposing the process into its functional elements and their interrelationships (see Lind, 1981, for one grammar of decomposition for thermodynamic processes). For example, a material inflow process to a reservoir can be decomposed into three functional elements: a source of material, a transport mechanism, and points of entry. Another part of decomposition is determining what if any are the alternatives for each process or functional element within a process (i.e. the redundancy and diversity in the system) and what is their interrelationship, such as mutually exclusive, simultaneously capable of effecting the goal, large vs small capability. For example, there may be multiple, mutually exclusive inflow processes to deliver the same material to the same destination that vary in how much or how quickly they can affect the goal commodity. Functional elements can continue to be decomposed into alternatives and models consisting of functional elements to whatever degree is required to capture how the system being analysed works. Thus, functions consist of processes, each of which has a model and alternatives, and the functional elements of a model can themselves be processes (subprocesses) which possess models and alternatives. Requirements can occur at any level of decomposition of a process—some value of a commodity may be needed in order for the entire process to function properly, for an alternative version of the process, for a functional element of the process's model, or for alternative versions of a functional element. Decomposing how processes effect goals is a kind of qualitative modeling (Bobrow, 1985) which in this case exists within a knowledge-representation network.

The goal–means network is a canonical description of the domain. It is a technically accurate description but not strictly normative because alternative, technically accurate decompositions can exist. For example, a material process can be described as an inflow–outflow balance which can be decomposed into alternative inflow and outflow subprocesses which consist of source–transport and transport–sink functional elements, respectively. If some of the outflow sinks double as sources for inflow, then an alternative description would be to decompose the process into multiple recirculation vs injection processes. Where multiple technically accurate descriptions apply, they define candidates for multiple or alternative user models of system function, for example, disturbances can change "system–environment" distinctions in partially decomposable systems (Moray, in press; Iwasaki & Simon, 1986).

The network that results from the goal-directed analysis has several interesting properties (Fig. 2 is a portion of the network for the primary system thermodynamics in a nuclear power plant). First, moving up through the network defines reasons and the consequences that can result from disturbances if they are not checked. Moving down through the network defines causal chains and maps the routes for analysis of correctable causes for disturbances (Rasmussen, 1983). Second, because the representation is not based on how the system fails (root-cause

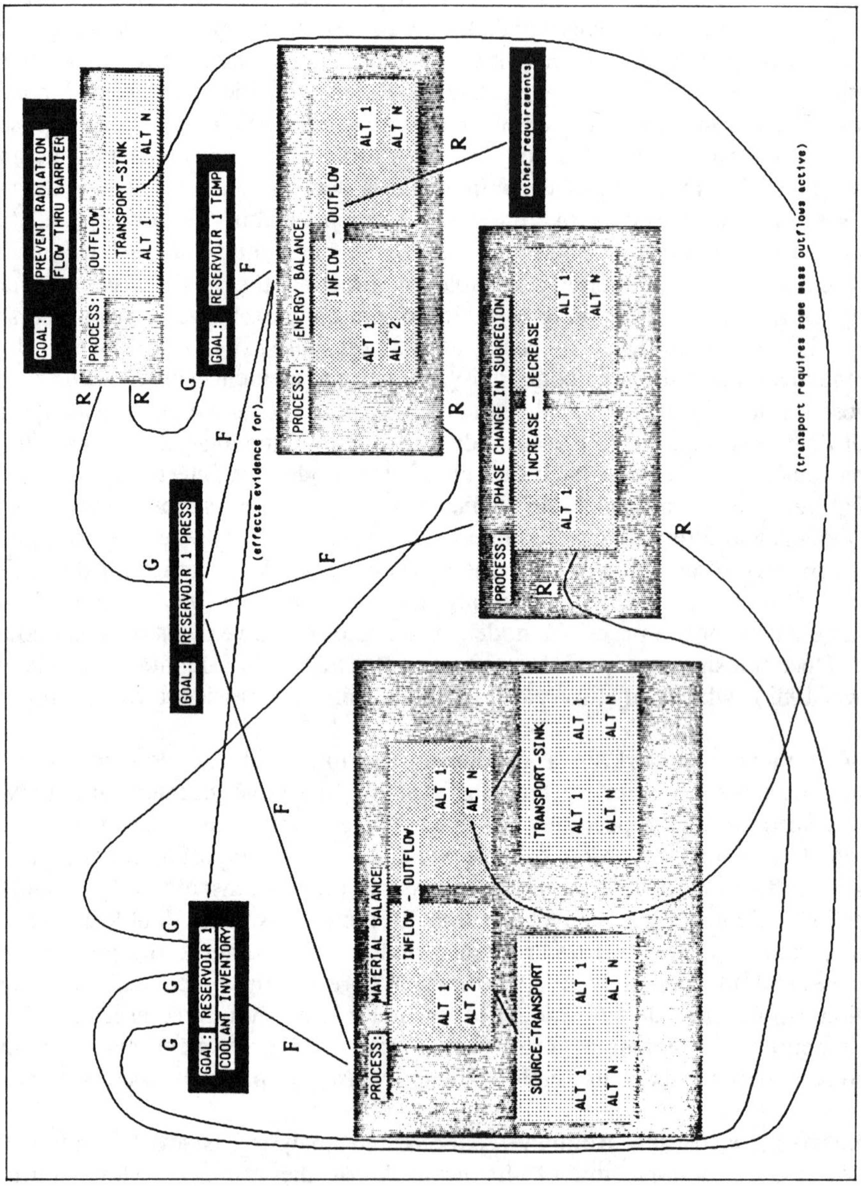

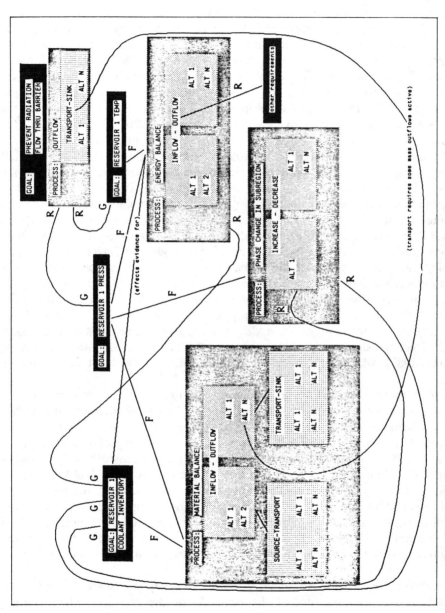

FIG. 2. Simplified example of goal–means network for one part (primary-system thermodynamics) of a nuclear power plant.

faults), but rather on how the system should function normally, it does not depend on complete enumeration of fault categories or possible incidents. It represents a fault in terms of the disturbances (departures from normal functioning) produced by the fault and how those disturbances propagate over time (a disturbance chain). Multiple failures are represented by their respective disturbance chains and by how those disturbances interact. Thus, particular scenarios including their temporal characteristics are represented in terms of the disturbances associated with the initiating fault and how those disturbances propagate over time and interact with the disturbances generated by other failures.

The network provides a framework to discover the kinds of problems that can arise, to define the kinds of information processing requirements, and to provide a conceptual foundation for investigations of actual problem-solving performance.

Cognitive demands

Now we have a framework with which to capture the kinds of cognitive demands and situations that can arise in a particular domain which the problem-solving agent must be able to cope with. At one level, cognitive demands involve collecting and integrating available evidence to answer the questions about plant state (evidential uncertainty, data sampling, situation assessment). A second layer of cognitive demands occurs across units in the goal–means structure—given an assessment of system state (the mapping of available evidence into unit/attribute states at some point in a scenario), what cognitive situations arise in the choice of a response including diagnosis, problem formulation (to choose the goal that should be focused on at that time in the problem-solving process), plan generation or selection, choice under uncertainty and risk, plan monitoring and adaptation. One important result of using the goal-directed problem space to identify the kinds of cognitive situations that can arise is that it helps the cognitive technologist to map the evolving problem-solving process, including possible trajectories (e.g. places where evidence indicates multiple possible interpretations) and error prone points (e.g. missing a state-dependent side effect), as it appears to problem-solvers in the situation.

Cognitive demands I: evidence utilization

One kind of demand is to determine the state of the system or parts of the system. There are evidence-processing requirements associated with each unit in the goal–means network that must be carried out to accurately specify the state of the system. These requirements can be determined by considering the monitoring and control activities that are needed to maintain goals. The particular questions that the person investigating the domain needs to answer are: what evidence can be used to determine: (*a*) *is a goal satisfied*? (*b*) *what constraints imposed by other goals are currently active*? (*c*) *is a process disturbed, which depends on the questions* (*i*) *should a process be active and* (*ii*) *is a process active*? (*d*) *can a process work if demanded*? (*e*) *how to initiate, tune, transfer or terminate processes*. Figure 3 shows how these cognitive activities map into the structure of the goal–means network.

The goal-directed framework specifies the *units* of description for situation assessment (independent of particular situations). The above categories specify the *attributes* of those units, e.g. a process may be inactive–available, inactive–

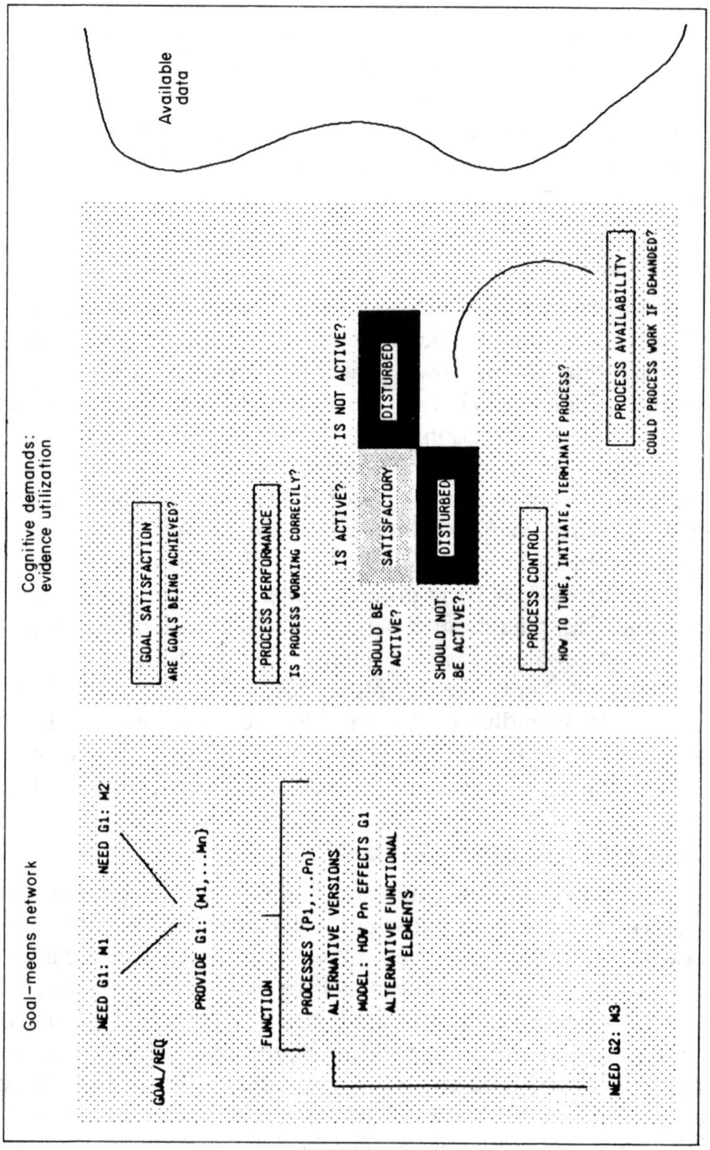
Goal–means network
Cognitive demands: evidence utilization
Available data
GOAL/REQ
NEED G1: M1
NEED G1: M2
PROVIDE G1: {M1,...Mn}
FUNCTION
PROCESSES {P1,...Pn}
ALTERNATIVE VERSIONS
MODEL: HOW Pn EFFECTS G1
ALTERNATIVE FUNCTIONAL ELEMENTS
NEED G2: M3
GOAL SATISFACTION
ARE GOALS BEING ACHIEVED?
PROCESS PERFORMANCE
IS PROCESS WORKING CORRECTLY?
IS ACTIVE?
IS NOT ACTIVE?
SHOULD BE ACTIVE?
SHOULD NOT BE ACTIVE?
SATISFACTORY
DISTURBED
DISTURBED
PROCESS CONTROL
HOW TO TUNE, INITIATE, TERMINATE PROCESS?
PROCESS AVAILABILITY
COULD PROCESS WORK IF DEMANDED?

FIG. 3.

unavailable, active–should be active, active performing correctly, etc. These unit:attribute categories define the *situation assessment* space; i.e. they specify the set of questions that must be answered or what judgements must be made in order to characterize system state. The plant data that are or could be available form the raw material from which answers to these questions can be built. The problem-solving agent must be able to *collect and integrate data to answer these questions about plant state.* For example, what parameters, targets, and limits indicate that a goal is satisfied or what data (e.g. tank levels, valve positions, pump status, flow rates) indicate that the function performed by a process is working correctly (e.g. a material flow process consisting of sources and transport mechanisms). In effect, a particular piece of plant data signals multiple messages or interpretations. At a high level, it signifies or activates a category in the situation assessment, e.g. $x > y$ signifies a violation of goal A; at another level, the datum signifies that, of the set of things that represent violations of goal A, the particular signal is $x > y$.

Uncertainty is one of the dimensions of the complexity of problem worlds. Uncertainty in how evidence testifies to the state of units of the goal–means structure occurs because a given datum may or may not be an accurate measurement/entry (e.g. sensor failure, mis-entry, overdue update on manual entry). Uncertainty also arises because a given datum may be evidence with respect to the state of several units in the goal–means network, in other words, there is a one to many data to unit mapping (Fig. 4). In this approach, this form of uncertainty is represented explicitly by multiple evidential links from datum to unit (or, from the point of view of units in the network, local functions for combining evidence; Cohen *et al.*, 1987). This is needed to be able to handle the flexibility with which experts can solve non-routine cases, to handle revisions of problem formulation especially in dynamic worlds, and to handle the possibility of multiple failures (non-numerical handling of uncertainty to control attention as in Pople, 1985 and Cohen & Gruber, 1985).

The problem-solving agent's ability to meet these evidence processing demands is a joint function of several factors. First, the cognitive demands imposed by the domain can vary—how much collection and integration of data is needed to answer the situation assessment questions, the kinds of integration required, the costs and risks associated with acquiring data, how the world can change over time (which affects, for example, data sampling rate requirements). Performance also depends on the kind of representation of the available data provided to the problem-solver (independent of different task demands, problem-solving performance is a function of the problem representation). Each of these factors in turns depends on the processing characteristics of the problem-solving agent. A complete cognitive analysis must address the interaction of each of these factors; failure to do so has been one source of failures in the history of decision support (see Woods *et al.*, 1986, for one example—the history of the alarm problem in process control and the various attempts to treat it). Here we are only focusing on how to map the first of these factors; see Woods (1984), Woods & Roth (in press) and Cole (1986) for examinations of how variations in problem representation effect human performance. Demands on the collection and integration of data are one potential source of performance difficulties on domain tasks, and it is one location in the cognitive landscape where support systems or decision automation may be deployed.

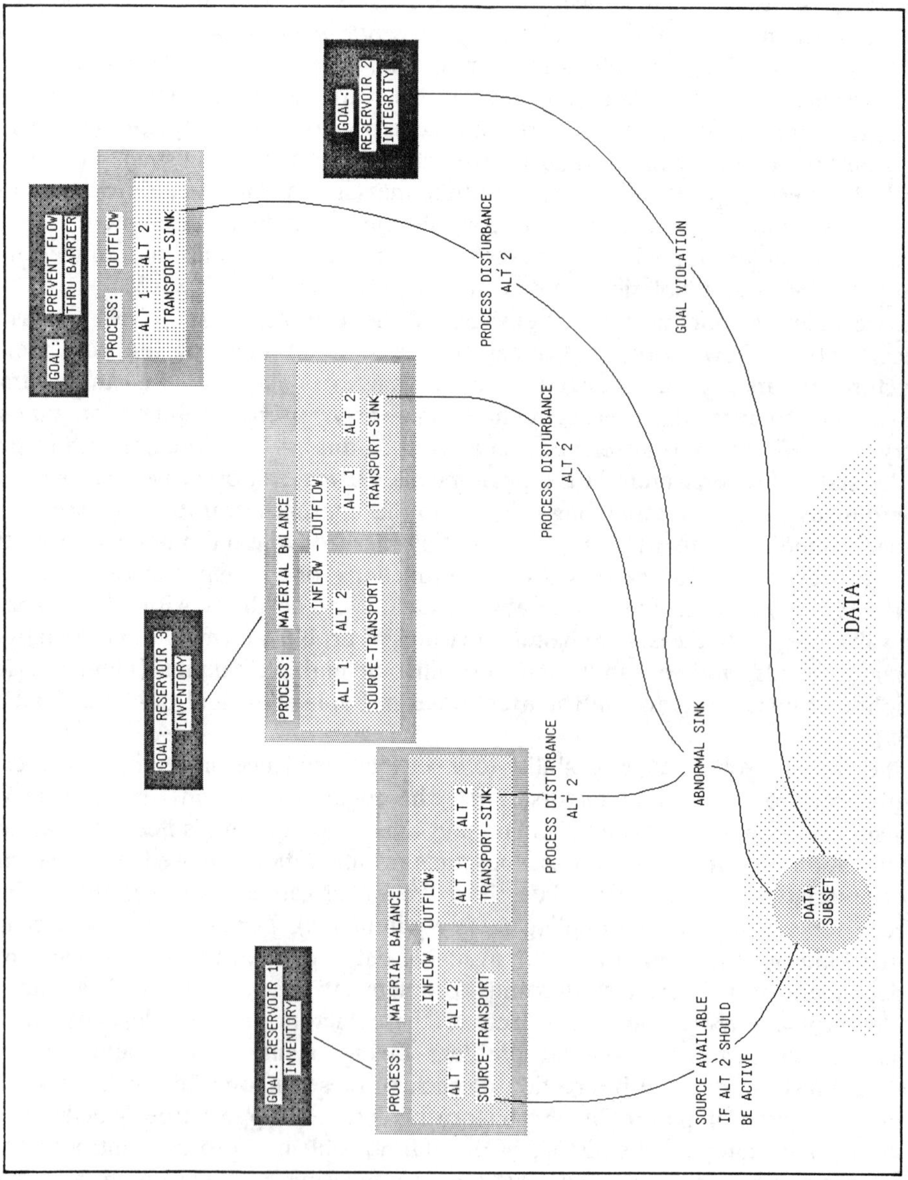

FIG. 4. Illustration of diagnosis in terms of multiple evidential links.

Cognitive demands II: pragmatic reasoning situations

How can we use the goal–means network plus evidence utilization demands as a problem space to specify the kinds of cognitive situations that can arise? Traditionally, cognitive situations are described in global categories like diagnosis and planning or in terms of elementary information-processing tasks (Card, Moran & Newell, 1983; Johnson & Payne, 1985). But the goal–means problem space offers a mechanism to describe at an intermediate level the kinds of situations the problem solver may confront. The result can be thought of as pragmatic reasoning situations, after Cheng, Holyoak, Nisbett & Oliver (1986) and Holland, Holyoak, Nisbett & Thagard (1986), in that the situations are abstracted from the specifics of the domain language and the reasoning involves knowledge of the things being reasoned about. The goal–means problem space and evidence-state mapping provide knowledge about the relationships between units (e.g. goal, inter-goal constraints, process decompositions, requirements) and about the state of units or relationships (especially about desired/actual state mismatches) that can be used to describe the reasoning situations that arise for particular incidents in particular worlds (c.f. Woods, in press).

For example, planning a response to a goal-of-interest should take into account the constraints imposed by related goals in order to avoid unintended effects (missing side effects is a typical error in complex systems, especially if the constraint is state dependent; Dörner, 1983). For example, does the problem-solver know that a requirement relation exists and is now relevant between units *A* and *B* (or a post-condition, goal, process alternative, inter-goal constraint, or other relationship)? If, for other reasons, *A* needs to be done, then *B* must first be satisfied; therefore, check if *B* is satisfied. If *B* is not satisfied, then *A* cannot or must not be done; therefore, do not even consider *A* with respect to the original reason you were interested in it or act to satisfy *B* so as to able to do *A*. If *B* is satisfied, then *A* can be done (e.g. Cheng & Holyoak, 1985). Similar reasoning processes can easily be seen for other relationships such as post-conditions where the problem-solver is biased towards one alternative over another because of the negative post-conditions associated with one of the paths.

To illustrate a few of the kinds of pragmatic reasoning situations that can arise and be described in terms of the goal–means space, consider one of the four dimensions that contribute to the complexity of a problem domain: the level of interconnections between parts. When interconnections are extensive, actions involving one part can often have side effects in addition to the intended effect; a single fault can have consequences for multiple parts; independent faults can have consequences for the the same part. Extensive interrelationships between parts also introduce complexity because they can produce "effects at a distance" (Boel & Daniellou, 1984, contains one example); i.e., a fault in one place can produce disturbances at other distant locations (given either a physical or functional distance metric). Automation is a potent source of effects at a distance. Highly coupled systems produce the opportunity for situations to arise where there are competing goals. Competing goals can take two forms. One is a choice involving goal sacrifice where the tradeoff is between repairing or replacing one means to achieve the primary goal (can I repair it? how much time is available to attempt repair?), and an alternative means that entails as post-conditions the sacrifice of other goals in order to meet the primary goal. A second case is multiple mutually constrained goals where two or more goals must be simultaneously satisfied in circumstances where the means to achieve each constrains how the others can be achieved.

When a large number of interconnected parts is combined with another dimension of complexity, dynamism, the disturbance management cognitive situation arises. In this situation the problem solver must cope with the consequences of disturbances in the short run as well as diagnoses and treat the underlying fault (get the plane into stable flight, then figure out how to restart the failed engine). The problem solver in this kind of problem situation first becomes aware of a disturbance (actual/desired state mismatch). The problem-solver has several possible strategic responses: attempt to adjust the disturbed process, find and correct, if possible, what produced the disturbance (diagnosis), respond to cope with the effects of the disturbance (if there is insufficient time or an inability to repair the affected process).

In disturbance management, various pragmatic reasoning patterns are relevant, e.g. if there are no pressing consequences of a disturbance and no more urgent disturbances on the agenda, then try to find a correctable cause. Note that knowledge about the time response of various processes is often an important element in disturbance management, e.g. the time available until a disturbance propagates from one part to another, responses to buy more time before consequences propagate, the time it will take to adjust or repair a process. Also, judgments about *prospective* solutions can play a significant role—will I be able to adjust or repair the process (implying knowledge of why it failed) before undesirable consequences occur.

Figure 5 illustrates a case of competing goals in terms of the goal–means structure that is abstracted from a real incident (U.S. Nuclear Regulatory Commission, 1985 and see Woods & Roth, 1986 for the cognitive analysis) and from human performance in simulated runs of the same incident (Woods *et al.*, 1982). One of two alternatives within a function was disturbed. The alternative process (B) needed to be activated to prevent consequence propagation with respect to Goal 1 (which would occur in about 30 min), but activation of the alternative would have a significant negative impact with respect to a second important goal. The goal competition was resolvable because Goal 2 was relatively less important than Goal 1 for these circumstances. In addition, instructions were available that specified that if Alternative *A* was lost, Alternative *B* should be implemented (implicitly the instructions specified that Goal 2 should be sacrificed in this circumstance). Another path open to the problem-solvers was to try to diagnosis and correct the disturbance in Alternative *A*. Actual problem-solvers chose to try to repair Alternative *A* because they judged that they would be able to do this within the time available (in part because they saw the disturbance as a mis-start of process *A* rather than as a failure). In addition, they needed to be prepared to adapt their response (i.e., switch to Alternative *B*) if conditions changed which increased the urgency of preventing consequence propagation or if repair efforts met with difficulties.

Some of the aspects of cognitive situations related to diagnosis of correctable causes are represented in the goal–means space in terms of working down through a function's decomposition and requirement links. Figure 6 illustrates the causal chain for a case where the problem-solver begins with a goal violation—reservoir level is

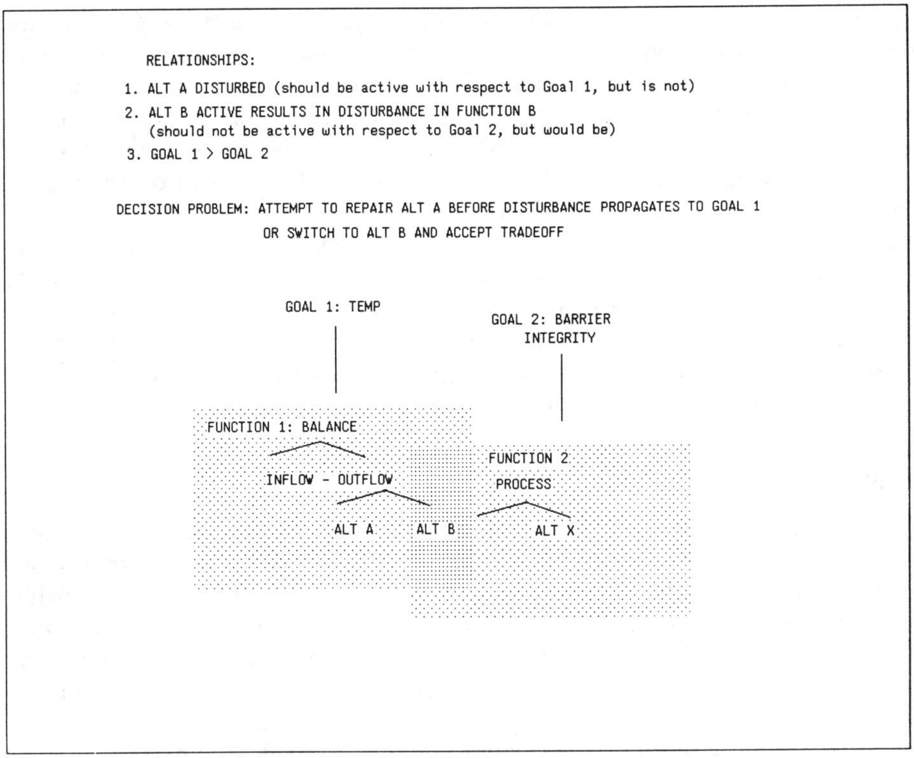

FIG. 5. Illustration of disturbance management cognitive situation.

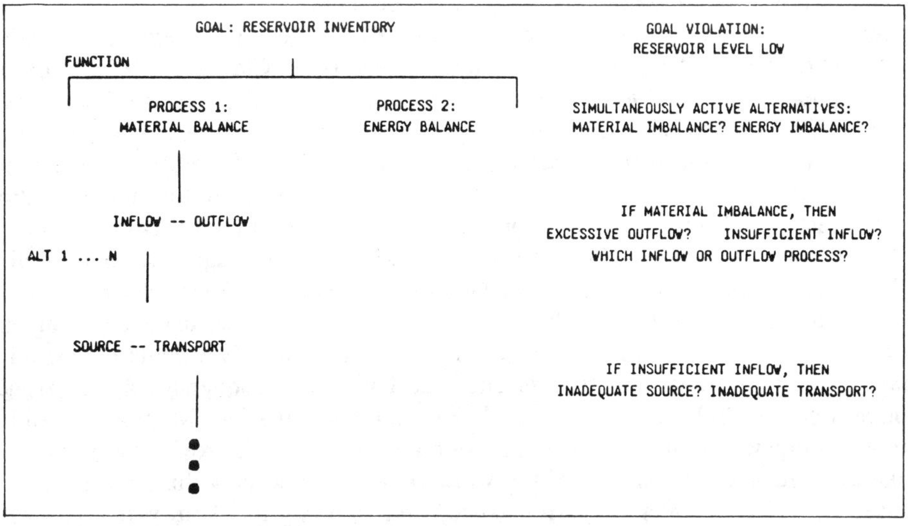

FIG. 6. Illustration of diagnosis in terms of goal–means structure.

low. The first step illustrates one difficult kind of diagnostic situation where multiple processes are simultaneously capable of effecting the goal. Thus, any one is capable of being the source of the disturbance. This situation is particularly difficult when multiple processes are active because the effect of one contributor partially occludes the effect of the other. For example, when one is identified as a contributor to the disturbance, the diagnostic search at that level can be terminated prematurely. Figure 6 continues a trace of the causal chain; the problem-solver terminates the search when a correctable item is located. This depends on the purpose of the problem-solver and on the action set available to him or her. Similarly, when there is an actual/desired state mismatch low in a causal chain, moving up through the goal–means space supports reconceptualizing the problem at a new level of abstraction to find alternative paths for goal satisfaction, for example, in plan adaptation (Rasmussen, 1985; Alterman, 1986).

Summary

We have briefly described a technique to map the cognitive demands imposed by the characteristics of a problem-solving world and illustrated some examples of its descriptive power. Analysing cognitive demands is one part of building a cognitive description of a problem-solving world. Understanding how representations of the world interact with different demands and with characteristics of the cognitive agents is another important part of a cognitive description. Mapping cognitive demands is part of a *problem-driven* approach to the application of computational power. In tool-driven approaches, knowledge acquisition focuses on describing domain knowledge in terms of the syntax of computational mechanisms, i.e. the language of implementation is used as a cognitive language. Semantic questions are displaced either to whomever selects the computational mechanisms or to the domain expert who enters knowledge. The alternative is to provide an umbrella structure of domain semantics that organizes and makes explicit what particular pieces of knowledge mean about problem-solving in the domain.

There is a clear trend in this direction in order to achieve more effective decision support and less brittle machine problem-solvers (e.g. Clancey, 1985; Coombs, 1986; Gruber & Cohen, 1986). Acquiring and using a domain semantics is essential for knowledge acquisition and engineering to be capable of avoiding potential errors and specifying performance boundaries when building "intelligent" machines (Woods *et al.*, in press). One tangible example of the need for and evolution towards semantic frameworks for applications of computational technology is the evolution from Mycin where large amounts of domain knowledge remained implicit in the computational mechanisms, to Neomycin (Clancey, 1983) which begins to make a domain semantics explicit, e.g. via support knowledge, to Langlotz, Shortliffe & Fagan (1986) who place rules in the context of fully articulated cognitive situations, e.g. a particular rule is shortcut based on frequency of occurrence through a choice under uncertainty and risk cognitive situation. Mapping cognitive demands is one way to acquire and use a domain semantics. It provides a framework to characterize what knowledge of the world is needed and how this knowledge can be used to achieve effective performance, and it helps to identify places where performance is more likely to go wrong (e.g. situations where state-dependent side effects become relevant or where a reasoning shortcut coded in a rule is not valid). Both of these types of information about problem-solving in a domain can be used to specify what cognitive tools are needed by cognitive agents and what kinds of computational mechanisms are capable of providing such tools.

Techniques for analysing cognitive demands not only help characterize a particular world, they also help to build a repertoire of general cognitive situations, that is, abstract and therefore transportable reasoning situations which are domain relevant because they use knowledge of the world being reasoned about (Cheng *et al.*, 1986). Thus, mapping cognitive demands is one of several efforts to develop a cognitive language of description that is independent of the language of the domain itself and the language of implementation technology (e.g. Gruber & Cohen, 1986). This independent cognitive language is essential not only to capture the cognitive demands imposed on a problem solver, but also to specify how capabilities for building cognitive tools can be used to provide effective decision support (Woods, 1986; Woods *et al.*, in press).

The approach to knowledge acquisition and cognitive task analysis which is described here emerged from the contribution of many people. The idea of Jens Rasmussen and Morten Lind on knowledge representation of complex, dynamic worlds provided the initial impetus. The approach continued to develop over the course of several in-depth applications to create effective problem-solving support systems by Westinghouse's Man–Machine Functional Design group; especially noteworthy are the contributions of James Easter, William Elm, Craig Watson, and John M. Gallagher, Jr. Many difficulties were overcome only because of many valuable discussions with and critiques by Emilie Roth.

References

ALTERMAN, R. (1986). An adaptive planner. In *Proceedings of the AAAI*. Los Altos, CA: Morgan Kaufman. pp. 65–69.

BOBROW, D. G. (Ed). (1985). *Qualitative Reasoning About Physical Systems*. Cambridge, Massachussetts. The MIT Press.

BOEL, M. & DANIELLOU, F. Elements of process control operator's reasoning: activity planning and system and process response times. In D. WHITFIELD Ed., *Ergonomic Problems in Process Operations*. Institute Chemical Engineering Symposium Series 90, 1984.

CARD, S. K., MORAN, T. P. & NEWELL, A. (1983). *The Psychology of Human–Computer Interaction*. Hillsdale, New Jersey: Lawrence Erlbaum Associates.

CHENG, P. W. & HOLYOAK, K. J. (1985). Pragmatic reasoning schemas. *Cognitive Psychology*, **17**, 391–416.

CHENG, P. W., HOLYOAK, K., NISBETT, R. & OLIVER, L. (1986). Pragmatic versus syntactic approaches to training deductive reasoning. *Cognitive Psychology*, **18**, 293–328.

CLANCEY, W. J. (1983). The epistemology of a rule-based expert system—a framework for explanation. *Artificial Intelligence*, **20**, 215–251.

CLANCEY, W. J. (1985). Heuristic classification. *Artificial Intelligence*, **27**, 289–350.

COHEN, P. & GRUBER, T. (1985). Managing Uncertainty. Technical report COINS 85-24, University of Massachusetts. (Also Pergamon Infotech State of the Art Report).

COHEN, P. *et al.* (1987). Management of uncertainty in medicine. *IEEE Conference on Computers and Communications*, IEEE, 1987. (Also University of Massachusetts COINS Technical Report 86-12).

COLE, W. G. (1986). Medical cognitive graphics. *Chi*'86 *Conference Proceedings*, ACM.

COOMBS, M. J. (1986). Artificial intelligence and cognitive technology: foundations and perspectives. In HOLLNAGEL, E., MANCINI, G. & WOODS, D. D. Eds, *Intelligent Decision Support in Process Environments*. New York: Springer–Verlag.

DAVIS, R. (1983) Reasoning from first principles in electronic troubleshooting. *International Journal of Man–Machine Studies,* **19,** 403–423.

DÖRNER, D. (1983). Heuristics and cognition in complex systems. In GRONER, R., GRONER, M. & BISCHOF, W. F. Eds., *Methods of Heuristics*. Erlbaum.

EMBREY, D. & HUMPHREYS, P. (1985). Support for decision making and problem solving in abnormal conditions in nuclear power plants. In *Knowledge Representation for Decision Support Systems*. Durham, England: International Federation for Information Processing.

GRUBER, T. & COHEN, P. (1986). Knowledge engineering tools at the architecture level. Technical report COINS 86-12. University of Massachusetts.

HOLLAND, J. H., HOLYOAK, K. J., NISBETT, R. E. & THAGARD, P. R. (1986) *Induction: Processes of Inference, Learning, and Discovery*. Cambridge, Massachusetts: MIT Press.

HOLLNAGEL, E. (1981). Report from the third NKA/KRU experiment: the performance of control engineers in the surveillance of a complex process. Technical Report N-14-81. Risø National Laboratory.

HOLLNAGEL, E. & WOODS, D. D. (1983). Cognitive systems engineering: new wine in new bottles. *International Journal of Man–Machine Studies,* **18,** 583–600.

IWASAKI, Y. & SIMON, H. A. (1986). Causality in device behavior. *Artificial Intelligence,* **29,** 3–32.

JOHNSON, E. & PAYNE, J. W. (1985). Effort and accuracy in choice. *Management Science,* **31,** 395–414.

LANGLOTZ, C., SHORTLIFFE, E. & FAGAN, L. (1986). Using decision theory to justify heuristics. *Proceedings of AAAI.* Los Altos, CA: Morgan Kaufman. pp. 215–219.

LIND, M. (1981). The use of flow models for automated plant diagnosis. In RASMUSSEN, J. & ROUSE, W. B. Eds., *Human Detection and Diagnosis of System Failures*. New York: Plenum Press.

MITCHELL, C. & MILLER, R. A. (1986). A discrete control model of operator function: a methodology for information display design. *IEEE Systems, Man, and Cybernetics,* **SMC-16,** 343–357.

MITCHELL, C. & SAISI, D. (1987). Use of model-based qualitative icons and adaptive windows in workstations for supervisory control systems. *IEEE Systems, Man, and Cybernetics.*

MORAY, N. (1987). Intelligent aids, mental models, and the theory of machines. *International Journal of Man–Machine Studies.* In press.

NEWELL, A. (1982). The knowledge level. *Artificial Intelligence,* **18,** 87–127.

PEARL, J., LEAL, A. & SALEH, J. (1982). GODDESS: a goal-directed decision structuring system. *IEEE Transactions on Pattern Analysis and Machine Intelligence,* **PAMI-4,** 250–262.

PEW, R. W. *et al.* (1986). Cockpit automation technology. Technical report 6133. BBN Laboratories Incorporated.

PEW, R. W., MILLER, D. C. & FEEHRER, C. E. (1981). *Evaluation of Proposed Control Room Improvements Through Analysis of Critical Operator Decisions.* Palo Alto, CA: Electric Power Research Institute (NP-1982).

POPLE, H. Jr (1985). Evolution of an expert system: from internist to caduceus. In DE LOTTO, I. & STEFANELLI, M. eds, *Artificial Intelligence in Medicine.* Elsevier Science Publishers B.V. (North-Holland).

RASMUSSEN, J. (1983). Skills, rules, and knowledge; signals, signs, and symbols; and other distinctions in human performance models. *IEEE Transactions on Systems, Man, and Cybernetics* **SMC-12,** 257–266.

RASMUSSEN, J. (1985). The role of hierarchical knowledge representation in decisionmaking and system management. *IEEE Transaciions on Systems, Man and Cybernetics,* **SMC-15,** 234–243.

RASMUSSEN, J. (1986). A cognitive engineering approach to the modelling of decision making *Technical Report Risø-M-2589.* Risø National Laboratory.

Sorkin, R. D. & Woods, D. D. (1985). Systems with human monitors: a signal detection analysis. *Human–Computer Interaction*, **1,** 49–75.

U.S. Nuclear Regulatory Commision. (1985). *Loss of Main and Auxillary Feedwater at the Davis–Besse Plant on June* 9, 1985. Springfield, VA: National Technical Information Service (NUREG-1154).

Warfield, J. N. (1973). Intent structures. *IEEE Transactions on Systems, Man, and Cybernetics,* **SMC-3,** 133–140.

Williams, M. D., Hollan, J. D., & Stevens, A. L. (1983). Human reasoning about a simple physical system: a first pass. In Gentner, D. & Stevens, A. S. Eds., *Mental Models.* Hillsdale, New Jersey: Lawrence Erlbaum Associates.

Wimsatt, W. C. (1972). Teleology and the logical structure of function statements. *Studies in the History and Philosophy of Science,* **3,** 1–80.

Woods, D. D. (1984). Visual momentum: a concept to improve the cognitive coupling of person and computer. *International Journal of Man–Machine Studies,* **21,** 229–244.

Woods, D. D. (1986). Paradigms for intelligent decision support. In Hollnagel, E., Mancini, G. & Woods, D. D. Eds, *Intelligent Decision Support in Process Environments.* New York: Springer–Verlag.

Woods, D. D., Elm, W. C. & Easter, J. R. (1986). The disturbance board concept for intelligent support of fault management tasks. *Proceedings of the International Topical Meeting on Advances in Human Factors in Nuclear Power.*

Woods, D. D. (1987). Coping with complexity: the psychology of human behavior in complex systems. In Goodstein, L. P., Andersen, H. B. & Olsen, S. E. Eds., *Mental Models, Tasks and Errors: A Collection of Essays to Celebrate Jens Rasmussen's* 60*th Birthday.* New York: Taylor Francis. In press.

Woods, D. D. & Roth, E. (1986). *The Role of Cognitive Modeling in Nuclear Power Plant Personnel Activities: A Feasibility Study.* Washington D.C.: U.S. Nuclear Regulatory Commission (NUREG-CR-4532).

Woods, D. D. & Roth, E. (1987). Cognitive Systems Engineering. In Helander, M. Ed., *Handbook of Human–Computer Interaction.* New York: Wiley. In press.

Woods, D. D., Roth, E. M. & Bennett, K. B. (1987). Explorations in joint human–machine cognitive systems. In Robertson, S. & Zachary, W. Eds., *Cognition, Computing and Cooperation.* Norwood, New Jersey: Ablex Publishing. In press.

Woods, D. D., Wise, J. A. & Hanes, L. F. (1982). *Evaluation of Safety Parameter Display Concepts.* Palo Alto, CA: Electric Power Research Institute (NP-2239).

Generic tasks for knowledge-based reasoning: the "right" level of abstraction for knowledge acquisition

TOM BYLANDER AND B. CHANDRASEKARAN

Laboratory for Artificial Intelligence Research, Department of Computer and Information Science, The Ohio State University, Columbus, Ohio 43210, U.S.A.

Our research strategy has been to identify *generic tasks*—basic combinations of knowledge structures and inference strategies that are powerful for solving certain kinds of problems. Our strategy is best understood by considering the "interaction problem", that representing knowledge for the purpose of solving some problem is strongly affected by the nature of the problem and by the inference strategy to be applied to the knowledge. The interaction problem implies that different knowledge-acquisition methodologies will be required for different kinds of reasoning, e.g. a different knowledge-acquisition methodology for each generic task. We illustrate this using the generic task of hierarchical classification. Our proposal and the interaction problem call into question many generally held beliefs about expert systems such as the belief that the knowledge base should be separated from the inference engine.

Introduction

Knowledge acquisition is the process that extracts knowledge from a source (e.g. a domain expert or textbook) and incorporates it into a knowledge-based system that solves some problem. Whether the knowledge is acquired by a knowledge engineer or by a program, ultimately the knowledge must be encoded in some knowledge-base representation. Consequently, knowledge acquisition cannot be separated from a broader theory of knowledge-based reasoning; a solution to knowledge acquisition must be compatible with a solution to the general problem of knowledge-based reasoning.

For some time now, we have been developing a *theory of generic tasks* that identifies several types of reasoning that knowledge-based systems perform and provides a overall framework for the design and implementation of such systems (Chandrasekaran 1983, 1984, 1986). In this paper, we present our theory as a way to exploit the "interaction problem". Because each generic task exploits it differently, each one should be associated with a different knowledge-acquisition methodology.

First, we pose and discuss the interaction problem. Next, we review our theory of generic tasks: the characteristics of a generic task and the generic tasks that have been identified so far. In view of the interaction problem, we propose our theory of generic tasks as a framework for identifying different knowledge-acquisition methodologies. We illustrate this using the generic task of hierarchical classification. Finally, we reflect on a number of beliefs that have driven much of the past research on knowledge acquisition and knowledge-based reasoning.

KNOWLEDGE-BASED SYSTEMS Vol. 1
ISBN 0-12-273251-0

The interaction problem

The interaction problem is this:

> Representing knowledge for the purpose of solving some problem is strongly affected by the nature of the problem and by the inference strategy to be applied to the knowledge.

In other words, how knowledge is represented has a close relationship to how knowledge is used to solve problems; knowledge is dependent on its use. The interaction problem is not a new notion. Minsky, in his famous frame proposal, argues that "factual and procedural contents must be more intimately connected to explain the apparent power and speed of mental activities" (Minsky, 1975, p. 211). Marr has noted that "how information is represented can greatly affect how easy it is to do different things with it" (Marr 1982, p. 21). Our argument takes a different perspective, that the problem and the inference strategy influence what knowledge is represented and how knowledge is encoded, i.e. knowledge will be represented to take advantage of what it will be used for.

The interaction problem, if true, has serious implications for how knowledge acquisition should be done. Because some knowledge representation must be the target of knowledge acquisition, knowledge-acquisition methodologies must take the interaction problem into account. Also, if different kinds of reasoning have different kinds of interactions, there is a need for a different knowledge-acquisition methodology for each kind of reasoning.

REASONS FOR THE INTERACTION PROBLEM

Figure 1 illustrates the situation that gives rise to the interaction problem. Starting with a problem and a domain expert (or some other source of knowledge), the goal of knowledge acquisition is to construct a knowledge-based system that solves the problem by the combination of a knowledge base and an inference strategy (a

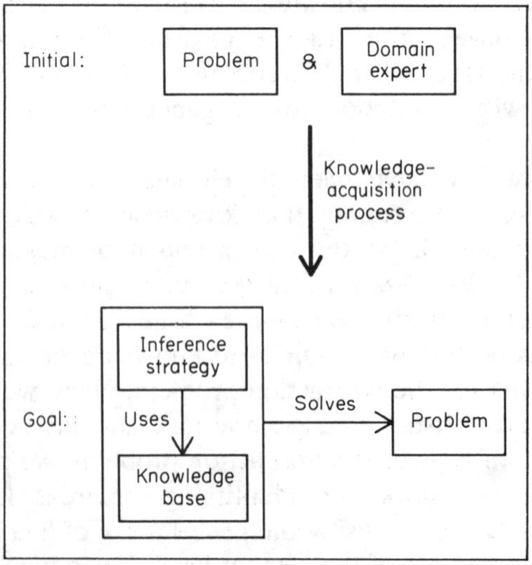

FIG. 1. Initial and goal states of the knowledge-acquisition process.

process to use or interpret the knowledge). There are two primary reasons why this leads to the interaction problem.

(1) Choice of knowledge. The knowledge acquisition process must *choose* what knowledge to ask for and what knowledge to encode. The choice is driven by the need to gain leverage on the problem by obtaining knowledge with high utility and to reduce complexity by avoiding or discarding knowledge with low utility. Not everything the domain expert knows has the same level of usefulness, and in any case, it is not feasible to acquire everything that the domain expert knows.

(2) Constraints of inference strategy. A knowledge representation requires some process that, given a description of a situation, can *use* (or *interpret*) the knowledge to make conclusions. It is this process which we call the "inference strategy" ("inference engine" is an equivalent phrase). The knowledge must be represented so that the inference strategy reaches appropriate conclusions (appropriate to the problem being solved) in a timely fashion. Consequently, the knowledge must be adapted to the inference strategy to ensure that certain inferences are made from the knowledge and not others. Also, give a choice of inference strategies, there will be an interaction between the strategy chosen and the form of knowledge.

EXAMPLES OF THE INTERACTION PROBLEM IN RULE-BASED REPRESENTATIONS

The idea of rules is to explicitly map situations to actions. Naturally then, the focus is on determining what conditions characterize the situations and what conclusions characterizes the actions. The result is that two different problems in the same domain can have different rules representing the "same" knowledge. For example in diagnosis, rules of the form "symptom→malfunction" will be implemented, while in prediction of symptoms, the rules will be in the form "malfunction→symptom". In each case, the knowledge will be adapted to the problem. One might argue that there is no difficulty with keeping both problems in mind, and so both kinds of rules can be in the same knowledge base. Of course, *given that one has already taken the interaction problem into account,* the knowledge base then will have rules appropriate for the problems to be solved.†

Another source of interaction is that special programming techniques are needed to encode problem-specific inference strategies. For example, R1 (McDermott, 1982), which is implemented in OPS5 (Forgy, 1981), performs a sequence of "design subtasks", each of which is implemented as a set of production rules. However, OPS5 has no construct equivalent to a subtask, so the grouping of rules and the sequencing from one set of rules to another are achieved by programming techniques. Clearly, the constraints of OPS5's production rule representation has had a significant effect on how R1's knowledge was encoded.

Different inference strategies for rules are also a source of interaction. If

† It is actually dangerous to have both kinds of rules in a knowledge base. Given some confidence in a malfunction, then some confidence in the symptoms it causes should be inferred. However, one shouldn't infer confidence in other malfunctions that cause the same symptoms.

EMYCIN's backward-chaining strategy is used, rules can combine with other rules to increase or decrease confidence in a given conclusion (van Melle, 1979). On the other hand, if OPS5's recognize-and-act strategy is used, only one rule at a time can be fired, so that situations must be matched to actions much more exactly. Also, the "context" must be carefully controlled to ensure that appropriate rules are considered. Note that the difference is not whether EMYCIN does forward- or backward-chaining, but that EMYCIN allows rules to act in parallel, while OPS5 applies rules in serial.

THE INTERACTION PROBLEM IN OTHER REPRESENTATIONS

Rule-based and logic-based representations are fairly similar with respect to the interaction problem. Like rules, logic provides for a direct way for drawing conclusions from situations. In the context of a specific problem, it is useful to encode only those propositions that can make problem-relevant conclusions. Logic-based representations are also like rules with respect to implementing special structures and dealing with different inference strategies. To implement R1 in predicate logic, for example, a subtask construct would also have to be implicitly programmed. Two different inference strategies for logic, such as PROLOG and resolution theorem proving, are quite different to use.

The emphasis in frame representations is on describing the conceptual structure of the domain. However, different problems might need quite different conceptual structures. For example, classificatory problem solving (Gomez & Chandrasekaran, 1981; Clancey, 1985) in general needs a generalization hierarchy (hypothesis-subhypothesis), while routine design (Brown & Chandrasekaran, 1986) in general needs a structural hierarchy (component-subcomponent). Of course, one of Minsky's original intentions was to express the interaction between knowledge and inference strategies. For example, the idea of attaching various kinds of information to a frame is for controlling how the frame will be used.

EXPLOITING THE INTERACTION PROBLEM

The interaction problem will not go away no matter what representation is chosen. Every knowledge-based system will be developed, debugged, and maintained so its knowledge works with its inference strategy and so its knowledge in combination with its inference strategy solves a certain problem. It is not feasible to undertake an exhaustive study of a domain to acquire any and all the knowledge associated with that domain. It is important to realize that knowledge-based systems are powerful only when selected portions of domain knowledge, appropriately interpreted, are needed to solve problems.

Instead of trying to lessen the impact of the interaction problem, our research strategy has been to *exploit* it. We claim that different representations can be exploited in different ways and are thus more applicable to certain kinds of problems than others. This is where our theory of generic tasks comes in. Our intent is to propose types of problem solving in which the representation and the inference strategy can be exploited to solve certain kinds of problems. For a particular domain and problem, our intent is to encode a selected portion of domain knowledge into an efficient and maintainable problem solving structure.

The proposal

Intuitively one would think that diagnosis in different domains would have certain types of reasoning in common, and that design in different domains would also have certain types of reasoning in common, but that diagnostic reasoning and design problem-solving will be generally speaking different. For example, diagnostic reasoning generally involves malfunction hierarchies, rule-out strategies, setting up a differential, etc., while design involves device/component hierarchies, plans to specify components, ordering of plans, etc. However, the formalisms (or equivalently the languages) that have been commonly used for knowledge-based systems do not capture these distinctions. Ideally, diagnostic knowledge should be represented by using the *vocabulary* that is appropriate for diagnosis, while design knowledge should have a vocabulary appropriate for design. Our approach to this problem has been to identify *generic tasks*—basic combinations of knowledge structures and inference strategies that are powerful for dealing for certain kinds of problems. The generic tasks provide a vocabulary for describing problems, as well as for designing knowledge-based systems that perform them.

CHARACTERIZATION OF A GENERIC TASK

Each generic task is characterized by information about the following:

(1) The type of *problem* (the type of input and the type of output). What is the function of the generic task? What is the generic task good for?
(2) The *representation* of knowledge. How should knowledge be organized and structured to accomplish the function of the generic task? In particular, what are the type of *concepts* that are involved in the generic task? What concepts are the input and output about? How is knowledge organized in terms of concepts? In essence, we adopt Minsky's idea of frames as a way to organize the problem solving process (Minsky, 1975).
(3) The *inference strategy* (process, problem-solving, control regime). What inference strategy can be applied to the knowledge to accomplish the function of the generic task? How does the inference strategy operate on concepts?

The phrase "generic task" is somewhat misleading. What we really mean is *an elementary generic combination of a problem, representation, and inference strategy about concepts.* The power of this proposal is that if a problem matches the function of a generic task, then the generic task provides a knowledge representation and an inference strategy that can be used to solve the problem. Figure 2 illustrates how this fits into knowledge acquisition. Identifying a generic task that is applicable to the problem is an intermediate goal of knowledge acquisition. The problem must match the type of problem that the generic task can solve. The generic task then specifies a type of strategy and a type (or form) of knowledge for solving the problem. Further stages of the knowledge-acquisition process (not illustrated in the figure) are to obtain domain knowledge and to particularize the inference strategy.

EXAMPLES OF GENERIC TASKS

Our group has identified several generic tasks. Here, we briefly describe the generic tasks of hierarchical classification (Gomez & Chandrasekaran, 1981) and object synthesis by plan selection and refinement (Brown & Chandrasekaran, 1986).

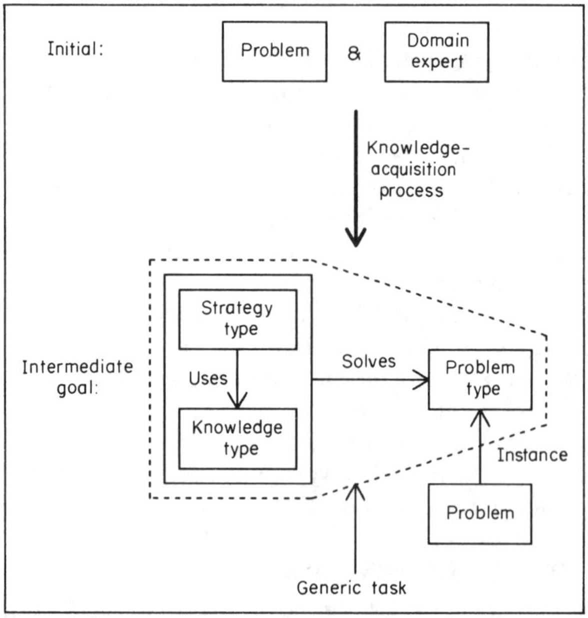

FIG. 2. Using generic tasks in knowledge acquisition.

Hierarchical classification

Problem: Given a description of a situation, determine what categories or hypotheses apply to the situation.

Representation: The hypotheses are organized as a classification hierarchy in which the children of a node represent subhypotheses of the parent. There must be knowledge for calculating the degree of certainty of each hypothesis.

Important concepts: Hypotheses.

Inference strategy: The establish–refine strategy specifies that when a hypothesis is confirmed or likely (the establish part), its subhypothesis should be considered (the refine part). Additional knowledge may specify how refinement is performed, e.g. to consider common hypotheses before rarer ones. If a hypothesis is rejected or ruled-out, then its subhypotheses are also ruled-out.

Examples: Diagnosis can often be done by hierarchical classification. In planning, it is often useful to classify a situation as a certain type, which then might suggest an appropriate plan. The diagnostic portion of MYCIN (Shortliffe, 1976) can be thought of as classifying a patient description into an infectious agent hierarchy. PROSPECTOR (Duda, Gaschnig & Hart, 1980) can be viewed as classifying a geological description into a type of formation. Hierarchical classification is similar to the refinement part of Clancey's heuristic classification (Clancey, 1985).

Object synthesis by plan selection and refinement

Problem: Design an object satisfying specifications. An object can be an abstract device, e.g. a plan or program.

Representation: The object is represented by a component hierarchy in which the children of a node represent components of the parent. For each node, there are plans that can be used to set parameters of the component and to specify additional constraints to be satisfied. There is additional knowledge for selecting the most appropriate plan and to recover from failed constraints.

Important concepts: The object and its components.

Inference strategy: To design an object, *plan selection and refinement* selects an appropriate plan, which, in turn, requires the design of subobjects in a specified order. When a failure occurs, failure handling knowledge is applied to make appropriate changes.

Examples: Routine design of devices and the synthesis of everyday plans can be performed using this generic task. The MOLGEN work of Friedland (Friedland, 1979) can be viewed in this way. Also R1's subtasks (McDermott, 1982) can be understood as design plans.

OTHER PROPERTIES OF GENERIC TASKS

Generic tasks have a number of other properties. First, there is not a one-to-one relationship between generic tasks and problems. A problem might match the function of more than one generic task, so that several strategies might be used to solve the problem, depending on the knowledge that is available. For example, a design problem might be solvable by hierarchical classification if there is only a limited number of possible designs. Also, generic tasks can be composed for more complex reasoning, i.e. one generic task problem solver might call upon another to solve a subproblem. Typically, knowledge-based systems must be analysed as using combinations of generic tasks, rather than just using a single one.

Second, we are interested in human-like problem-solving, which, we assume, is about concepts and is non-quantitative. This leads us to look for ways that knowledge can be distributed over the concepts of a domain. Distribution of knowledge opens the possibility for parallel processing, e.g. refinement of a hypothesis in hierarchical classification can be done in parallel. The criterion of non-quantitative problem solving generally excludes normative methods such as linear programming and Bayesian optimization.

Third, a complete description of a generic task should include not only the items mentioned above, but also how it can be used for explanation, learning, and teaching. In this view, things like explanation and learning are not separate generic tasks, but are processes to be integrated into a generic task.

Fourth, each generic task can be associated with a programming language that embodies the kind of knowledge and inference strategy that the generic task specifies. For example, CSRL is a language developed for hierarchical classification (Bylander & Mittal, 1986), and DSPL, a language for object synthesis using plan selection and refinement (Brown, 1985). The languages of the generic tasks should be useful for guiding and mediating knowledge acquisition, which implies that each language should be useful for: pointing out what knowledge needs to be obtained, decomposing problems into subproblems, combining information in decision-making (especially uncertain information), debugging knowledge, allowing incremental expansion of the knowledge base, and providing process guidance to implementors.†

† This list is adapted from (Bradshaw, 1986).

OTHER GENERIC TASKS

Other generic tasks that have been identified include knowledge-directed information retrieval (Mittal, Chandrasekaran & Sticklen, 1984), abductive assembly of explanatory hypotheses (Josephson, Chandrasekaran & Smith, 1984), hypothesis matching (Chandrasekaran, Mittal & Smith, 1982), and state abstraction (Chandrasekaran, 1983). More detail on the overall framework can be found in Chandrasekaran (1986).

Exploiting hierarchical classification

Each generic task exploits domain knowledge differently; it calls for knowledge in a specific form that can be applied in a specific way. Because the knowledge-acquisition methodology must be able to extract and select the appropriate knowledge, each generic task calls for a different knowledge acquisition methodology. For illustration we consider the generic task of hierarchical classification (HC). In HC, the emphasis is on obtaining the classification hierarchy that contains the hypotheses that are relevant to the problem and can be used with the establish–refine strategy. This section does not provide a complete knowledge-acquisition methodology for HC, but outlines a number of considerations that a methodology must take into account. Additional guidelines for using HC can be found elsewhere (Mittal, 1980; Bylander & Smith, 1985).

DETERMINING HYPOTHESES OF INTEREST

HC is useful for determining the hypotheses that apply to a situation. An important step then is to decide upon the hypotheses that the problem-solver should potentially output. For example in diagnosis, the potential malfunctions of the object should be considered. The goal here is to determine the specific categories that should be produced, so if a general category is considered (e.g. "something is wrong with X"), then more specific categories should be generated (e.g. by asking "What types of problems can occur with X?"). Determining the usefulness of a category is discussed below.

ANALYSING COMMONALITIES AMONG HYPOTHESES

Once a collection of classificatory hypotheses have been identified, one needs to determine the commonalities among the hypotheses. These commonalities become potential candidates for mid-hierarchy hypotheses in the classification hierarchy. The easiest example to handle is when one hypothesis is clearly a subhypothesis of another, i.e. it asserts a more specific category. In general, two hypotheses may have commonalities along the following lines:

Definitional—The two hypotheses share a definitional attribute, e.g. hepatitis and cirrosis are liver diseases. Rain and snow are forms of precipitation;
Appearance—The two hypotheses are recognized using common pieces of evidence. Both cholestasis and hemolytic anemia have jaundice as a common symptom. Wet grass is symptomatic of both rain and dew;
Planning—The two hypotheses are associated with similar plans of action. Both

the common cold and allergies are reasons to take plenty of facial tissue with you. Either lightning or strong winds are good reasons for staying inside.

The ideal hypothesis asserts some definitional attribute over all its subhypotheses, has an appearance common to all its subhypotheses, and also provides constraints on the plans associated with its subhypotheses.

In general, the hierarchy should follow a definitional decomposition whenever possible. However, there are cases where appearance is an important consideration. For example, the Dublin–Johnson syndrome is a benign hereditary disorder that mimics key symptoms of cholestasis (jaundice, conjugated hyperbilirubinemia—high amounts of conjugated bilirubin in the blood). Because it *looks* so much like cholestasis, it is most useful to make it a subhypothesis of cholestasis.

ASSESSING EVIDENCE FOR OR AGAINST HYPOTHESES

The above two steps should generate a large number of hypotheses. However, not all of them will be useful for HC, i.e. there is a need to select a classification hierarchy that can be used to exploit the establish–refine strategy, getting rid of any intermediate hypothesis do not provide additional problem-solving power. Because the language we have used for HC, CSRL, requires a classification *tree* (Bylander & Mittal, 1986), we have become familiar with some of the strategies for evaluating hypotheses. However, the following questions are relevant whether a tree or tangled hierarchy is used.

Are there sufficient criteria to distinguish the hypothesis from other hypotheses? In other words, does this hypothesis have a different appearance from other hypotheses?

Is there evidence that distinguishes the hypotheses from its siblings? Because the establish–refine strategy does not consider a hypotheses unless its parent (or one of its parents in a tangled hierarchy) is relevant, evidence that distinguishes the hypothesis from its siblings is especially important.

Is the evidence normally available? Evidence for or against an hypothesis is not very useful if it is not likely to be available to the system when it is running. For example in medical diagnosis, some tests are relatively risky, expensive, or time-consuming to perform, so it is best to use hypotheses that rely on outward signs and symptoms and generally available laboratory data.

We have generally used another generic task, hypothesis matching, for mapping evidence to confidence values in hypotheses (Chandrasekaran *et al.*, 1982). However, we do not want to complicate the central issue by considering combinations of generic tasks. Examples of how hypothesis matching can be exploited are provided in Sticklen, Chandrasekaran & Smith (1985) and Bylander & Mittal, (1986).

DEBUGGING HYPOTHESES

An important part of knowledge acquisition is being able to find out what knowledge was incorrect or left out when something goes wrong. In HC, the following problems can occur:

Missing hypothesis—add the hypothesis to the classification hierarchy;

Wrong confidence value—debug the knowledge that produces the confidence value. Sticklen *et al.* (1985) describes how hypothesis matching can be debugged. The problems below assume that the confidence values are reasonable in view of the evidence considered;

Relevant hypothesis not considered—a hypothesis is not considered if one of its ancestors is not refined.† There are two possible problems with the ancestor.

There is not enough evidence to support the ancestor. To resolve this problem, one needs to find more evidence for the ancestor, lower the threshold for refining the ancestor, or implement more suitable ancestors, i.e. find better hypotheses for the establish–refine strategy.

The hypothesis is not definitionally a subhypothesis of the ancestor. In this case, the solution is to implement more suitable ancestors.

Irrelevant hypothesis considered—A hypothesis is considered only if one of its parents was refined.‡ Two causes of this problem are similar to the previous problems—when there is not enough evidence to oppose the parent or the hypothesis is not definitionally a subhypothesis of the parent. Similar fixes apply to these cases.

Another possible cause is that the establish–refine strategy being used is too simple. Sometimes a hypothesis should not be considered even if its parent is established. For example, if one of the hypothesis's siblings is confirmed, and the hypothesis is incompatible with its siblings, then the hypothesis should not be considered. The solution here is to adapt the establish–refine strategy to take this additional information into account. It should be noted that this problem is not a defect of establish–refine. Instead, it shows that establish–refine is really a *family* of strategies. The CSRL language, for example, provides a default establish–refine strategy and allows other establish–refine strategies to be defined.

KNOWLEDGE ACQUISITION FOR HIERARCHICAL CLASSIFICATION

The point of HC is to determine the hypotheses that describe a situation. The point of knowledge acquisition for hierarchical classification is to obtain the knowledge (the classification hierarchy) so that HC can be effectively performed. That is, knowledge acquisition needs to exploit the interactions between the representation, inference strategy, and the problem. Exploiting HC means the construction of a classification hierarchy that contains the hypotheses to be considered and that allows the establish–refine strategy to efficiently search the hypotheses. Thus a knowledge acquisition methodology for HC needs to evaluate each hypothesis for its relevance as a potential output and in view of the evidence that can support or oppose it.

A re-examination of past beliefs

Some generally held beliefs about knowledge-based systems need to be re-examined in light of the interaction problem and our proposal to exploit it. These beliefs have

† This statement is true only for a classification tree. For a tangled hierarchy, a hypothesis is not considered if every path from the root node to the hypothesis has a node that was not refined.

‡ It is possible that some other problem solver might directly ask a classifier to consider a hypothesis. This problem would be then attributable to the other problem solver, not the classifier.

served the first generation of knowledge-based systems well, especially in stimulating much research and discussion. However, we believe it is the time to reconsider them.

BELIEF 1: KNOWLEDGE SHOULD BE UNIFORMLY REPRESENTED AND CONTROLLED

This belief denies the interaction problem and implies that there is nothing to be gained by using different representations to solve different problems. Our experience is that when the problems of a domain match the generic tasks, the generic tasks provide explicit and powerful structures for understanding and organizing domain knowledge.

BELIEF 2: THE KNOWLEDGE BASE SHOULD BE SEPARATED FROM THE INFERENCE ENGINE

This belief denies that the inference strategy affects how knowledge is represented. However, its real effect has been to force implementors to implicitly encode inference strategies within the knowledge base. Both MYCIN, whose diagnostic portion is best understood as HC, and R1, which is best understood as routine design, show that this separation is artificial.

BELIEF 3: CONTROL KNOWLEDGE SHOULD BE ENCODED AS METARULES

Although metarules address the problem of how to have multiple, explicit strategies in a rule-based system, the metarule approach ignores other aspects of the interaction problem. The "separation of control knowledge from domain knowledge" promotes the view that domain knowledge can be represented independent of its use, i.e. that different sets of metarules can be applied as needed. However, given a clear strategy (whether metarules or inference engine) and a problem to be performed, the domain knowledge will be adapted to interact with the strategy to solve the problem.

BELIEF 4: THE ONTOLOGY OF A DOMAIN SHOULD BE STUDIED BEFORE CONSIDERING HOW TO PROCESS IT

We believe that ontology should not be performed just for its own sake, but in view of the problems that need to be done. For example, to apply HC to a domain, there is a need to focus on the hypothesis space and evaluate hypotheses. Although other knowledge structures (e.g. component hierarchies, causal networks) may be useful for other generic tasks, if HC is going to be performed, then knowledge acquisition should concentrate on those aspects of the domain that are relevant to HC. This is not to say that a domain should not be analysed to identify what generic tasks are appropriate; however, this kind of domain analysis does not require an exhaustive ontology of the domain.

BELIEF 5: CORRECT REASONING IS A CRITICAL GOAL FOR KNOWLEDGE-BASED SYSTEMS

Everything else being equal, being correct is better than being incorrect. However, an emphasis on correctness detracts from more critical issues. One of those issues is developing an understanding of the appropriate strategies to be applied to a problem. For example, there has been much research and debate about normative methods for calculating uncertainty. The reasoning problem, though, is not how to precisely calculate uncertainty, but how to avoid doing so. In diagnosis, for example, there is much more to be gained by using abduction (assembling composite hypotheses to account for symptoms), then by independently calculating the degree of certainty of each hypothesis to several decimal places of accuracy.

BELIEF 6: COMPLETENESS OF INFERENCE IS A CRITICAL GOAL FOR KNOWLEDGE-BASED SYSTEMS

Everything else being equal, being complete is better than being incomplete, but an emphasis on completeness ignores the fact that certain kinds of inferences will be more important than others for a particular problem. For example in our description of HC, we did not mention that when a subhypothesis is confirmed, one can infer that its ancestors are also confirmed. This is not because we believe that a HC problem solver should never perform this inference, but because other inferences are the crucial aspects of HC: refinement of a hypothesis if it is likely and pruning of its subhypotheses when it is ruled out.

BELIEF 7: A REPRESENTATION THAT COMBINES RULES, LOGIC, FRAMES, ETC. IS WHAT IS NEEDED

Such representations appear to be a good compromise since they let you represent knowledge in the "paradigm" of your choice. Unfortunately, this is, at best, only an interim solution until something better is found. None of the individual representations fully address the interaction problem, nor do they distinguish between different types of reasoning.

Generic tasks at the "right" level of abstraction

The first generation of research into knowledge-based systems has conducted an extensive search for a "holy grail" of representation, in which knowledge could be represented free of assumptions of how it would be used. For any particular problem, though, certain kinds of inferences and certain pieces of knowledge will be critical to the problem, and consequently, domain knowledge needs to be organized so those inferences are performed efficiently. This is how the interaction problem arises, and why it will never go away. Instead of futilely trying to avoid it, the interaction problem needs to be studied and understood so that methods of exploiting it can be discovered and applied.

Our theory of generic tasks is an attempt to provide the "right" level of abstraction for this and other problems of knowledge-based reasoning. Each generic task provides a knowledge structure in which knowledge can be organized at a conceptual level. In hierarchical classification, the concepts are hypotheses organized as a classification hierarchy. Each generic task identifies a combination of a problem definition, representation, and inference strategy that exploits the interaction problem. We have shown how the generic task of hierarchical classification can be associated with a knowledge-acquisition methodology that takes advantage of the interactions between domain knowledge and the establish-refine strategy.

Research supported by Air Force Office of Scientific Research, grant 82-0255, and Defense Advanced Research Projects Agency, RADC Contract F30602-85-C-0010

References

BRADSHAW, J. M. (1986). Presentation at the *Knowledge Acquisition for Knowledge-Based Systems Workshop,* Banff, Canada.

BROWN, D. C. (1985). Capturing mechanical design knowledge. *Proceedings of the 1985 ASME International Computer in Engineering Conference,* Boston.

BROWN, D. C. & CHANDRASEKARAN, B. (1986). Knowledge and control for a mechanical design expert system. *Computer,* **19,** 92–100.

BYLANDER, T. & MITTAL, S. (1986). CSRL: A language for classificatory problem solving and uncertainty handling. *AI Magazine,* **7,** 66–77.

BYLANDER, T. & SMITH, J. W. (1985). Mapping medical knowledge into conceptual structures. *Proceedings of Expert System in Government Symposium. IEEE Computer Society,* McLean, Virginia, pp. 503–511.

CHANDRASEKARAN, B. (1983). Towards a taxonomy of problem solving types. *AI Magazine,* **4,** 9–17.

CHANDRASEKARAN, B. (1984). Expert systems: matching techniques to tasks. In *Artificial Intelligence Applications for Business.* Norwood, New Jersey: Ablex, pp. 116–132.

CHANDRASEKARAN, B. (1986). Generic tasks in knowledge-based reasoning: high-level building blocks for expert system design. *IEEE Expert* **1,** 23–30.

CHANDRASEKARAN, B., MITTAL, S. & SMITH, J. W. (1982). Reasoning with uncertain knowledge: the MDX approach. *Proceedings of the Congress of American Medical Informatics Association.* San Francisco: AMIA, pp. 335–339.

CLANCEY, W. J. (1985). Heuristic classification. *Artificial Intelligence,* **27,** 289–350.

DUDA, R. O., GASCHNIG, J. G. & HART, P. E. (1980). Model design in the prospector consultant system for mineral exploration. In *Expert Systems in the Microelectronic Age.* Edinburgh University Press, pp. 153–167.

FORGY, C. L. (1981). Technical Report CMU-CS-81-135, *OPS5 Users Manual.* Carnegie–Mellon University.

FRIEDLAND, P. (1979). Knowledge-based experiment design in molecular genetics. *Ph.D. thesis,* Computer Science Department, Stanford University.

GOMEZ, F. & CHANDRASEKARAN, B. (1981). Knowledge organization and distribution for medical diagnosis. *IEEE Transactions on Systems, Man and Cybernetics,* **SMC-11,** 34–42.

JOSEPHSON, J. R., CHANDRASEKARAN, B. & SMITH, J. W. (1984). Assembling the best explanation. *Proceedings of the IEEE Workshop on Principles of Knowledge-Based Systems.* IEEE Computer Society, Denver, pp. 185–190.

MARR, D. (1982). *Vision.* W. H. Freeman.

MCDERMOTT, J. (1982). R1: A rule-based configurer of computer systems. *Artificial Intelligence* **19,** 39–88.

VAN MELLE, W. (1979). A domain independent production-rule system for consultation programs. *Proceedings of the Sixth International Conference on Artificial Intelligence.* Tokyo, pp. 923–925.

MINSKY, M. (1975). A framework for representing knowledge. *The Psychology of Computer Vision.* McGraw–Hill, p. 211–277.

MITTAL, S. (1980). Design of a distributed medical diagnosis and database system. *Ph.D. thesis,* Department of Computer and Information Science, The Ohio State University.

MITTAL, S., CHANDRASEKARAN, B. & STICKLEN, J. (1984). Patrec: a knowledge-directed database for a diagnostic expert system. *Computer* **17,** 51–58.

SHORTLIFFE, E. H. (1976). *Computer-Based Medical Consultations: MYCIN.* New York: Elsevier.

STICKLEN, J., CHANDRASEKARAN, B. & SMITH, J. W. (1985). MDX-MYCIN: the MDX paradigm applied to the MYCIN domain. *Computers and Mathematics with Applications,* **11,** 527–539.

Training Knowledge Engineers

The Knowledge Acquisition Grid: a method for training knowledge engineers

MARIANNE LAFRANCE
Department of Psychology, Boston College, Chestnut Hill, MA 02167, U.S.A.

This paper describes the *Knowledge Acquisition Grid,* developed to assist knowledge engineers in the manual transfer of expertise. The *Grid* is used in a knowledge-acquisition module which itself is part of a larger program designed to train people in knowledge engineering techniques offered by Digital Equipment Corporation. The *Grid* describes a two-dimensional space in which five forms of expert knowledge and six basic types of interview questions constitute the horizontal and vertical dimensions respectively. Description of the rationale, dimensions, components, and strategy for use of the *Grid* in the knowledge-acquisition component of building an expert system is provided along with discussion of the need for greater attention in general to the social psychology of expert interviewing.

Introduction

Knowledge-based systems require the kind of specialized and extensive know-how usually available only in the heads of a small coterie of experts. To date, the gaining of this expertise has proven to be one of the great challenges in building expert systems. This paper describes one component of a training program in Knowledge Acquisition designed to provide knowledge engineers with the conceptual and practical means with which to more effectively conduct the manual transfer of expertise.

BACKGROUND

Started in 1982, the original 14-week course in building knowledge-based systems covered a range of available Artificial Intelligence tools but no material specifically focussed on how to deal with the complexities of acquiring knowledge from the experts. In response to this perceived gap, a 1 and 1/2 day module in "Expert Interviewing" was created with the aim to introduce knowledge engineers to in-depth interviewing by drawing on theory and research in psychology, ethnography, non-verbal communication, cognitive science, and the sociology of specialized interviewing.

The guiding assumptions for the design were twofold: first, that experts who are in the position to know the most may be the least able to articulate what they know; and second, that available methods could be adapted and developed for the expressed purpose of eliciting this particular kind of *tacit knowledge.* The module has since been expanded to reflect ongoing work in knowledge-based systems. Now called "Knowledge Acquisition", the segment has expanded and become an integral part of the now 7-week overall program in the art and science of putting expert systems together.

KNOWLEDGE-BASED SYSTEMS Vol. 1
ISBN 0-12-273251-0

FOCUS ON MANUAL KNOWLEDGE ACQUISITION

Although the course includes material on automated knowledge acquisition, it relies on a firm footing in manual interviewing methods. There are a number of reasons for this emphasis. First, manual knowledge acquisition is the most routinely used and broadly useful method of acquiring expertise. Manual methods can be applied to any knowledge domain. Second, even when automated knowledge-acquisition features become practically feasible for some aspects of the process, manual methods will still be called upon to complement and cover content not accessible by purely automated means. Third, manual interviewing methods are uniquely good at taking into account considerable expert-to-expert and domain-to-domain differences. A final reason for focusing on manual methods of gathering expertise is that as they are used in an increasing range and variety of settings, they will suggest promising directions for how expertise transfer can be more effectively automated.

Training objectives

The training module in knowledge acquisition is designed to increase knowledge-acquisition effectiveness by exposing participants to a variety of issues and tools. More specifically the overall objectives are as follows:

(1) an enhanced appreciation for knowledge acquisition as an integral process in building expert systems;

(2) an understanding of the social psychological nature of the expert interview;

(3) facility in identifying the prevailing myths about knowledge acquisition and in dealing with the realities;

(4) competence in using specialized interviewing techniques to enhance the expert's involvement and the knowledge engineer's comprehension;

(5) the ability to recognize different kinds of expert knowledge;

(6) a repertoire of different types of questions to access different types of knowledge;

(7) skill in using the Knowledge Acquisition Grid;

(8) awareness of the cycles involved in knowledge acquisition;

(9) acquaintance with new developments in the automation of knowledge acquisition.

NEED FOR TRAINING IN KNOWLEDGE ACQUISITION

Expert systems are valuable to the extent to which they are knowledge-rich, but the need for the knowledge has not been matched by effective means for drawing it out. Expert knowledge is not easily captured. The process is time-consuming, painstaking and complicated. In fact, knowledge acquisition has been called the critical bottleneck problem in the development of knowledge-based systems (Feigenbaum, 1980).

In acknowledgement of this bottleneck, one response has been to increase efficiency by developing means for automating the transfer of expertise. Another tack has been to recognize that several issues contribute to the bottleneck and that methods must be developed to deal with each of them. A number of the important ones are: (1) lack of understanding about the nature of expertise; (2) poor manual interviewing skills; and (3) a limited repertoire of questioning strategies. Each of these are described briefly below.

Nature of expertise
Knowledge engineers who assume that the problem of knowledge acquisition is essentially quantitative are likely to bypass important know-how. Work in cognitive psychology has shown, for example, that experts differ from novices not only in the amount they know, but also in how their knowledge is represented and bundled (Chi, Glaser & Rees, 1982; Chiesi, Spilich & Voss, 1979, Larkin, McDermott, Simon & Simon, 1980; Milojkovic, 1982).

A conception of know-how based on qualitative aspects has important implications for people concerned with acquiring knowledge from experts. For example, there is need to recognize that expert knowledge is complex. The Forms of Knowledge Dimension of the Knowledge Acquisition Grid acknowledges this and so articulates five different kinds of expert knowledge, each with its own structure and characteristics and each enclosing a subset of the expert's expertise. These will be described more fully below in the section entitled, The Knowledge Acquisition Grid.

Poor manual interviewing skills
The process of knowledge acquisition usually demands an extended and intense collaboration between domain expert and knowledge engineer. To say that this requires interpersonal competence and communication skills severely understates what is involved. Experts and knowledge engineers alike have different experiences and goals as well as preferred ways of thinking and problem-solving. If these are ignored, the knowledge engineer's understanding of a particular domain may be incomplete or significantly misconstrued.

Experts cannot be simply hooked-up and drained of everything they know. An approach that takes account of the psychological nature of expertise and the interpersonal quality of acquiring and representing it will have more validity than one that gives little weight to these important social psychological issues.

Limited questioning strategies
Designing more proficient knowledge acquisition depends in part on expanding ways to tap into human expertise. It is insufficient merely to have the patience to ask "why" of an expert 400 times in a row. A enlarged and more targeted repertoire of questions and probes is required. For example, recent research in cognitive and social psychology has shown that different types of questions elicit different kinds of answers (Clark, 1985; Loftus, 1975). The Types of Questions Dimension of the Knowledge Acquisition Grid articulates six distinct kinds of questions directed at different aspects of expert knowledge. These also will be described in the section on The Knowledge Acquisition Grid.

In sum, these bottlenecks in knowledge acquisition can be substantially relieved through the application of a manual method that comprehends the nature of the acquisition process and is adept at grasping various kinds of expertise using a diversified set of questions.

THE KNOWLEDGE ACQUISITION GRID

The Knowledge Acquisition Grid was created to deal with the realization that expertise resists single category compartmentalization and that no single question can elicit all the required information.

FORMS OF KNOWLEDGE

QUESTION TYPES	LAYOUTS	STORIES	SCRIPTS	METAPHORS	RULES-OF-THUMB
Grand Tour					
Cataloging Categories					
Ascertaining Attributes					
Determining Inter-connections					
Seeking Advice					
Cross-checking					

FIG. 1. The Knowledge Acquisition Grid.

Structure of the Knowledge Acquisition Grid

The Knowledge Acquisition Grid shown in Fig. 1 organizes expert knowledge and knowledge engineer questions as separate but interacting dimensions. The first dimension represented by the columns in the Grid, describes the *Forms of Knowledge* in which an expert's know-how is stored. The second dimension, represented by the rows in the Grid, describes the *Types of Questions* that are available to knowledge engineers directed to making it explicit.

DIMENSION 1: FORMS OF KNOWLEDGE

Since expert knowledge is complex, expert-system builders need a way to recognize and access material that is distributed in memory and implicit in character. The *Forms of Knowledge* dimension describes five distinct categories. A brief description of each form is as follows:

Layouts

Layouts incorporate the expert's "map" of the task, including an understanding of its boundaries, organization, and basic classifications. Layouts subsume and give coherence to the expert's facts and heuristics by specifying the goals to which they aim, and the criteria used to characterize the problem at hand. By getting access to how the expert sets the task and organizes current information in light of prior knowledge and the present context, the knowledge engineer is better able to understand how to frame the problem.

Stories

Stories represent the classic cases and typical examples culled from the expert's long experience with the problem domain. Stories can be of a number of types in addition to the now familiar *Talking Aloud Protocol* (Waterman & Newell, 1971). For example, *Explanatory Stories* are those in which the expert, seeking to account for some puzzling situation, narrates a set of events that unfold in such a way as to lead up to and produce the phenomenon in question. A *Diagnostic Story/ Prescriptive Story* describes some phenomenon in a way that shows what was wrong and what needed fixing.

Scripts

Scripts give the expert's sequential and procedural knowledge of the domain. The basic elements of a script are its roles, standard props or objects needed to carried out the actions, a standard sequence of scenes wherein one act enables the next, and some normal results from performing the activity successfully (Abelson, 1981). To know an expert's scripts is to have a temporal chart of critical actions, and to be able to understand each action in terms of the prior knowledge required to perform it.

Metaphors

Metaphors encapsulate the expert's alternative images of the task, each of which includes unique features, constraints, and options. Metaphors describe one thing by reference to another apparently dissimilar thing so that the first is understood more completely than if the comparison had not been made. Their advantage is being able to present an idea which can later be reconstructed and embellished through probes directed to the expert by the knowledge engineer.

Rules-of-thumb

Rules-of-thumb provide the myriad tactics and heuristics for interpreting and dealing with the array of circumstances encountered in carrying out the task. A rule-of-thumb encapsulates tacit knowledge about which conditions warrant which actions, and about how to gather data on and assess current conditions. Rules-of-thumb are concrete, implementable strategies of minor to moderate scope which can single out and define as issues those specific, limited conditions for which they serve as the most complete strategy.

Rationale for multiple forms of knowledge

Each of these Forms of Knowledge is a different way in which know-how can be represented in the mind of the expert. Together, the five forms suggest that expertise in a domain is not encoded in the expert's mind in a single bundle, but that varying experiences with the problem lead to different mental representations.

Support for this idea comes from research in cognitive psychology that shows that learning is a segmented process, and that the various segments are stored in different parts of the memory (Anderson, 1980). Psychologists hypothesize that breaking up of information in this way allows people to make better generalizations and more useful predictions. Similarly, Minsky's society-of-mind theory proposes that intelligent action emerges from the interactions of many small systems operating within an evolving overall administrative structure (Minsky, 1986).

The Forms of Knowledge dimension thus sees expertise as made up of a number of smaller subsystems, each with its own character and function. There are, in addition, a number of practical advantages to seeing expert know-how as multi-modal. In the first place, the Forms of Knowledge provide a way during knowledge acquisition to catalogue a domain's real complexity. In the spirit of getting a handle on a domain, there is sometimes danger that that which initially does not seem to fit may be left out. The five kinds of knowledge provide the means to tap a domain's particular intricacies. Secondly, the Forms of Knowledge encourages broad coverage. In seeing knowledge as multi-modal, the knowledge engineer is more likely to check whether other modes of expressing the expertise will turn up relevant but previously unarticulated know-how. Third, each Form of Knowledge reflects a different but practical slant on the problem domain. Layouts, Stories, Scripts, Metaphors, and Rules-of Thumb implicitly prescribe how to deal with different features of the task.

Next, the five Forms of Knowledge have the advantage in allowing understanding of the expert's domain in his or her own words. This gives the knowledge engineer a way to stay close to concepts of a domain while at the same time having the means to organize them. Finally, the five Forms of Knowledge provide a buffer between acquiring the knowledge from the expert and transforming into a particular representation. This allows the expertise to be understood first on its own terms, a stage which is important for subsequently being able to select appropriate representations for it.

One caveat should be mentioned here: even though expert knowledge is generally multi-modal, not all Forms of Knowledge are evident in all knowledge domains, nor are they used equivalently by all experts. Nevertheless, knowledge engineers should recognize the potential in each area, and not assume that unfamiliar forms are necessarily unproductive ones.

DIMENSION 2: TYPES OF QUESTIONS

The process of constructing a knowledge base requires the knowledge engineer to have a number of question types. Survey researchers, for example, have shown that using different question formats optimizes the information received. Alternatively, the use of only one type of question severely limits the kind of answers that can be obtained. The Knowledge Acquisition Grid presents five *Types of Questions*. A brief description of each type with an example is given below:

Grand Tour questions

Grand Tour questions cast a wide net over the domain in order to understand the boundaries as the expert sees them. The material sought includes an overview of the expert's perspective, goals, organization, and classifications. Sub-areas within the expert's domain are comparably pursued by means of *Mini-Tour* queries. An example of a Grand Tour question is: "Could you describe the kinds of things that schedulers do? Please do not edit things out of your description, even things you think may not be important". This particular phrasing of the question is designed to elicit the expert's *Layout* description but could as equally be directed at eliciting a Story or Metaphor.

Cataloging the categories

The expected outcome of these questions is an organized taxonomy of the expert's terms and concepts. An example of a Cataloging the Categories Question is: "When you gave me an overview of your job, you talked about 'schedulers'. Are there different types of schedulers? Are schedulers a subtype of some other kind of job?" This type of question might follow a *Layout* response to a *Grand Tour Question.*

Ascertaining the attributes

These questions aim to discover the distinguishing features and range of possible values of the expert's concepts. An example of an ascertaining the attributes question is: "You've described a number of types of scheduling situations that you've encountered. I wonder whether you could now take the first two that you mentioned, and describe some ways in which these two are similar to each other but different from the third example that you gave". As stated, this particular question is a follow-up to a *Stories* description.

Determining the interconnections
These questions directed at uncovering the relations among the concepts in the domain. Of particular interest is the existence of a causal model for the whole domain or parts of it. An example of Determining the Interconnections Question is: "In describing the routine set of steps for scheduling an order you said that checking the request date occurs before anything else. Why is that the case?" This specific question is directed at previously obtained *Scripts* information.

Seeking advice
These questions are designed to reveal the expert's recommendations and hence strategies for how to deal with a variety of conditions such as how to determine current conditions and which conditions warrant which actions. An example of a Seeking Advice Question would be: "You've compared scheduling to playing a board game; from your experience with playing board games what advice could you give on doing scheduling?" This type of question would be in response to a previously volunteered *Metaphor*.

Cross-checking questions
These questions are designed to validate and examine the limits on previously obtained information. Cross-checking questions actually consist of five subtypes including, the *Naive Question, Playing Devil's Advocate, Posing Hypothetical Situations,* asking *How Sure Are You*? and *Seeking the Exception.* An example of a Naive Question is "Bear with me while I ask what may appear to be a naive question. Could you tell me why orders need to be scheduled?" An example of a Devil's Advocate Question would be: "Let me play devil's advocate in response to your story about the need to set priorities on scheduling multiple orders. What if you didn't set priorities?"

Rationale for multiple-question types
Research indicates that questions variously restrict and direct the kinds of answers that can be obtained. For example, there are two broad types of questions: questions that ask the recipient to provide the questioned item, and second, utterances that require the hearer to simply agree or disagree with the content of the question. *Wh-questions,* those beginning with the words, *who, whose, which, what, why* and *how,* ask that the question variable be supplied and hence provide considerable leeway for the recipient to respond. The second type, *yes/no questions* call on the hearer merely to concur or not with the proposition put forward by the questioner (Woodbury, 1984).

Within the *wh-question* category itself, there are differences in the specificity of the information solicited. Questions beginning with *what, why,* and *how* are open-ended whereas questions beginning with the words, *who, where, when,* and *which* appear to require answers that are as specific as the hearer is able to make them.

In addition, subtle variations in the wording of questions has also been shown to affect how hearers respond to them. For example, research on eye-witness testimony has revealed that witnesses to a car-accident report that the car was going significantly faster if the question was phrased: "How fast was the car going when it *crashed* into the wall?" than if it was put, "How fast was the car going when it *ran* into the wall?". A comparable rewording in another study showed the same effect. In this case, survey respondents report more headaches if the question is phrased: "Do you get headaches *frequently* and if so, how often?" than if they are asked instead, "Do you get headaches *occasionally* and if so, how often?" (Loftus, 1979).

Studies such as these show that replies are significantly effected by how questions are phrased. Other research on questions has focussed on asking whether some question strategies are more effective than others in eliciting full and valid disclosure. The theoretical notion here is that the memory trace is composed of several features and hence the effectiveness of a question or retrieval cue is related to the amount of feature overlap with the event being reported. In other words, some questions pick up elements that are neglected when other kinds of questions are asked. A second outcome of this research is the recognition that there are probably several retrieval paths to the encoded event, so that information that is not readily accessible by one retrieval cue might be available with a different cue (Tulving, 1974).

The Knowledge Acquisition Grid is based on this theoretical framework. The matrix formed by forms of knowledge and types of questions was designed to increased the feature overlap between the encoding of the expertise by the expert and the subsequent retrieval of it by a knowledge engineer in a knowledge acquisition context. The multiple forms of both knowledge and questions encourage using many retrieval paths.

There are then a number of advantages to being able to call upon a number of different question types. The six types of questions in the Grid are designed to be seen as a package, the total contents of which are geared to producing a comprehensive survey of the expert's knowledge. Secondly, the questions enable the knowledge engineer to approach the domain from a number of directions, thus increasing the chances of revealing important material overlooked when other routes are taken. A third advantage to having access to several question types is their respective attention to different features. Each question is aimed at uncovering different aspects of the expertise. For example, questions from Cataloging the Categories focus on getting hold of the inventory of the expert's domain. Seeking Advice questions aim to disclose the expert's methods for hitting upon a solution.

The six types of questions also vary in their level of specificity which, in combination, have the advantage of pulling out material varying in detail and composition. Grand Tour questions seek to draw out broad and comprehensive descriptions in order to grasp the scope of the problem domain, whereas Cross-Checking questions aim to verify fine detail. A fifth advantage to utilizing different types of questions is the ability to contain knowledge engineer bias. The range of questions should counteract the tendency of knowledge engineers to favor one type of question over another at the expense of a more comprehensive grasp of the relevant expertise.

Sixth, having access to a number of different types of questions guards against unwarranted assumptions by both knowledge engineer and expert. In asking a different type of question about the same material, the knowledge engineer may find that he or she has incorrectly assumed certain connections or inclusions. Alternative questions may also encourage the expert to be more explicit about things previously assumed to be obvious. There are also a number of advantages having to do with doing well by the expert. For example, having access to a number of different types of questions provides the knowledge engineer with opportunities to change direction or pace or content, and thus sustain the expert's interest and motivation. Multiple questions also indirectly convey to the expert that the knowledge engineer respects the range and complexity of what he or she knows. Finally, multiple questions provide cross-validation without cross-examination. By being able to draw on a range of queries, knowledge engineers can come back repeatedly to the same content area while avoiding the appearance of redundancy or skepticism.

STRATEGY FOR USING THE KNOWLEDGE ACQUISITION GRID

The five *Forms of Knowledge* and the six *Types of Questions* in the Grid do not exhaust either dimension. Rather each dimension is intended to provide the knowledge engineer with a wide-angled lens to see more of the relevant expertise. Moreover, the categories within each dimension overlap to some degree. For example, a Story about a particular situation may contain within it a Metaphor or Rule-of-Thumb. The important feature of the matrix nature of the Grid is not category independence but domain scope.

The Grid can be used both passively, in providing assistance in decoding expert's replies, and actively in deliberating evoking particular knowledge content. Each of these will be taken up in turn followed by some suggestions for structuring alternative paths through the Grid.

Decoding experts' replies

The Grid provides an explicit depiction of the reciprocal relationship that can exist between knowledge-acquisition queries and experts' descriptions. Questions do not only elicit content but can follow it as well. The Grid provides a way of dealing with replies which do not seem to be tied to a question as phrased. Consequently, a practiced knowledge acquisitor can make sense of the answers provided. The Grid provides a tool for appreciating and identifying the different forms in which expertise comes.

One of the marks of the good knowledge acquisitor is the ability to decode answers of all types and specially those which appear not to be a reply to the immediately preceding question. More specifically, knowledge engineers who use the Grid develop the habit of asking themselves after *each* reply: "To which question is this an answer?" and "Of what knowledge form is this an example?"

Evoking particular knowledge forms

The Grid provides the means for deliberately evoking a particular kind of response from the expert. At times the expert may be asked for a Metaphor to open up a new line of investigation on the problem domain. Or a Story about an atypical situation may be solicited in order to get a different slant on the domain after a protocol analysis has been done on a typical case.

Particular knowledge forms are evoked as a way to call up what is typically tacit knowledge. Since experts are frequently hard pressed to describe what they know, direct "why" or "how" questions are as frequently unsuccessful at uncovering important information: but if they can be given a mental handle, such as a request for a particular kind of experience with the domain, then previously implicit material can be made explicit.

Alternative paths through the Grid

Each dimension of the Grid has its own internal order. In the Types of Questions dimension, open-ended queries are at the top, while increasingly more specific, directed questions are located at the bottom. The Forms of Knowledge Dimension is somewhat less linear, but the more declarative kinds of knowledge are represented at the left, and the more procedural kinds of know-how are on the right side of the grid.

One way then of using the Grid is to start at the upper left and move back and forth in a more or less systematic manner to the lower right. Such an orientation begins with a broad investigation of the factual aspects of the expertise, and concludes with a focussed attention on specific heuristics.

This tactic is not, however, the only way to use the Grid. It can also be used in a more cyclic fashion. The knowledge engineer moves from general queries to specific ones and back to the general, and from the declarative to the procedural and back to the declarative in the knowledge mode.

How the Grid is used depends ultimately on the particular needs of different expert systems teams; there is no single best way to deploy it. The value of the Grid is its multi-modal depiction of expert knowledge and the multi-form investigative means to get a hold of it.

Summary

Skill with the Grid comes with practice. In the knowledge-acquisition training module, participants learn how to use it through exposure to videotape examples as well as group exercises and guided feedback. Moreover, participants learn to use the Grid in the context of learning about other aspects of the knowledge acquisition process, including strategic, technical, and organizational concerns.

Manual knowledge acquisition can be made systematic. It can also be learned. The Knowledge acquisition training module has been developed to demonstrate both these properties. Traditionally, manual knowledge acquisition has been seen as putting in lots of unproductive hours. Experience with the Grid in particular has shown that greater sophistication with manual interviewing techniques can lead to time well spent and to greater overall knowledge-acquisition proficiency.

References

ABELSON, R. P. (1981). The psychological status of the script concept. *American Psychologist,* **36,** 715–729.

ANDERSON, J. R. (1980). *Cognitive Psychology and its Implications*. San Francisco, California: Freeman.

CHI, M. T. H., GLASER, R. & REES, E. (1982). Expertise in problem solving. In STERNBERG, R. Ed., *Advances in the Psychology of Human Intelligence,* Vol. 1. Hillsdale, New Jersey: Lawrence Erlbaum, pp. 7–75.

CHIESI, H. L., SPILICH, G. J. & VOSS, J. F. (1979). Acquisition of domain-related information in relation to high and low domain knowledge. *Journal of Verbal Learning and Verbal Behavior,* **18,** 257–273.

CLARK, H. H. (1985). Language use and language users. In LINDZEY, G. & ARONSON, E. Eds, *The Handbook of Social Psychology,* 3rd edn, Vol. 2. Reading, Massachusetts: Addison–Wesley, pp. 179–232.

FEIGENBAUM, E. A. (1980). *Knowledge Engineering: The Applied Side of Artificial Intelligence.* Stanford, California: Stanford University, Heuristic Programming Project.

LARKIN, J., MCDERMOTT, J., SIMON, D. F. & SIMON, H. (1980). Expert and novice performance in solving physics problems. *Science,* **208,** 1335–1342.

LOFTUS, E. F. (1975). Leading questions and the eyewitness report. *Cognitive Psychology,* **7,** 560–572.

LOFTUS, E. F. (1979). *Eyewitness Testimony.* Cambridge, Massachusetts: Harvard University Press.

MILOJKOVIC, J. D. (1982). Chess imagery in novices and master. *Journal of Mental Imagery,* **6,** 125–144.

MINSKY, M. (1986). *Society of Mind.* New York: Simon and Schuster.

TULVING, E. (1974). Cue-dependent forgetting. *American Scientist,* **62,** 74–82.

WATERMAN, D. A. & NEWELL, A. (1971). Protocol analysis as a task for artificial intelligence. *Artificial Intelligence,* **2,** 285–318.

WOODBURY, H. (1984). The strategic use of questions in court. *Semiotica,* **48,** 197–228.

Modelling human expertise in knowledge engineering: some preliminary observations

DAVID C. LITTMAN

Cognition and Programming Project, Department of Computer Science, Yale University, New Haven, CT 06520, U.S.A.

This paper reports the results of an empirical analysis of the knowledge engineering behavior of six persons with extensive experience in artificial intelligence (AI). The six persons were given the task of designing an AI program and were videotaped while they did so; during their 2–3 h design sessions, they were asked to talk aloud about what they were doing and why they were doing it. The paper identifies several recurrent behaviors common to all the AI designers. For example, several components of the designers' goal structures are identified, as is the importance of focussing on a "touchstone", or key issue, around which much of the designer's behavior revolves. Several potential implications of the research for the design of knowledge engineering tools are explored.

1. Introduction: motivation, and goals

Knowledge transfer from expert humans to machines is slow, difficult, and prone to error. In order to automate the knowledge transfer process, knowledge engineers will have to write intelligent knowledge-acquisition programs that are easily as complex as the domain programs they want to produce. One potentially useful approach to acquiring information that might assist developers of automated knowledge acquisition tools may be to:

study human expertise in knowledge acquisition as a skill in its own right

and

attempt to write programs that can reproduce it.

That is, it may be productive to view the problem of automated knowledge engineering as a *knowledge engineering problem itself*! Our goal, as "meta knowledge-engineers", would then be to identify the knowledge and skills that human knowledge engineers use when they perform the task of identifying an expert's underlying model of a domain. We would then engineer our knowledge about knowledge engineering into knowledge-acquisition programs. It seems plausible to suppose that, if we could understand the skills and knowledge that knowledge engineers use when they build models of someone's reasoning in a domain (perhaps their own!), we could contribute to the goal of understanding how to build general automated knowledge-acquisition tools. Ultimately, we would like to be able to write AI programs that can, themselves, write general knowledge engineering programs.

In this paper, we present preliminary analyses of videotaped protocols of six AI

KNOWLEDGE-BASED SYSTEMS Vol. 1
ISBN 0-12-273251-0

researchers who were given a transcript of a fragment of tutorial interaction between a tutor and a student in which the tutor was teaching the student some basic principles of large weather systems. The AI researchers were given the following task:

> Write "interesting" AI computer programs that simulate the interactions of the tutor and the student.

We gave our AI designers the task of writing simulation programs for two main reasons. First, building simulation programs is often a useful initial step in producing performance-oriented knowledge-based systems. Second, by giving our AI designers the goal of producing simulation programs, we forced them to focus their attention on the actual domain knowledge and rules of interaction that the tutor and student used. Forcing our AI designers to focus on the actual domain knowledge and tutorial knowledge appeared to have the effect of preventing them from looking for some AI "hacks" that would reproduce the tutorial interaction presented in the transcript of the tutorial dialogue without formulating a plausible representation of the student's and the tutor's knowledge.

Our goal in analysing the protocols of the AI researchers as they designed their programs to simulate the interaction of the tutor and the student is guided by this hope:

> We hope that what we learn through systematic empirical study and simulation of human knowledge engineers will assist us in the knowledge transfer endeavor. We further hope that, by understanding the knowledge and skills used by AI researchers to build AI programs, we will be able to build AI programs ourselves that will behave as AI researchers do when they are writing AI programs.

We realize that the observations presented in this paper are preliminary. In addition, it may seem that much of what we observed in our subjects has been noted in books and papers such as Hayes-Roth, Waterman & Lenat (1983) and Chandrasakaran (1985), or is part of the tacit knowledge of members of the AI community; both are true. The intended contribution of this work, however, is threefold:

to suggest that knowledge engineering may itself be a reasonable domain for knowledge engineering;
to begin to provide an *empirical basis* for constructing intelligent knowledge engineering programs;
to suggest that the videotape protocol methodology may be a useful technique for assisting us in systematizing our empirical knowledge of knowledge engineering.

The preliminary conclusions from this work read like a bad-news good-news story. The bad news is that the amount of knowledge used by human knowledge engineers is enormous: building programs to simulate human knowledge engineers will be a formidable task. The good news comes in three parts:

the program design behavior of different human knowledge engineers appears to be similar;
the design behavior of our AI designers appears to be consistent with models of

the design process developed by researchers studying the cognitive underpinnings of software design;†

protocol methodology appears to provide useful insights into program design processes of knowledge engineers.

This paper is thus intended to illustrate the potential value of the videotape protocol methodology for identifying some of the consistent, empirical aspects of the behavior of *human* knowledge engineers that may facilitate our work as we confront the task of building intelligent *machine* knowledge engineers. In Section 2 we describe the stimulus materials and methods we used in this experiment. In Section 3, we describe seven of the main themes we discovered in the behavior of our AI designers and present empirical data from the protocols of their problem-solving interviews. Finally, in Section 4, we present some implications, limitations, and future directions for our work.

2. Methods

2.1. SUBJECTS

The six subjects were experienced designers of AI programs. Each AI designer was an advanced student in the Yale Artificial Intelligence program. Each subject had at least six years of programming experience and had written at least one AI system containing more than 5000 lines of code. Most subjects had written several such AI systems and had worked on significantly larger AI systems.

2.2. STIMULUS MATERIALS AND PROCEDURE

Each subject was given a transcript of a fragment of dialogue between a tutor and a student and asked to design an AI program that would simulate the behavior of the tutor and the student. During the fragment of dialogue, the tutor teaches the student about certain aspects of large weather systems. The fragment of dialogue, which is included as Appendix 1, contains 10 interchanges between tutor and student.‡ We chose this fragment of tutorial dialogue because the participants have quite different goals, as well as very different knowledge. The task appeared to engage the interest of all subjects: all stated that the task was challenging, though they felt a longer transcript would have been useful. Each subject was videotaped while solving the AI design problem; the interviewer encouraged subjects to talk aloud during their interviews.

3. Central themes in AI designers' performance

In this section we present brief descriptions and examples of each of the following seven themes that appeared to be central to our AI designers' behavior:

1. the importance of the knowledge engineer's goal structure;

† cf. Adelson & Soloway (1984).

‡ This transcript is a slightly modified version of a transcript presented by Stevens, Collins & Goldin (1982). Misconceptions in students' understanding. In SLEEMAN, D. & BROWN, J. S. Eds, *Intelligent Tutoring Systems*.

2. the importance of world knowledge;
3. the selection of a general representational schema;
4. causal simulation of the domain;
5. the identification of heuristics;
6. model testing strategies and progressive refinement;
7. focusing on a "touchstone".

Each of the six subjects in our study exhibited several of these themes; we now briefly consider each theme in turn.

3.1. IMPORTANCE OF THE KNOWLEDGE ENGINEER'S GOALS

Each of our AI designers appeared to have a goal structure that guided the process of designing the simulation program. Five of the most important goals of the AI designers are:

Identify the domain knowledge and tutorial knowledge: subject 2 was most explicit about his need to have clearly in mind the knowledge that his program would have to represent. He was the most cautious of the designers: rather than jumping right in and selecting a representation for the knowledge, he spent most of his time just understanding the knowledge his program would have to manipulate. *Subject 1, Subject 2, Subject 3* and *Subject 4* all spent considerable effort understanding the tutorial knowledge of the tutor so that they could understand how the tutor constructed responses to the student.

Use a consistent representation: Subject 1 was most explicit about his enforcement of the constraint of using a consistent representation for the knowledge in his program. *Subject 1* realized that a common representation for the knowledge of the tutor and the student would facilitate the communication between the two parties in his simulation of the tutorial interaction. Interestingly, the heuristic of using a common representation led *Subject 1* to ignore the possible desirability of using *different* representations for the tutor and the student; thus, he did not question the psychological reality of using the same basic representation.

Identify and solve key problems. All of our subjects identified key issues that their programs would have to address. Some key issues were problems of representing knowledge; other key issues related to how to organize the control structure of the program.

Use stepwise refinement. Each of the subjects recognized the need to develop their designs so that the degree of specificity of each of the components of the program stayed "in synch". *Subject 3* was most articulate about this issue when he noted that he would design a *modular program* and said that he would "go back and forth" between the modules as he fleshed out the design. The strategy of going back and forth apparently served to prevent designers from making commitments to one aspect of the design that would make other aspects difficult to design.

Identify the abstract structure of the problem: Subject 3 explicitly used the method of abstraction to help him design his program. As we illustrate in section 3.5, *Subject 3* abstracted away from the domain the *pattern of the tutorial interaction.*

These five components of the goal structures of our AI designers were important aspects of their design processes. A tentative implication for knowledge engineering design tools of the importance of the designer's goal structure may be that such tools should provide the designer with a mechanism for directly expressing and fulfilling his or her goals.

3.2. IMPORTANCE OF WORLD KNOWLEDGE

The importance of world knowledge for the process of constructing intelligent knowledge-based systems cannot be overemphasized. We repeatedly observed our subjects appealing to their world knowledge as they designed their programs. Our subjects appealed to world knowledge in two main ways.

Appeal to domain knowledge

In order to understand what the student understood, and how the tutor generated responses to the student, all subjects referred extensively to their knowledge of the weather domain. In addition, almost all subjects regretted not knowing *more* than they did know about the weather domain. For example, *Subject 1,* who had extensive training in physics, noted several times the limitations of his knowledge about weather systems. On one occasion he simply did not know the reason for the form of the tutor's response to the student. *Subject 1*'s puzzlement over the goal the tutor was trying to achieve with the reply to the student occasioned this comment:

> "Obviously my model of weather is off. It's a confounding factor. The first thing I'd do in building (this) AI system would be to get out some books on weather and *understand what is really going on*".

This quotation by *Subject 1* illustrates the importance to knowledge engineers, and to programs which would simulate their behavior, of a reasonably powerful model of the domain in which they are engineering knowledge.

Appeal to general world knowledge

Our subjects frequently appealed to world knowledge that was not related directly to the weather domain. For example, one of our subjects noted that transcripts of dialogue sometimes failed to capture potentially useful information, such as the motivation and mood of the participants. Another subject pointed out that different tutors might have very different styles for interacting with students; equally, some subjects observed that motivational differences of students might require their tutorial programs to reason about differences in students' learning style. Such differences in tutoring and learning styles have an impact on the design of the AI program since they would either have to be simulated or ignored altogether.

3.3. SELECTION OF GENERAL REPRESENTATIONAL SCHEMA

Early in the process of constructing their AI programs, both *Subject 1* and *Subject 3* settled on general frameworks for conceptualizing the rules and knowledge of the weather domain. For example, *Subject 1* decided, very early on in his analysis of the transcript of the weather tutoring session, that he would use a frame-slot notation for the weather domain knowledge of *both* the student and the tutor. Approximately 2 min after beginning his task, *Subject 1* identified his representation:

> "If there's a climate *frame,* one of its *slots* is going to have to be temperature".

Subject 1's statement is consistent with his subsequent analysis of the requirements of his program's representation of the weather domain. He continued to pursue the applicability of the frame-slot representation, identifying the kinds of knowledge the frames would have to represent as well as heuristics for filling the slots with values. Thus, *Subject 1*'s early decision to use a frame-based representation for the weather domain knowledge played an important role in the subsequent design work of his AI program.

Subject 3 initially settled on a *causal-chain* representation for the knowledge in the weather domain,† and this had a strong impact on his subsequent work. While *Subject 3* recognized that his initial causal-chain representation of weather domain knowledge was not entirely adequate, he based most of the machinery for the tutor's question-generation behavior directly on the causal-chain representation. For example, the causal-chain representation would permit the tutor to work *back* along causal chains to help students explain how observed effects arose (e.g. precipitation) and to work *forward* along causal chains to help students learn to predict what would happen given a particular state of the world (e.g. water-burdened clouds running up against a mountain.)

Thus, while *Subject 1* and *Subject 3* selected different initial representations for weather domain knowledge, their early selections of representations strongly affected their subsequent designs of their programs.

3.4. CAUSAL SIMULATION OF THE DOMAIN

In building their representations of the weather domain, our subjects used naive physical process theory to mentally simulate the causality in the weather domain as they tried to understand what the student understood and what the tutor was trying to teach. The following quotation illustrates *Subject 1*'s use of mental simulation to understand the effect of the Japanese Current on the climate in the northwest United States: *Subject 1* performed this reasoning when he was trying to understand why the tutor asked the student how the Japanese Current affects rainfall in Washington and Oregon.

> "I'm inferring that the Japanese Current is cold. If I make (this) assumption about the Current's path, then the Current starts out in Japan, which is fairly far north, and goes even farther north. So, it *probably gets very cold when it comes down the coast and probably makes Washington and Oregon a fairly cold place*".

Though *Subject 1* made the incorrect assumption that the Japanese Current is cold, this quotation illustrates the use of simulation to reason about causality as part of the process of designing the behavior of the model of the student and the tutor. While some subjects performed simulation more than others, every subject used simulation of the domain during their program design behavior.‡

3.5. IDENTIFICATION OF HEURISTICS

Subjects identified two kinds of heuristic reasoning that their programs would have to perform. First, subjects identified several heuristics for reasoning in the weather

† A causal-chain representation relates causes in the weather domain (e.g., air masses cooling) to their effects (e.g., releasing moisture as precipitation).

‡ See Littman & Soloway (in prep) for a more extensive treatment of mental simulation in design and understanding of programs.

domain; second they identified several program control heuristics that the tutor and the student used to control the tutorial session.

Domain heuristics
Subject 1 interpreted the student's behavior in terms of his frame representation for the student's knowledge about climate. *Subject 1* proposed that the slots of the student's climate frames were filled by applying heuristics which could include default methods for filling slots.

> "The second heuristic is that the slots have metrics associated with them with distributions on the values and a heuristic for selecting a value out of that distribution based on the "is-part-of" relations for geographic areas".

Heuristics for controlling program behavior
Subject 1 was very interested in understanding how the tutor constructed responses to the student, since he viewed this as the primary aspect of his program's control mechanism. After reading and analysing several interchanges between the student and the tutor, *Subject 1* said:

> "The interviewer is definitely building this question (to the student) out of the student's responses".

Then, after seeing a few more interchanges, *Subject 1* had a flash of insight and stated the tutor's heuristic for constructing replies to the student:

> "If the student comes up with a false statement, the statement is immediately contradicted *in an interesting way*. It's (the tutor's response) not just "NO". . . . (the response) relates back to a previous question of the interviewer; *the correct response* (to that question) *is what the interviewer says now*".

Thus, *Subject 1* identified two kinds of heuristic reasoning for his program to perform, one related to the domain, the other related to the control structure of his program. It is interesting to note that both domain and program control heuristics were based on the frame-slot representation *Subject 1* had selected: *Subject 1* designed the domain heuristics to *fill* slots in the frame; he designed the program control heuristics to *use* information in the frames.

Subject 3 also focused on the heuristics used by the tutor and student to construct their responses. *Subject 3* abstracted the transcript of the tutorial interaction by categorizing each of the tutor's and each of the student's statements; the designer's categorization of the statements in the transcript produced a foundation for a schematic description of the types of utterances the tutor and student made. Using the schematic description of the student's and tutor's utterances, *Subject 3* then attempted to identify heuristics the tutor and the student used to produce a response of each type.

3.6. MODEL TESTING AND PROGRESSIVE REFINEMENT

All of our subjects tested their evolving models of the tutor and student at various stages in the design process. On the basis of their tests of their evolving models, they refined and corrected their program designs.

Model testing
All subjects noted the importance of testing their evolving models of their programs. Most subjects used a variant of the "predict-and-verify" method. Here is how *Subject 1* expressed his strategy:

> "So I guess I am playing this game, *predicting what the interviewer is trying to say*".

The important observation to make about the skills of progressive refinement and testing the adequacy of the model is that both require a great deal of heuristic knowledge. For example, how did our designers know:

which components of the design should be tested?
when to test components of the design?
how to test different components of the design?
how to interpret the results of the tests?†

The last point, interpreting the results of the test of the model, points out the importance to our subjects of debugging their designs. For example, *Subject 1* spent a great deal of time resolving contradictions between his representation of the tutor's knowledge of the Japanese Current and his factual knowledge about the fate of moisture in bodies of air that are cooling. The following quotation illustrates *Subject 1*'s concern with identifying the sources of contradictions. He said this:

> "*I get a contradiction* because I know that if things cool off they lose their moisture. So I know there is a lot of rainfall in Washington and Oregon. But I am thinking that if the (Japanese) Current is coming down from the North it will already have lost its precipitation so I get a contradiction".

In fact, *Subject 1* had made a faulty assumption about the Japanese Current (it is actually a warm body of water) and built the faulty assumption into his representation of the tutor's knowledge. *Subject 1*'s faulty representation of the tutor's knowledge of the Japanese Current led him to a contradiction between what he believed about the Japanese Current (that it is cold) and the affirmative response of the tutor to the student's assertion that it is probably rainy in the Pacific Northwest *because of the effect of the Japanese Current on the precipitation patterns. Subject 1* attempted to resolve this contradiction several times, putting it aside each time until some new piece of evidence became relevant to its resolution. The behavior of "putting aside" the contradiction, until some new information became relevant to its resolution, is consistent with the model of software design proposed by Adelson & Soloway (1984) in which "demons" help expert software designers turn their attention to old issues when new information becomes relevant to them.

Progressive refinement
Our subjects began their AI tasks with only a very general idea of how their programs should behave. After an initial phase of identifying important attributes of the student's and the tutor's knowledge and reasoning, they decided on a potential representation for the knowledge and reasoning. As our subjects filled in their

† The credit assignment monster raises its head here.

designs, they refined their initial representations by a three-step process:

1. derive predictions about the behavior of the tutor and the student from their program designs;
2. evaluate the predictions with the transcript of the tutorial session;
3. alter the model to be consistent with the transcript.

Our subjects knew that their programs would have to be designed in a "propose-test-modify" cycle. *Subject 3,* for example, was fully aware that his first attempt to design the program would be primarily a "throwaway". *Subject 3*'s recognition that much of his initial program would be a throwaway is apparent in this comment:

> "It's always a problem: How far to expand something you *know* you are going to throw away".

3.7. FOCUSING ON A "TOUCHSTONE"

After studying each of the statements of the tutor and the student, our AI designers had a general idea of the knowledge the AI program would have to include and the rules it would have to use to manipulate the knowledge effectively. Following this initial assessment of the requirements of the program, several of our AI designers identified a tutorial interaction, or a response, that they considered to be crucial for their programs. *Subject 2,* for example, decided that the third interaction, in which the tutor asks the student how the Japanese Current affects rainfall in the Pacific Northwest, was crucial for his program. He said that if he could understand how to generate the tutor's question, and how to generate the student's response, then he would understand:

- how to represent the domain knowledge;
- how to represent the tutor's strategies for asking questions;
- how to represent the student's strategies for building informative responses.

A touchstone is valuable because it provides a test for the essence of something. The skill of selecting touchstones for testing knowledge-based programs is important; it permits the designer of the program *to identify what would count as an effective solution to the problem.* If we understood better how knowledge engineers know "what counts as a good solution", we may be able to design support tools to assist them in developing such solutions.

4. Implications, limitations, and future directions

4.1. IMPLICATIONS

The current work suggests four main implications:

The means by which expert knowledge engineers extract mental models of domain experts is, itself, *heavily knowledge based.*

A great deal of the behavior of experienced knowledge engineers appears to be based on *heuristic classification* of problem types (Clancey, 1985). For example, the AI designers in our study who had some experience studying or designing tutoring systems made a commitment to a general representation schema earlier in their design processes than the AI designers who had never worked in the domain of tutoring systems. The experience of the AI designers who had worked in the domain of tutoring systems appeared to permit them to classify the problem type quite readily and to select a representational system based on the problem type. Thus, it would be very useful to study the organization of our knowledge engineer's knowledge.

The process of empirically studying the methods that knowledge engineers use to extract mental models of domain experts is a potentially useful enterprise.

The videotape protocol methodology appears to provide a potentially useful tool for studying the process of designing knowledge-engineered programs.

Each of these implications must be explored by future research before it will be possible to evaluate them fully.

4.2. LIMITATIONS

The information presented in this paper does *not* constitute even an attempt to formulate an initial empirical theory of knowledge engineering. Rather, it attempts to show that describing consistent aspects of the behavior of knowledge engineers can suggest interesting avenues to pursue toward the goal of developing computational theories of how knowledge engineers do their work. The ideas presented in this note are limited for three main reasons:

the subject sample is small;

the domain in which the AI designers were asked to work is limited: a more adequate empirical descriptive theory of knowledge engineering should study behavior of knowledge engineers in several domains;

the results reported here are just a first pass at analysing the extensive protocol data collected from our subjects' design sessions.

Even with these limitations firmly in mind, it seems plausible to imagine that a more detailed empirical study of the behavior of knowledge engineers could provide useful insights into ways of devising intelligent, knowledge-based knowledge engineering tools.

4.3. FUTURE DIRECTIONS

In the future, we plan to address several issues.

More detailed analysis of the data

The protocol data we collected are very rich. The goal in performing further analyses will be to identify:

the actual knowledge used by our subjects;

the goal structure of our AI designers that controlled their design behavior;

the heuristics they used to achieve their goals.

Understand how designers develop their models

Our AI programmers were in the position of trying to build models of the weather domain and of the reasoning of the tutor and student. It seems plausible that building models is a very general attribute of all AI programming. We intend to explore this issue in great detail, with both the current data as well as others, to identify the kinds of models AI programmers use and build as well as how they build them.

Develop a process model
In this paper we have identified just a few of the salient behaviors of our designers. As we identify more of the knowledge and heuristics used by our designers, we plan to develop a process model of their behavior.

Develop a simulation
As the process model of our designers' behavior becomes fleshed out, we will develop a simulation model of their behavior. Clearly, we have only just begun to explore this potentially fruitful domain of problem solving.

I would like to thank Elliot Soloway for his unflagging support of my work. James Spohrer and Andrew Liles made very useful comments on drafts of this paper; William Clancey offered useful criticisms as well. I would especially like to extend my appreciation to my subjects.

References

ADELSON, B. & SOLOWAY, E. (1984). A model of software design. *Yale Computer Science Department Technical Report* #342, 1984, Department of Computer Science, Yale University, New Haven, Connecticut.

CHANDRASAKARAN, B. (1985). Generic tasks in expert system design and their role in explanation of problem solving. *Technical Report, July, 1985,* Department of Computer and Information Science, Laboratory for Artificial Intelligence, The Ohio State University, Columbus, OH.

CLANCEY, W. (1985). Heuristic classification. *Stanford University Computer Science Department Research Report* #STAN-CS-85-1066.

HAYES-ROTH, F., WATERMAN, D. & LENAT, D. eds. (1983). *Building Expert Systems.* Reading, Massachusetts: Addison–Wesley.

LITTMAN, D. & SOLOWAY, E. Mental simulation of programs: Some data and some theory. Department of Computer Science, Yale University, New Haven, Connecticut.

STEVENS, A., COLLINS, A. & GOLDIN, S. (1982). Misconceptions in students' understanding. In SLEEMAN, D. & BROWN, J. S. Eds, *Intelligent Tutoring Systems.* London: Academic Press.

Appendix 1. Stimulus transcript

The following transcript is the actual transcript shown to our subjects.

tutor: 1. Do you know what the climate is like in Washington and Oregon?
student: 1. Is it cold?

tutor: 2. Can you guess about the rainfall there?
student: 2. Normal, I guess.

tutor: 3. How do you think the Japan Current along the coast affects rainfall there?
student: 3. It's probably rainy.

tutor: 4. Yes. There are rain forests in Washington.
student: 4. Then the Japanese Current is warm.

tutor: 5. Right.
student: 5. And the wind blows in from the sea.

tutor: 6. What happens to the moist air blown over Washington and Oregon by the winds?
student: 6. It condenses and rains because the moist air cools and cannot hold the water. It cools when the wind blows it and it lowers from the sky.

tutor: 7. What happens to the temperature of the moist air as it rises?
student: 7. The moist air gets warm.

tutor: 8. No, it cools when it rises.
student: 8. So the mountains also cool the moist air when they come into contact with the moist air.

tutor: 9. No. The contact with a solid mass does not cool off moist air.
student: 9. Because it is not big enough.

tutor: 10. Right.
student: 10. So then it is because the mountains cause the air mass to rise?

Discourse and Protocol Analysis

Knowledge elicitation using discourse analysis

N. J. Belkin and H. M. Brooks

The School of Communication, Information, and Library Studies, Rutgers University, New Brunswick, N.J. 08903 U.S.A.

P.J. Daniels

Admiralty Research Establishment, Ministry of Defence, Queens Road, Teddington, Middx. TW11 0L9, U.K.

This paper is concerned with the use of discourse analysis and observation to elicit expert knowledge. In particular, we describe the use of these techniques to acquire knowledge about expert problem solving in an information provision environment. Our method of analysis has been to make audio-recordings of real-life information interactions between users (the clients) and human intermediaries (the experts) in document retrieval situations. These tapes have then been transcribed and analysed utterance-by-utterance in the following ways: assigning utterances to one of the prespecified functional categories; identifying the specific purposes of each utterance; determining the knowledge required to perform each utterance; grouping utterances into functional and focus-based sequences. The long-term goal of the project is to develop an intelligent document retrieval system based on a distributed expert, blackboard architecture.

1. Introduction

We are concerned with the problem of designing intelligent automated interfaces to mediate between people who feel they require information, and the (usually) computer-based knowledge resources which might contain information which could be of use to them. These three elements: user; intermediary; and knowledge resource, and the relations among them, constitute the general information system which is our focus of attention. The intermediary and knowledge resource elements together constitute the *information provision mechanism* (IPM). Such systems arise in, for instance, social security benefits offices, student advisory services and bibliographic retrieval systems. At present, almost all such information systems require a human intermediary. We assume that any automated interface will need to perform at least some of the functions that human intermediaries perform (as well as being capable of recognising situations when human intermediaries are necessary); such an interface would perforce be intelligent.

Human intermediaries are required in such systems for a number of reasons. Generally speaking, the users in such systems are unfamiliar with the contents, structure and access mechanisms of the data base, nor should they be required to be, for they are at best intermittent participants in any one such system. The intermediary's role in this respect is to use his/her knowledge to choose the appropriate data base(s), to structure the search appropriately to the data base, and to formulate and apply the relevant query. These tasks might seem somewhat

KNOWLEDGE-BASED SYSTEMS Vol. 1
ISBN 0-12-273251-0

mechanical and requiring relatively little creative "intelligence". However, the knowledge required for this performance is substantial, and the reasoning processes complex. More important, however, is that these tasks depend upon a specification of the user's information requirements that is unlikely to be forthcoming.

The typical situation in information systems, and indeed in many decision support systems, is that the users are unable to specify the information which might be useful in managing the problem which prompted them to enter the system. That is, they may recognise that their state of knowledge with respect to their problem or task is *anomalous* (see Belkin, 1980), but are usually unable to say what it will take to resolve the anomaly. Thus it appears that much of the human intermediary's task in the *information interaction* between user and intermediary is concerned with the building up of models of the user, the user's situation and the user's problem, which can be used as the basis for the subsequent tasks of database interrogation. This kind of activity is without doubt highly complex, requiring a wide range of learning capabilities and sophisticated reasoning mechanisms, within the context of a highly motivated and complicated interaction with the user. The advisory systems considered by Coombs and Alty (1984), although concerned with a slightly different problem domain, exhibit the same level of complexity, and the same type of interaction pattern.

In order to achieve our goal of an intelligent interface in the information system, we will need to discover and specify the necessary and sufficient *functions* of the intermediary, the *knowledge* necessary for performing those functions, and the *dialogue structure* in which those functions are performed. For implementation, we will need to use these results to determine **representational schemes** for knowledge, **retrieval strategies** and **general system architecture**, including **interactions** among the functions. These tasks cannot be accomplished by asking intermediaries and users what they are doing, or by requiring them to state their rules of action, because of the nature of the situation in which they find themselves. Rather, it appears that the most reasonable approach is a detailed, multi-faceted observation and analysis of the information system in action.

The knowledge we wish to elicit relates to the goals as specified. That is, we wish to know what intermediaries do, what knowledge they need and how they apply that knowledge. We need to discover how they model users and problems, what their interaction strategies are, and responses to the user. We believe, indeed, that contrary to usual advice on expert system construction, it will be necessary to spend a great deal of time and energy developing a realistic model and deep understanding of the situation before attempting implementation.

Our primary method of acquiring (or eliciting) the knowledge we require has been through functional discourse analysis of real information interactions, between human users and human intermediaries in operational information systems. This technique was based initially on a set of functions for information interaction derived from an abstract analysis of the information situation, and from our experience and observation of information interaction (Belkin, Seeger & Wersig, 1983) and by a simulation (Belkin, Hennings & Seeger, 1984), but the functions have been emprically confirmed through the discourse analysis itself.

The identification of these functions has led to a proposal for a "distributed expert" architecture for the IPM (Belkin, Seeger & Wersig, 1983), and the general notion of functional distribution and blackboard architecture, and some of the specific functions themselves, have been also suggested independently by Croft (1985) and by Fox (1987). In the remainder of this paper, we explain why we chose discourse analysis as our method of knowledge elicitation, and describe the results this method has achieved.

2. Methods of knowledge elicitation

Buchanan (1983) defines knowledge acquisition as:

> the transfer and transformation of problem solving expertise from some knowledge source to a program. Potential sources of knowledge include human experts, textbooks, databases and even one's own experience.

What appears on the surface to be a relatively straightforward task turns out in practice to be extremely difficult, time consuming and complex (Buchanan, 1982; Welbank, 1983). Within the field of Artificial Intelligence there is little in the way of methods or techniques that could be used to facilitate knowledge elicitation. Attention has been turned, therefore, to other disciplines where knowledge transfer is of interest, e.g. cognitive psychology.

The main techniques for knowledge elicitation can be summarized as follows:

(a) **Interviewing the expert:** either informally or making use of structured interviewing techniques.
(b) **Verbal protocol analysis:** that is, analysing recordings of experts thinking aloud as they carry out a task.
(c) **Observational studies:** observing and recording the behaviour of the experts as they work on real problems, in their normal working environment, in as unobtrusive a way as possible.

A review of Knowledge Acquisition methods is presented by Welbank (1983). Gammack and Young (1985) propose a number of "psychological" techniques including Personal Construct Theory (Kelly, 1955), and concept sorting. The use of Personal Construct Theory to acquire knowledge for expert system construction has also been investigated by Shaw (e.g. 1984).

Interviewing the expert is a commonly used technique. It has a number of disadvantages though, and may not be useful in areas where experts have particular difficulty in articulating their knowledge (Welbank, 1983). The use of verbal protocol analysis is also very common (e.g., Kuipers & Kassirer, 1983). Frequently the task given to the expert is a standard case study or an artificial problem. A comprehensive survey of the use of verbal protocol analysis has been carried out by Ericsson and Simon (1980). Verbal protocol analysis has also been used to investigate various aspects of the functional behaviour of librarians (Ingwersen, 1982). Observational techniques are probably the least used because they tend to be extremely time consuming and require complex, indepth analysis (Welbank, 1983). However, they do have the advantage that they can be used to discover what the expert actually does and are useful for extracting information about the role of the expert, ordering of tasks, etc. (Kidd, 1985*b*). One example of the use of observational techniques in the expert system context is the investigation into student–advisor interactions carried out by Coombs and Alty, with a view to developing an expert system advisor (Alty & Coombs, 1980; Coombs & Alty, 1984).

Current thinking tends towards the use of multiple techniques, with certain methods seen as more appropriate for particular types of knowledge (Gammack & Young, 1985; Kidd, 1985*b*). in other contexts, this has been done for some time e.g. the detailed study of librarian–user carried out by Ingwersen and Kaae (1980) made use both of verbal protocol analysis and of interviews. Perhaps the most important factor is that whatever technique is used, it should elicit knowledge at the level of granularity required for that particular expert system application. That is, the more detailed and complex the knowledge to be elicited, the more detailed and complex the analysis must be to extract that knowledge (Clancey, 1983; Kidd & Cooper, 1985; Kidd, 1985*a*).

3. Discourse analysis for knowledge elicitation

3.1. INTRODUCTION

The long-term aim of this project is to produce an intelligent information system interface which will simulate the functional performance of a good human intermediary. There are a number of possible methods which could be used to achieve our objective. Our approach has been to employ the methods which seemed best suited to developing a particular specification, given the problem-solving environment under investigation. These methods include:

(a) functional analysis of real interactions between human experts (search intermediaries) and clients (information system users);
(b) interviews with experts;
(c) system simulation, using humans to play the roles of various system functions;
(d) classification of problem and knowledge types.

The nature of the problem domain and the expert-client interaction has meant that most of our efforts have concentrated on the first of these methods, i.e. functional analysis of expert–client interactions. This method produces an extremely detailed analysis of what is, in any case, a highly complex situation, in a way that is as unconstrained as possible by pre-conceived ideas. Results from methods (c) and (d) have been reported in Belkin, Hennings and Seeger (1984) and in Stinton (1984).

3.2. GENERAL METHODOLOGY

The general method in most of our research has been to collect data consisting of natural language human–human interactions. The collection of data took place at two academic online information retrieval services at London University with four different trained intermediaries, who carry out searches of biliographic databases. The users of these services are primarily postgraduate students, but also include academic staff, and other types of researchers, for example, from industry and research centres. Having obtained the permission of both participants, audio-recordings were made of a number of presearch interviews, from the point that the user entered the service until the beginning of the online search, when the intermediary logged onto the system. These interviews were collected for several projects within the overall research program. These general analytic techniques for all the projects was similar: here we report on the methods used on one specific project involving a core set of seven interactions.

The interviews were transcribed from the tapes according to a specific format. Initially we used a transcription protocol based on that devised by G. Jefferson (Sacks, Schegloff & Jefferson, 1974). This protocol preserves paralinguistic features and speech overlaps. Although the interactions were relatively easy to transcribe using this method, for the purpose of the analysis the transcripts proved too cumbersome. In particular, the flow of the discourse was difficult to follow because of the way in which the protocol represented the large number of overlapping speech sequences. A method of transcription which preserved the time-line sequence of the discourse was gradually developed (Brooks & Belkin, 1983; Price, 1983; Daniels, Brooks & Belkin, 1985; Brooks, 1986*b*). A sample from one of the interactions, transcribed according to the time-line protocol, is given in Fig. 1. The speech of each participant is typed on alternate lines, with the participant indicated at the beginning of each line. All text is typed in lower case except for proper names. Punctuation marks are not used to indicate grammatical sentences or part sentences, but breath pauses and true pauses are indicated by commas and periods in parentheses respectively. Each transcription was checked by another member of the team.

The transcript was then divided into utterances, which are the units of analysis, and this division was again checked by another person (Daniels, Brooks & Belkin, 1985). An utterance can be defined as a speech sequence by one participant during the conversation. It may or may not comprise of complete grammatical entities, and may be terminated by a contribution made by the other participant. If the contribution of one participant takes the conversational turn, the previous speech

I three countries but (,) are (,) you prepared to read about (,) I dunno (,) ya know
U uh mm

I anywhere else in the world (,) if:: if it's
U yeah (,) if it's related to the questions I'm asking

I yeah (,) if it's about community education (,) but primarily in Africa (.) and:: (.) then
U uh mm mm

I those three countries
U then I'm looking at (.) history as well because these three (...) the

I uh mm
U reason I chose them is because they've got a history of community education (,) which

I = intermediary, U = user, (,) = breath pause (.) = 0·5 s pause

FIG. 1. Sample transcript from interview no. 4.

I so (,) community schools /82a now there are quite alot of (,) headings (,) that (...) are (,)
U yeh I see /83

I (,) headings (,) that (...) are (,) begin with community /82b I think perhaps if we just
U

I them all quickly /82c you'd be interested in something that was just generally about
U

I community (,) and community action and presumably attitudes (.....) /82d it would
U yeh (,) yes /84

I = intermediary, U = user, (,) = breath pause, (.) = 0·5 s pause, /84 = utterance number 84, /82a = part utterance (a)

FIG. 2. Sample transcript, subdivided into utterances, from interview no. 4.

sequence is regarded as a completed utterance (Brooks & Belkin, 1983; Price, 1983). An example of a transcript divided into utterances is given in Fig. 2. The seven interview transcripts together comprise some 2000 individual utterances.

The transcripts were then subdivided into connected groups of utterances or foci. The focus of a dialogue (Grosz, 1978) can be said to highlight that part of the mutual knowledge of the participants relevant at any given point in a dialogue, by grouping together those concepts or themes that are in the focus of attention. The current focus is likely to dictate the structure of the discourse, and the topics to which reference can be made at that given moment. In the analysis of our seven interviews, shifts of focus were generally indicated by the occurrence of "frame" words (Sinclair & Coulthard, 1975), often accompanied by pauses of varying duration. Examples of frame words include:

Right OK (.) right (..) (Interview 190684HBA)
now the next thing is to get (,) our strategy (Interview 190684HBA)

An extract from one of the transcripts showing a shift in focus is given in Fig. 3.

Each interview was analysed both at the focus and at the utterance level. The purpose of the analysis was to identify:

(a) the tasks and functions being carried out at each level;
(b) the knowledge required to perform the tasks and functions; and
(c) interactions between functions and relations among foci.

The function of a focus or utterance is the general information interaction goal, under which several tasks can be subsumed. By function of an utterance, we mean to indicate **why** the participant said it; that is, what the participant intended to gain or communicate by it, what model-building activity it was relevant to, and so on. A single utterance may serve more than one function. Tasks are the specific goals of an utterance or part utterance.

In several successive studies we tested the adequacy of pre-determined functional

I	**now** I gather you're (cough.) excuse me (.) you're a visitor (.) um /3 yes are you
U	yes I am /4
I	part of the university or /5 Ya (,) um I I just (,) we
U	well I teach at a Canadian university /6
I	ask you this because I—it's awful to bring up charges straight away (laugh.) but just so
U	
I	that you know (,) you that its a ten pound and it's (inaud ...)/7 **right OK (,) right**
U	yes /8
I	(laugh ...) we've got that out of the way (laugh ...) what's what's the subject of your
U	(laugh) (laugh ..)
I	query /9 right (.....) any anything particularly (,) specific /11
U	greek turkish relations /10

I = intermediary, U = user, /8 = utterance number 8, (.) = 0·5 s pause; frame word marking focus shift is in bold.

FIG. 3. Extract of transcript from interview 190684HBA demonstrating a shift in focus.

patterns by attempting to categorise all of the utterances in a number of dialogues using several coding schemes. Each iteration used the previous results to add, delete or modify the definition of specific functions until the classification stabilized on the functions indicated in Table 1. Having analysed the interactions in this way, it was possible to obtain time-line displays of their functional structures over the course of the dialogue.

Two of the core set of interviews were then subjected to separate analyses by three different people, each of whom attempted to identify and categorise the tasks performed by the utterances made by both parties during the dialogue. Each of the remaining five interviews was analysed by at least two people independently. The results were cumulated for all seven interactions to produce an inventory of tasks (or subfunctions).

Not only must the functions necessary to achieve a successful information interaction be specified, it is also necessary to identify and specify the knowledge resources associated with the functions, since they will require some knowledge in order to achieve their goals. Eliciting the knowledge required to carry out each task and function was carried out by detailed analysis of each utterance and enquiring what knowledge the intermediary required to make each of the utterances.

To verify the subfunctions and knowledge resources identified for the User Model function, interviews were arranged with two experienced intermediaries at the search services from which the original data were collected. A questionnaire was devised which first asked the intermediaries to comment on the appropriateness and

TABLE 1
The functions of an intelligent interface for document retrieval (after Belkin, Seeger & Wersig, 1983)

Name of function	Description
Problem State (PS)	Determine position of user in problem treatment process e.g. formulating problem, problem well specified etc.
Problem Mode (PM)	Determine appropriate mechanism capability e.g. document retrieval
User Model (UM)	Generate description of user type, goals, knowledge etc. e.g. graduate student, dissertation etc.
Problem Description (PD)	Generate description of problem type, topic, structure, subject etc.
Dialogue Mode (DM)	Determine appropriate dialogue type and level for situation e.g., menu
Retrieval Strategy (RS)	Choose and apply appropriate retrieval strategies to the knowledge resource
Response Generator (RG)	Determine propositional structure of response to user
Explanation (EX)	Describe mechanism operation, restrictions etc. to user as appropriate
Input Analyst (IA)	Convert input from user into structures usable by functional experts
Output Generator (OG)	Convert propositional response to form appropriate to user, dialogue mode etc.

accuracy of the five User Model subfunctions, and second asked for further information, confirmation, etc. of various problem areas. Each question was accompanied by a concrete example in the form of an excerpt from a transcript or a played-back recording. The information gained in these interviews confirmed the User Model analysis and clarified the problem areas.

Discourse analysis of the human–human interactions was also used to identify and define a problem structure associated with the informtion system. This identification and specification of the problem structure depended crucially on the analysis of the discourse into foci, on the specification of the goal of each focus and on an investigation of the factors initiating a shift in focus (Daniels, Brooks & Belkin, 1985).

4. Results

4.1. INTRODUCTION

Whilst the main emphasis of this paper is on the methodology we are using to develop the design and specification of an intelligent, distributed expert-system based interface, we would like to indicate to what effect these methods have been employed. Although the results were obtained primarily within a document retrieval environment, they also seem to apply in other sorts of advisory interction we have studied.

4.2. ADEQUACY OF THE FUNCTIONS

Functional analysis of the interviews has elicited a common set of functions which appear to be the minimum required for an intelligent interface in the document retrieval environment, and perhaps for information systems in general. These are the functions specified in Table 1, which were finally determined through repeated coding of interactions and modification of the classification.

4.3. TASK ANALYSIS AND SUBFUNCTION IDENTIFICATION

The functions identified for an intelligent information system (Table 1) are defined at rather a general level. To implement such a system, for each function, it is necessary to specify the tasks concerned with the accomplishment of that function, the knowledge resources required by that function in order to achieve its goal and the information contributed to and acquired from the other functions. Most of our efforts have concentrated on the specification of the User Model (Daniels, 1986), Problem Description and Retrieval Strategies functions (Brooks, 1986*a*, 1986*b*).

A sample transcript, focus 1 from interview 120684HBA, with its accompanying task analysis, is presented in Fig. 4.

I emm (.) what's the (.) the problem /1 mm hm /3
U ok I'm just beginning (.) a research project /2 (.) err (.) I'm

I yeah /5
U at LSE /4a (...) and um (...) working in the Geography Department /4b and I'm

I
U err (.) doing a (.) thesis (.) beginning a thesis on err (.) forestry (.) and err the

I interesting /6 oh that's interesting
U impact of recreation (.) of conflicts of recreation and forestry /4c

I (.) mm (.5sec) (inaud........) /7
U and err (...) one of the things I'm aiming to do (.)

I
U eventually is to look at the cost benefits of different err (.) management schemes

I mm
U for recreation (...) /8

I = intermediary, /1 = utterance number 1, U = user, (.) = 1/2 sec pause.

FIG. 4. Transcript of focus 1, interview 12068HBA.

The task analysis for each utterance in Fig. 4 was determined as follows:

1: This is an open question from the intermediary directed towards **problem description**, in some form, from the user.
2: The user's response to the intermediary's opening question is to describe the context of the problem, and the user's position within the problem. This utterance seems concerned with establishing the **state** of the user's **problem**.
4a: In this part utterance, the user describes his **status**, and the college he is attending.
4b: This is concerned with further user description including **user status**.
4c: This a composite, containing information about the user's **goals**, a re-iteration of the **state** of the problem, and finally an initial description of the user's **research**.
8: The last utterance in this focus involves further description by the user of his **research topic**.

TABLE 2

The tasks of information interactions in an academic online IR setting (after Daniels, Brooks & Belkin, 1985)

Name of task	Description
1 CAPAB	Explain the capabilities of the system to the user
2 UGOAL	Determine the user's goals
3 USER	Determine the status of the user
4 KNOW	Determine the user's knowledge of the field
5 IRS	Determine the user's familiarity and knowledge of the information retrieval system
6 BACK	Determine relevant aspects of the user's background
7 PREV	Determine user's previous reference activities
8 PREVNON	Determine the user's previous non reference activities in connection with this problem
9 PDIM	Determine the problem dimension, i.e. the temporal state of the user's problem
10 SUBJ	Define the subject area or background to the search
11 SLIT	Determine the formal characteristics of the subject literature
12 RES	Specify the contents of the user's research
13 TOPIC	Specify the topic of the search
14 DOCS	Determine the content or description of the documents the user would like to retrieve
15 TERMS	Select terms for searching
16 QUERY	Formulate the query
17 STRAT	Evolve the search strategy: how the query will be implemented
18 DB	Select the databases to be searched
19 OUT	Select the output requirements
20 EXPL	Bring user's knowledge of IR systems to the minimum level necessary for the user to be able to co-operate effectively
21 DISP	Literal display of some aspect of the system
22 INFORM	Explain the intermediary's intentions to the user
23 PLAN	Specify the plan of the interview, structure the activity, what should be done next
24 MATCH	Compare models that participants hold

TABLE 3
Subtask analysis for focus 1, interview 120684HBA

Utterance	Speaker	Task
1	Intermediary	RES
2	User	PDIM
3	Intermediary	—
4a	User	USER
4b	User	USER
4c	User	UGOAL . . . RES
5	Intermediary	—
6	Intermediary	—
7	Intermediary	—
8	User	RES

Through iterative analysis of this type, 24 tasks, exhaustive of the utterances in our seven interactions were identified (Table 2). Table 3 gives the task analysis for the section of transcript in Fig. 4. (Utterances 3, 5, 6, and 7 are phatic and therefore not coded.)

The majority of the 24 tasks we have been able to identify in the interviews are concerned with the accomplishment of some particular function. Thus we grouped the tasks according to their related functions and consider them as **subfunctions** of the respective functions (see Table 4).

TABLE 4
*Functions and subfunctions for information interaction**

Functions	Subfunctions	Functions	Subfunctions
Problem Mode (PM)	CAPAB	Retrieval Strategies (RS)	TERMS
			QUERY
User Model (UM)	UGOAL		STRAT
	USER		DB
	KNOW		
	IRS	Response Generator (RG)	OUT
	BACK		
		Explain (EX)	EXPL
Problem State (PS)	PREV		INFORM
	PREVNON		DISP
	PDIM		
		Meta-Goals	MATCH
Problem Description (PD)	SUBJ		PLAN
	RES		
	TOPIC		
	DOCS		
	SLIT		

* Acronyms explained in Table 2. The tasks of information interactions in an academic online IR setting.

4.4 FUNCTION SPECIFICATION

Having identified the tasks associated with the functions, we attempted to identify the knowledge resources which the intermediary needs to access in order to accomplish the functions during the information interaction. Utterances concerned with User Model and Problem Description subgoals were analysed in depth. Inferences were made about the knowedge which appeared to be used.

User model knowledge

In the following extract, the intermediary is using her knowledge of ways in which users can be classified by status and which of these categories are important in the current exchange. In this example, the intermediary knows that users can be categorised according to whether they belong to London University or are visitors and therefore, that information about this aspect needs to be elicited. (Extract taken from interview 190684HBA.)

I: now I gather you're . . . a visitor/3 yes are you part of the university or /5

Problem description knowledges

(Extract taken from interview no. 4.)

I: yeah now does it mean when people talk of community education do they mean . . . primary, secondary, vocational, technical, and universities, or d-do they really mean only a certain level or type /22

The intermediary's question about the level of schooling implies some internal knowledge by the intermediary about the meta-structure of the subject domain, (i.e. knowing that the topics of research projects in this area tend to be limited to a particular educational level). This knowledge of the domain meta-structure provides a framework within which to question the user about her research.

Retrieval strategy knowledge

(Extract taken from interview 120684HBA.)

I: Um (,) there's a social sciences citation index (.....) /95a um (.) what we've got to do there is concentrate on just title words /98

Here the intermediary makes a decision about the search tactic needed (title-word search) based on her knowledge of the record structure of a particular database.

The types of knowledge needed by the User Modelling function tend to be categorical in nature. The specific knowledge used embodies a finite, restricted range of possibilities and tends to be qualitative rather than quantitative. In contrast, the knowledge needed by the Problem Description function appears much more varied and to be closely connected with the subject domain of the search topic. The knowledge elicited by the analysis for these two functions is described in more detail in Brooks, Belkin and Daniels (1985); Daniels (1986); and Brooks (1986*a*, 1986*b*).

4.5. TEMPORAL ORDERING AND INTERACTION AMONG FUNCTIONS

The pattern of interaction between the tasks associated with the User Model and Problem Description functions were analysed. Information contributed by, or donated to, other external functions was noted. These interactions were diagrammed on a focus-by-focus basis. A brief example of such a map is given in Tables 5 and 6 [For further examples see Brooks, Belkin & Daniels (1985) and Brooks (1986*b*).]

There appears to be no obvious single order in which these functions are, or should be, performed, but rather that any sequence is circumstance driven. The complexity of the interactions and the data-driven nature of the sequencing can be seen in the analysis of the foci of interview 190684HBA given in Table 7. The

TABLE 5
Interaction map for Problem Description, focus 5, interview 290684KSA

TOPIC	RES	SUBJ	DOCS	I/O
			Key-paper-2 delusional thinking and perceptual disordorder →	RS[TERMS] select terms from title of key documents
		Current theory: delusional thinking is an attempt to find organization in the mind. The mind becomes disorganized because of strange perceptions		
	Perception ↤			
Perception ↤				

TABLE 6
Interaction map for Retrieval Strategy, focus 5, interview 290684KSA

DB	TERMS	QUERY	STRAT	I/O
	CONSIDER ← - - - - - delusional thinking perceptual disorder			PD[DOCS] key-paper-2 title
	↓			
	OCCURRENCE ← - - - - - delusional thinking ← - - - - -		- - - ┐	PD[SUBJ] PD[RES]
	↓		⋮	
	SYNONYMS ← - - - - - delusional thinking		- - - ┘	UM[KNOW] users knowledge of terminology

TABLE 7

An analysis of the foci in interview 190684HB giving the function and main task for each focus

Focus	Function and task	Description
1	UM [USER]	Establish the status of the user—is he from the university or an external client, is he faculty?
2	PD [TOPIC]	Develop an initial description of the search topic
3	RG [OUTPUT]	Establish whether the user wants a particular type of document e.g. magazine, newspapers
4	PS [PREV]	Find out details about the user's previous search
5	EXPLAIN	Explain to the user the difficulties of restricting to UK journals only
6	RS [DB]	Explore the range of possible databases to search
7	PD [TOPIC]	Define the topic further-particularly with respect to historical period
8	EXPLAIN	Explain the difficulties of searching for historical period
9	PD [TOPIC]	Discuss the importance of historical period. Determine topics which are *not* of interest to the user
10	PD [SLIT]	Find out how much literature exist on various aspects of the topic
11	PD [TOPIC]	Further description of the topic—the types of relations between Turkey and Greece
12	RS [DB]	Discuss which databases it might be appropriate to search. Read down catalog as far as Middle East Index
13	PD [SUBJ]	Find out if Middle East conflicts also include Greek–Turkish conflicts
14	RS [DB]	Discuss suitable databases further
15	RS [TERM]	Select free text terms to express search topics
16	EXPLAIN	Explain the search procedure and what will happen online
17	PM [CAPAB]	Discuss whether databases exist which are concerned with UK publications only and where indexes to UK dissertations can be found
18	RS [QUERY]	Formulate a query for restricting output to UK publications
19	UM [GOAL]	User describes his research goals as an explanation of why he is interested in UK material on the topic only

acronym for the function of each focus is given together with the main functional task being carried out.

Comparison of the functional structure of individual interviews has enabled us to indicate the extent to which model-building activities are carried out by both parties. It was also possible to demonstrate the significance of the Problem Description, Problem State and User Modelling functions to the success or failure of an interaction (Brooks & Belkin, 1983; Price, 1983).

4.4. DIALOGUE STRUCTURE

As the previous section demonstrates, the pattern of interaction between user and intermediary is highly complex, and the sequencing of functions appears to be highly data driven. This has been a general result of all our analyses, and has led us to conclude that there is no fixed order in which the functions ought to be performed. Previous work (Belkin & Windel, 1984), however, suggested that a general problem structure for the information system situation could provide a means for driving and guiding a human–computer dialogue in this domain. We have taken this suggestion further by specifying a problem structure for the information retrieval situation based upon the functions, the general and specific goals of the situation, and the focus shifts in the interactions we have analysed (Daniels, Brooks & Belkin, 1985). We do not discuss this issue further here, but note that this approach allows one to represent, and take into account, individual aspects of an interaction, whilst retaining a commonality of form. That is, the foci of the interaction can be seen as concentrating the participant's attention on the accomplishment of a particular goal, with shifts of focus moving attention from goal to goal across the problem structure.

5. Discussion

As Welbank (1983) says, interviews alone cannot produce a detailed context of *real* behaviour; rather, they need to be backed up by observational studies. Our methodology concentrates on observational techniques supported by interviews. We define our methods as observational since they involve recording the expert at work on real problems in her normal environment. Verbal protocol analysis, on the other hand, requires the expert to verbalise her thoughts and actions as she carries out what is often an artificial task. Although this method has been used to study librarian-user interactions (Ingwersen & Kaae, 1980), it would have been inappropriate in our situation. We considered that interviews with intermediaries would not be productive as the principal method of knowledge elicitation either. Experts typically are unable to articulate their problem-solving expertise and this is particularly true for information provision environments where the tasks and goals of intermediaries are not well-formulated, well-defined or static. Unlike medical diagnosis for example, there is no metalanguage for describing problem-solving procedures.

We view the information interaction between user and intermediary as consisting to a large extent of model building by both participants within a cooperative dialogue. Therefore, the knowledge to be elicited must concern the way in which the participants use discourse to carry out their tasks. It is necessary then to analyse the recorded observations at the discourse level. This is the level of granularity required. The analysis that we have carried out goes beyond an analysis of broad concepts and beyond a simple functional analysis. Discourse analysis provides a "microlevel" of representation which will allow complete, complex and accurate interface implementation.

Using the methods outlined, we can also take account of the co-operative nature of the interaction, e.g. user inputs and contributions. Moreover, we have been able to outline a dialogue structure which could drive the computer half of such a dialogue. This will provide the basis for the development of a true mixed-initiative discourse component for the interface. Kidd & Cooper (1985) have noted that one of the main problems with current expert systems is their rigidly constrained system-oriented dialogues.

We have also been able to propose a representation for the user model and the problem description. The user model representation consists of frames embodying information about the user's status, goals, level of knowledge, previous information system experience and general background (Daniels, 1986). It is suggested (Brooks, 1986*a*) that the problem description is represented as several layers of interconnected partitioned semantic networks (Hendrix, 1979).

From the methodological point of view, we have taken great care with intercoder reliability and cross-checking. These are common procedures in psychological and sociological research but we have found very little evidence of their use in the knowledge engineering literature.

Finally, it should be noted that this is an extremely time-consuming method of analysis. The whole process, from recording, through transcription, to the analyses, involved a team of researchers over many months. On the other hand, it has provided us with sufficient detail to begin to implement an interface which will exhibit complex behaviour and be capable of sustaining mixed initiative discourse.

6. Conclusion

The outcome of this project has not merely been to identify and specify the functions required for an intelligent interface for document retrieval systems. In addition, we feel that contributions have been made to the following:

(a) Knowledge elicitation methodology: Detailed discourse analysis of recorded interactions between expert and client has not been, up to now, well explored as a method for eliciting expert knowledge.
(b) The application of expert system techniques to complex domains: Many expert systems have been confined to very restricted, well defined domains. The application we are investigating, however, involves multi-dimensional problem-solving in a complex domain. This has implications for methodology and system architecture.
(c) Cognitive modelling: We regard the knowledge elication task as serving not only to develop an operational system but also to further our understanding of what human intermediaries do and how they do it. Interaction between human intermediaries and users involves complex cognitive modelling. Thus our research has offered insight into the development and use of multiple models by more than one party.

In conclusion, we believe that knowledge elicitation is of prime importance to expert system development and should not be regarded as secondary to the implementation of an operational system. Unfortunately eliciting knowledge at a detailed and comprehensive level is time-consuming and mitigates against the immediate implementation of expert systems.

References

ALTY, J. L. & COMBS, M. J. (1980). Face-to-face guidance of university computer users—I: a study of advisory services. *International Journal of Man–Machine Studies,* **12,** 390–406.

BELKIN, N. J. (1980). Anomalous states of knowledge as a basis for information retrieval. *Canadian Journal of Information Science,* **5,** 133–143.

BELKIN, N. J., HENNINGS, R.-D. & SEEGER, T. (1984). Simulation of a distributed expert-based information provision mechanism. *Information Technology: Research, Development, Applications,* **3,** 122–141.

BELKIN, N. J., SEEGER, T. & WERSIG, G. (1983). Distributed expert problem treatment as a model of information system analysis and design. *Journal of Information Science,* **5,** 153–167.

BELKIN, N. J. & WINDEL, G. (1984). Using MONSTRAT for the analysis of information interaction. In: DIETSCHMANN, H. J. Ed. *Representation and Exchange of Knowledge as a Basis of Information Processing.* Amsterdam: Elsevier Science. pp. 359–382.

BROOKS, H. M. (1986*a*). Developing and using problem descriptions. In: BROOKES, B. C. Ed. *IRFIS 6: Intelligent information systems for the information society.* Frascati, Italy, Sept., 1985. Amsterdam, North Holland, 1986, pp. 141–161.

BROOKS, H. M. (1986*b*). An intelligent interface for document retrieval systems: Developing the Problem Description and Retrieval Strategy components. Ph.D. thesis. London, England. The City University, Department of Information Science; 1986.

BROOKS, H. M. & BELKIN, N. J. (1983). Using discourse analysis for the design of information retrieval mechanisms. In: *Research and Development in Information Retrieval. Proceedings of the Sixth Annual International ACM SIGIR Conference, Washington, D.C., 1983.* New York: ACM. pp. 31–47.

BROOKS, H. M., BELKIN, N. J., DANIELS, P. J. (1985). Problem descriptions and user models: developing an intelligent interface for document retrieval systems. In: *Informatics 8: Advances in Intelligent Retrieval.* London: ASLIB. pp. 191–214.

BUCHANAN, B. G. (1982). New research in expert systems. In: HAYES, J. E., MICHIE, D. AND PAO, Y. H. Eds. *Machine Intelligence 10.* Edinburgh: Edinburgh University Press. pp. 269–299.

BUCHANAN, B. G., *ET AL.* (1983). Constructing an expert system. In: HAYES-ROTH, F. & WATERMAN, D. A. Eds. *Building Expert Systems.* Reading, MA: Addison-Wesley. pp. 127–168.

CLANCEY, W. J. (1983). The epistemology of rule-based expert systems: a framework for explanation. *Artificial Intelligence,* **20,** 215–251.

COOMBS, M. J. & ALTY, J. L. (1984). Expert systems: an alternative paradigm. *International Journal of Man–Machine Studies,* **20,** 21–43.

COOMBS, M. J. & ALTY, J. L. (1980). Face-to-face guidance of university computer users—II: characterizing advisory interactions. *International Journal of Man–Machine Studies,* **12,** 407–429.

CROFT, W. B. (1985). An expert assistant for a document retrival system. In: *RIAO 85. Actes* of the conference: Recherche d'Informations Assistee par Ordinateur, Grenoble, France, 18–20 March, 1985. Grenoble: I.M.A.G. pp. 131–149.

DANIELS, P. J. (1986). The user modelling function of an intelligent interface for document retrieval systems. In: BROOKES, B. C. Ed. *IRFIS 6: Intelligent Information Systems for the Information Society.* Frascati, Italy, Sept. 1985. Amsterdam, North Holland, 1986. pp. 162–176.

DANIELS, P. J., BROOKS, H. M. & BELKIN, N. J. (1985). Using problem structures for driving human-computer dialogues. In: *RIAO 85: Actes* of the conference: Recherche d'Informations Assistee par Ordinateur, Grenoble, France, 18–20 March, 1985, Grenoble: I.M.A.G. pp. 131–149.

ERICSSON, K. A. & SIMON, H. A. (1980). Verbal reports as data. *Psychological Review,* **87**(3), 215–251.

FOX, E. A. (1987). Development of the CODER system: a test bed for artificial intelligence methods in information retrieval. *Information Processing and Management,* **23**(9), 341–366.

GAMMACK, J. G. & YOUNG, R. M. (1985). Psychological techniques for eliciting expert knowledge. In: BRAMER, M. A. Ed. *Research and Development in Expert Systems. Proceedings* of the 4th Technical Conference of the B.C.S. Specialist Group on Expert Systems, University of Warwick, 18–20 December 1984. Cambridge: Cambridge University Press. pp. 105–112.

GROSZ, B. J. (1978). Discourse knowledge. In: Walker, D. E. Ed. *Understanding Spoken Language.* New York: Elsevier-North Holland. pp. 229–346.

HENDRIX, G. (1979). Encoding knowledge in partitioned networks. In: FINDLER, N. V. Ed. *Associative Networks: Representation and Use of Knowledge in Computers.* New York: Academic Press. pp. 305–326.

INGWERSEN, P. (1982). Search procedures in the library analysed from the cognitive point of view. *Journal of Documentation,* **38**(3), 165–191.

INGWERSEN, P. & KAAE, S. (1980). User–librarian negotiations and information search procedures in public libraries: analysis of verbal protocols. *Final Research Report, DB-TEK*-50. Copenhagen: Royal Danish School of Librarianship.

KELLY, G. A. (1955). *The Psychology of Personal Constructs.* New York: Norton.

KIDD, A. L. (1985*a*). The consultative role of an expert system. In: JOHNSON, P. & COOK, S. Eds. *People and Computers: Designing the Interface. Proceedings* of the B.C.S. HCI Specialist Group Conference, University of East Anglia, 17–20 Sept. 1985. Cambridge: Cambridge University Press. pp. 248–254.

KIDD, A. L. (1985*b*). Knowledge elicitation. B.C.S. Specialist Group on Expert Systems Lecture, London, October, 1985.

KIDD, A. L. & COOPER, M. P. (1985). Man–machine interface issues in the construction and use of an expert system. *International Journal of Man–Machine Studies,* **22,** 105–112.

KUIPERS, B. & KASSIRER, J. P. (1983). How to discover a knowledge representation for causal reasoning by studying an expert physician. In *Proceedings of the 8th International Joint Conference on Artificial Intelligence,* Karlsruhe, 18–20 Aug. 1983. pp. 49–56.

PRICE, L. E. T. (1983). Functional and satisfaction analyses of information interaction dialogues. M.Sc. thesis, Department of Information Science, The City University, London.

SACKS, H., SCHEGLOFF, E. & JEFFERSON, G. (1974). A simplest systematics for the organization of turn-taking in conversation. *Language,* **50,** 1974, 731–734.

SHAW, M. L. G. (1984). Knowledge engineering for expert systems. In: *Interact '84. Proceedings* of the IFIP Task Group on Human–Computer Interaction, Imperial College, London, 4–7 Sept. 1984. pp. 328–332.

SINCLAIR, J. MCH. & COULTHARD, R. M. (1975). *Towards an Analysis of Discourse. The English Used by Teachers and Pupils.* Oxford: Oxford University Press.

STINTON, C. (1984). Intermediary knowledge resources in information interaction. M.Sc. thesis. Department of Information Science, The City University, London.

WELBANK, M. (1983). A review of knowledge acquisition techniques for expert systems. *British Telecom Research Laboratories Report, December 1983.* Martlesham Heath: Martlesham Consultancy Services.

Specification of expertise

PAUL E. JOHNSON

Department of Management Sciences, University of Minnesota, Minneapolis, Minnesota, U.S.A.

IMRAN ZUALKERNAN

Department of Computer Science, University of Minnesota, Minneapolis, Minnesota, U.S.A.

SHARON GARBER

3M, St Paul, Minnesota, U.S.A.

This paper describes a framework for representing the expertise required to perform a task. The framework is based upon a procedure employed in research conducted by cognitive scientists and psychologists in order to understand human problem-solving and decision-making processes. The procedure consists of giving individuals real or simulated tasks to perform and asking them to "think aloud" while they work. The comments such individuals make are referred to as verbal protocols, and the techniques used to analyse these comments are termed, collectively, protocol analysis.

The primary motivation for our paper is to address a fundamental difficulty in knowledge-acquisition methods. This difficulty is that the process of human thinking we wish to understand is not available to direct observation. In the case of the human expert, this difficulty is compounded by the fact that the expertise of interest is typically not reportable, due to the compilation of knowledge which results from extensive practice in a domain of problem-solving activity (Anderson, 1981).

One result of the lack of awareness on the part of human experts is the use of what is often called rapid prototyping in the development of an expert system. Through interaction with a system prototype, the expert often displays expertise not discoverable by other means. While we do not disagree with the basic notion that human experts can reveal knowledge and skill in the performance of a specific task (including prototype debugging), we believe that building a prototype system early in the knowledge-acquisition process may result in a commitment to a specific model of thinking (inference process) that does not adequately represent the expertise in question.

A system prototype is a specification of the requirements (e.g. Ramamoorthy, Prakash, Tsai & Usuda, 1984) for a computational model. In this paper we shall propose an alternative means of creating such a specification for an expert system, which is based upon the use of human problem-solving data. The virtue of our approach, we believe, is that it permits the construction of a system prototype itself to be based upon a specification of expertise.

In the balance of the paper we shall attempt to do several things. First, we will describe, briefly, the background of protocol analysis methods as these have been employed in the psychological investigations of human problem-solving, in order to make clear both the strengths and weaknesses of this method of understanding

KNOWLEDGE-BASED SYSTEMS Vol. 1
ISBN 0-12-273251-0

human thinking. Second, we shall propose a definition of expertise that has evolved from work done at the University of Minnesota over the past several years on the nature of expert thinking. Third, we shall describe a framework for developing a representation of expertise based upon the use of protocol analysis techniques. Finally, we shall present examples of the use of this framework.

Protocol analysis

Historically, protocol analysis derives from the attempt by psychologists at the turn of the century to use introspective methods in order to gain an understanding of human mental processes. Such introspective methods typically require intensive training on the part of subjects who then attempt to search their conscious memory for the basis of some task performance. Such efforts fell into disfavour among American psychologists when it was claimed that they distorted the process they were attempting to describe.

In Europe, psychologists used versions of the classic introspective method to augment more standard observational techniques for assessing an individual's thought process. The use of such methods has generally been considered informative by these investigators, particularly in the discovery phase of research where the attempt is made to develop a theory that can later be subjected to more standard verification experiments.

In the 1970s, American psychology moved away from behaviorism and toward mentalism and information-processing models of thought. Investigators such as Newell and Simon (1972) began to use a first cousin of the traditional introspective method, called concurrent protocol analysis, to examine problem-solving behavior. Based upon previous work by deGroot (1965), Newell and Simon asked chess players to think aloud while solving chess-game situations. The technique used by Newell and Simon required that subjects think aloud without attempting to rationalize their problem-solving activities, which these investigators regarded as an essential weakness of the earlier introspective methods. The concurrent protocol technique removed some of the earlier criticism and was responsible for much of the development of the theory of human information processing.

Our objective in the present paper is to show how the techniques developed by early investigators such as Newell and Simon can be extended to develop a specification of human expertise. In the next section we present a definition of expertise that has resulted from our attempts to apply the techniques of protocol analysis to the study of expert thinking in a variety of problem-solving domains.

Definition of expertise

Expertise is a kind of operative knowledge. It is characterized by generativity, or the ability to act in new situations, and by power, or the capacity to achieve problem solutions. Expertise is a kind of knowledge, and not a property of the behavior we observe as individuals perform tasks.†

† Expertise as we have defined it here is an instance of what Newell (1981) has termed a knowledge level concept; a theory of expertise also seems to us similar in many respects to what Marr (1982) has termed a computational theory.

One way to represent expertise for problem-solving is as a set of requirements that must be satisfied in order to solve problems in a given domain. If, for example, our goal is to solve quadratic equations, we can represent what is required to perform this task by means of the quadratic formula. For an equation of the form: $ax^2 + bx + c = 0$, the quadratic formula $[-b \pm (b**2 - 4ac)^{1/2}]/2a$ is a statement of what needs to be computed to achieve a solution.

The quadratic formula is a rule that has the capability of generating a certain behaviour, namely, solving quadratic equations. The formula displays a certain competency, i.e. it can be used to solve quadratic equations for integer, real and complex numbers. To be able to use this formula, however, requires certain abilities, for example, the ability to divide, multiply, subtract etc. If the quadratic equations involve complex numbers, it is also necessary to know how to do complex arithmetic. The abilities required to use the quadratic formula are realized by what Simon (1980) has termed the problem solver's inner environment. Addition, for instance, can be implemented in a number of ways, depending upon the internal-processing architecture of the problem-solver performing addition.

In the above example, expertise in solving quadratic equations can be represented by the quadratic formula, plus the abilities needed to use this formula. But this description is still incomplete. In addition to a statement of rules, and the abilities needed to apply these rules, we need to know the goals of the problem solver, and how these goals are related to the use of abilities. In our example it is important to know, for instance, that the discriminant $(b**2 - 4ac)$ is a good determiner of which goals to follow, i.e: (1) if the discriminant < 0, invoke the abilities for complex arithmetic; (2) if the discriminant $>= 0$, invoke the abilities for real arithmetic; and (3) if the discriminant $= 0$, the ability to perform a square root is not required.

The statements about discriminants are examples of rules that define goals for achieving a problem solution. The application of these rules is invariant with respect to the resources used to implement the abilities; that is to say, it does not matter how one does real arithmetic, rule 2 will still hold true. The rules represent a portion of what needs to be computed in order to solve quadratic equations.

Expertise, then, is a kind of knowledge that is used to perform a task. A part of the representation of this knowledge is the set of abilities which is required to realize the solution of problems to which the expertise applies. These abilities are themselves realized in different ways, depending on the artifact solving the problem.

Expertise can also be thought of abstractly as the knowledge that enables a mapping from some problem space to a solution space. Hence, any representation of expertise must include the inputs a problem solver accepts and the outputs it produces. Specifying what inputs are valid is important because it determines what kinds of features (data) need to be extracted from the environment for problem solving to occur. Specifying what outputs are valid is important because the outputs identify potential problem solutions.

To create a specification of expertise we begin with the assumption that any problem-solving process must have two components, a problem solver and a problem. The problem solver starts with little or no information, gathers what is needed, and proceeds by setting goals and executing actions to satisfy these goals. The problem solver is further assumed to have some criterion for determining when a solution is achieved and what is needed to achieve this solution. Given these assumptions, our representation (specification) of expertise in problem solving has five parts:

(1) possible solutions and components of solutions to a problem or class of problems;
(2) relevant information or data needed for solving such problems;
(3) goals that are required to achieve intermediate as well as final problem solutions;
(4) permissible ways of moving between intermediate problem solutions, including the data that must be processed to make such transitions;
(5) abilities required to make the transitions between potential problem solutions.

We now turn to a description of a framework for obtaining information from the expert, and for developing, from this information, a specification of expertise.

Framework

We claim that a specification of experties can be based upon inferences made from a record of problem-solving activity. Protocol is a record or trace of problem solving that consists of the verbal and motoric activity generated by an artifact while in the act of solving a problem. Protocol is analogous to the output of a program that has the same behavior as the problem solver. However, protocol is more than a trace of a program, it contains explicit statements of what the problem solver is trying to do.

Our objective in knowledge acquisition is to construct an artifact (e.g. an expert system) that solves a class of problems which is currently solved by an expert (or experts). To construct this artifact we need a specification of its requirements. This specification outlines what needs to be computed to solve that class of problems.

There will be a number of artifacts that can achieve the same performance in a variety of ways. The expert's method works because it is adapted to the capabilities of the human information-processing system and the demands of the problem-solving task. Since we may implement our specification on a variety of processors, we seek a description that does not depend on a particular processing architecture. The purpose of knowledge acquisition, in this case, is not to learn *how to solve a problem,* but rather to discover *what is required* to solve a problem.

Protocol provides a record of problem-solving activity. We propose to use this record to develop a specification of the requirements for any artifact that attempts to solve the same problem as the expert giving the protocol. Thus, given a class of problems, and a protocol record from experts solving these problems, our task is to determine a method for transforming information in this record into a specification of expertise.

We now state our framework for creating a specification of expertise as follows:

(1) The expert can be viewed as a processor that has the capability of producing a certain problem-solving behavior using expertise. The task of knowledge acquisition is to determine this expertise.
(2) The expert has developed a set of actions and abilities that are necessary to realize this expertise.
(3) Although we cannot observe the expertise directly, we can observe the invocation of the expert's actions and abilities in a record of problem-solving behavior.

(4) Since we can observe the invocation of actions and abilities by the expert, we can develop some representation of the expertise.
(5) A statement of the expertise required to perform a task serves as a specification of the requirements for a computer program that is designed to perform the task.

We next present a specific methodology for collecting and analysing protocol data to arrive at a specification of expertise.

Methodology

We begin with a record of the behavior generated by an expert. This record must be analysed in order to build a specification of the expertise required to solve the problem upon which the record is based. In the work we shall describe here, a protocol is assumed to consist of the verbal thinking-aloud comments generated by an expert who is instructed to say whatever he/she thinks of while solving a problem. The purpose of data analysis is to transform a given protocol into a form that can be used to build a specification of expertise. In the analysis presented here, this transformation has two parts: a syntactic analysis and a semantic analysis.

The syntactic analysis is based upon the identification of three categories of behavior in the protocol record: operations, episodes, and data cues:

Operations
Operations are primitive activities of problem solving that do not depend upon a particular context. Examples of operations include collecting data, doing a computation, etc. Syntactically, operations are indicated by verbs. A coder with some knowledge of the domain is usually better at identifying operations than someone who has no domain knowledge and only the notion of verbs to go on. We have found that someone who is less expert in a field often makes a good coder since he or she is familiar with basic vocabulary, but does not tend to extrapolate beyond the available information.

Episodes
In the protocol data, there are patterns of operations that are repeated within and across different problems. Such patterns are called episodes. These patterns exist together, in time, due to some necessity of performing them in a certain order. The simplest pattern is a sequence of operations.

Data cues
Data cues are operands. They comprise the data processed by the problem-solving operations. Syntactically, data cues are indicated by domain nouns.

Whereas a syntactic analysis allows the protocol record to be partitioned into separate categories of behavior, it is the semantic analysis that forms the basis for a representation of expertise. Our model of semantic analysis assumes that the problem space is a fundamental construct of problem solving (Newell, 1979). We define a problem space as a set of problem states, and we assume that problems are solved by moving from one problem state to another. We assert that the expertise required to solve problems can be characterized by the semantic categories of actions, abilities, goals, conditions, strategies and solutions. Each semantic category is derived from the syntactic analysis described above.

Actions
Actions are the means of making transitions between problem states. They are derived from operations performed by the problem solver. Actions are the semantic primitives of problem solving.

Goals
Goals represent desired states of problem solving that an expert is trying to achieve. They are derived from problem-solving episodes. Satisfying goals produces new problem states.

Abilities
Abilities represent capacities to perform the actions needed to achieve goals. For example, to solve a quadratic equation using the quadratic formula one needs the ability to perform arithmetic operations.

Conditions
Conditions are problem states defined by new data cues. New data cues can be further subdivided into environmental data and internal data. Environmental data correspond to the expert seeking some information from the problem domain. Internal data correspond to data produced via some internal computation. To identify problem conditions, it is necessary to keep track of when new information is articulated in the protocol.

Solutions
Solutions are end states of problem solving. They coincide with the achievement of goals. Solutions are identified by examining states of problem solving in which the expert is attempting to perform actions to achieve goals that are defined as solutions by the vocabulary of the domain.

Strategies
Strategies are permissible ways of moving between problem states. They are characterized by the data that must be processed in order to make transitions between problem states, plus the goals and abilities relevant to making these transitions.

REPRESENTATION OF EXPERTISE

The objective of the analysis proposed here is to formulate a specification for expertise based upon a coded record of problem-solving behavior. Syntactically, we shall define this specification as consisting of a set of components. The first component is a "circle" (see Fig. 1). Each circle has a unique string as its identifier and represents a solution or part of a solution. Each circle can itself have levels of specification. The second component is a set of directed relationships between circles. These relationships are represented by arrows. The arrows are of two kinds:

arrows between two circles;
arrows starting from a symbol represented by = =.

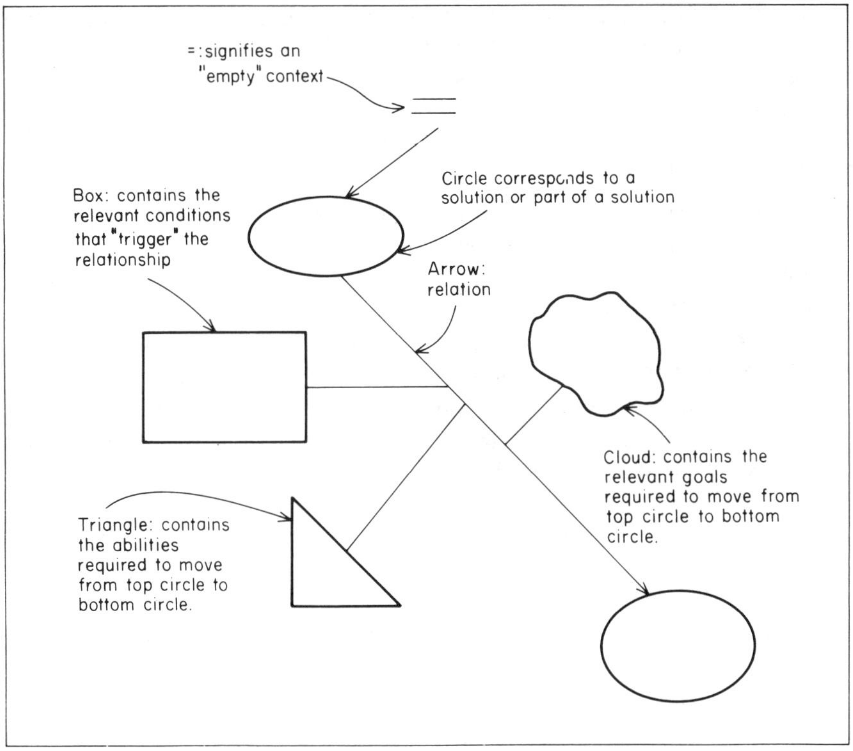

FIG. 1. Graphical specification of expertise.

The third component is a "box" associated with each arrow. Each box includes a set of possible conditions. The fourth component is a "cloud", which is also associated with each arrow. Each cloud contains a set of goals. Finally, each arrow also has associated with it a fifth component called a "triangle". Each triangle includes a set of problem-solving abilities.

Each circle identifies a possible solution. Relationships between circles determine directed "pathways" of problem solving. These, in turn, determine ways of moving between possible solutions. The goals in the cloud associated with each relationship are the goals that are satisfied in "travelling" on that relationship or "pathway". The conditions in the box associated with the relationship represent the set of possible "triggers" that activate the goals. The triangle identifies the abilities that are required in order to achieve the problem-solving goals. The union of conditions in all the boxes is the input specification. The set of all possible subsets of circles (at all levels) defines the output.

We illustrate the above formalism with an example based upon our previous discussion of the quadratic equation. A part of the expertise encoded in the quadratic formula is its ability to characterize a solution as either complex or real. We shall map that part of expertise in the quadratic formula that is responsible for distinguishing between complex and real solutions.

The expertise we are looking for is that which splits the possible solutions into complex and real. This expertise is encoded in the square-root symbol. The particular use of this symbol, and its semantics in the quadratic formula, define the expertise for splitting the output into two parts. We can further refine the output to reflect the ability to specify equal roots. We specify the requirements for expertise in this case as shown in Fig. 2.

PROTOCOL ANALYSIS

To create a specification from coded protocol records, we make the assumption that expertise reflects adaptation to the demands of a class of problem-solving tasks. This adaptation is expressed in one or more states of problem solving that recur across problem conditions. Because these stable states serve to define basic sequences of problem-solving activities, we shall refer to them as problem sequencers.

Problem sequencers are hypotheses about what stable states are necessary for achieving a problem solution. For example, one important recurring activity in problem solving is to propose a possible solution. Another important recurring activity is to evaluate each proposed solution. So "propose" and "evaluate" are two

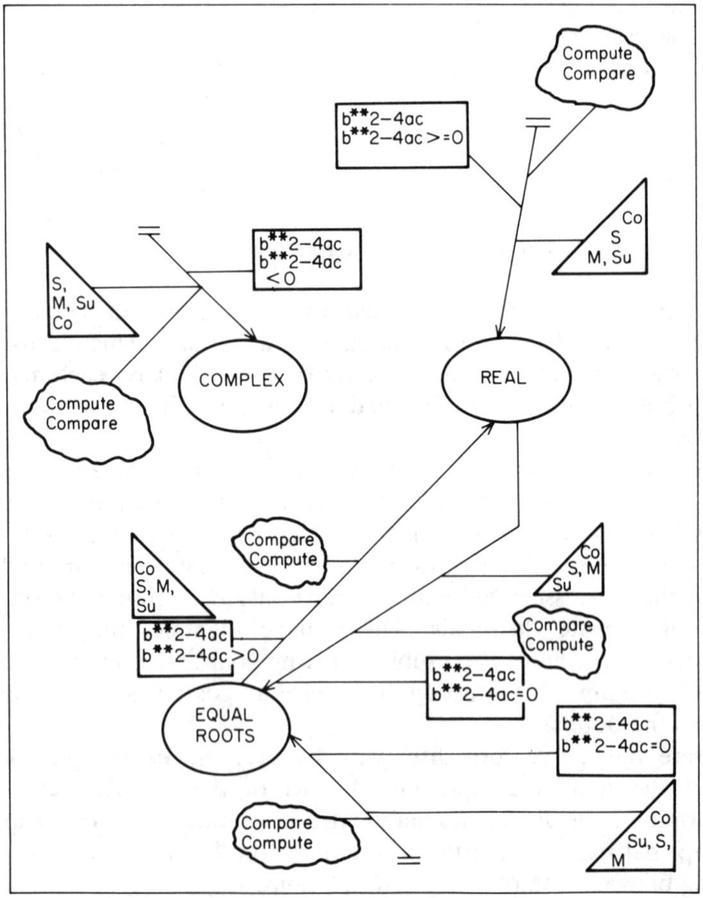

FIG. 2. Specification of a part of expertise in solving quadratic equations. s = square; su = subtract; M = multiply; Co = compare with 0.

problem sequencers that may reflect the expertise used in achieving a final problem solution. More complicated sequencers are possible depending on the task (see below).

An essential property of a problem sequencer is that it divides the protocol record into temporal intervals. A temporal interval is bounded by places (conditions) in the protocol record where a given sequencer occurs. The sequencer and the associated goals and abilities falling in the temporal interval between two problem conditions determine the relationship between the solutions or parts of solutions associated with the conditions. Once a temporal interval is determined by a given problem sequencer, a relationship is created such that the ends of the interval are the circles associated with the problem conditions. All the goals falling in the interval are included in the cloud associated with that relation. This, in turn, uniquely determines the assignment of conditions to boxes. If there is more than one goal in an interval, then the box contains the union of those goals.

We now describe an application of the above methodology in specific work currently in progress at the University of Minnesota.

Case study: industrial experimental design

The problem we wish to consider is often referred to as off-line quality control. It can be described briefly as follows. A client comes to an industrial statistician for help in designing an experiment to study a process. The client is typically an engineer or a scientist who is interested in improving particular aspects of a system. For an engineer, the problem might be finding the optimum paramaters for a process to create some product. The experiment must be statistically sound and must also satisfy certain cost constraints given by the client.

The objective of the statistician's proposed design in an industrial setting is to improve the performance of some process. This process can be characterized in terms of an experimental unit that flows through the system over time. The experimental unit is the material to which a treatment is applied in a single trial. The experimental unit may be a typewriter ribbon, a lump of cookie dough, a plot of land, a manufacturing plant or it may be a group of patients in a hospital or class of drugs. The statistician considers certain properties of the experimental unit, such as availability, size, and representativeness of the objects in the experiment.

The statistician is interested in various aspects of the process. These include time and cost of running the experiment, ease in changing factor levels, and blocking information. For example, differences in operators or in amounts of raw materials or in temperature and humidity conditions during the course of the experiment are important blocking considerations in designing an efficient experiment.

At the termination of the process, the measurements made on the experimental unit constitute the values of the response variable or variables. The statistician knows which characteristics to select and how these are measured.

METHOD

We studied two statisticians solving a series of industrial design problems. Each problem was based upon an actual encounter between an industrial statistician and a client. Our investigation consisted of the following:

Selection of test problems
We were interested in representing the expertise for formulating experimental designs in a manufacturing environment. To accomplish this objective, we discussed the problem area with experts and selected actual problems in a specific manufacturing domain.

Selection of experts
Two experts were chosen to solve problems. The first expert (E_1) was a University Professor with considerable industrial experience. The second expert (E_2) was an industrial statistician in a large manufacturing company. The experts were given identical problems and verbal protocols were collected during problem solving. These protocols were analysed as follows.

Analysis of data
We identified (and subsequently labelled) the following problem-solving goals which were, in turn, derived from episodes in the protocols of the two experts.

(1) Understand product/process
(2) Relate factors to response
(3) Propose design
(4) Probe for additional information
(5) Evaluate design
(6) Understand experimental conditions
(7) Calculate sample size
(8) Consider blocking
(9) Reduce factors/levels of factors
(10) Consider interactions
(11) Consider experimental error

Two types of analysis were performed in order to develop a specification of expertise for the experimental design task. The first of these, termed a frequency analysis, identifies potentially important goals. The second, termed a temporal analysis, examines alternative problem sequencers and divides the protocol record into intervals for purposes of identifying the components of a specification of the expertise required to perform the task.

Frequency analysis. Since we are concerned about the "what" rather than the "how" of problem solving, one convenient way of representing our data is through simple frequency counts. For instance, we have a record of our two experts solving the same problem and arriving at similar results. The first question we wish to ask, in this instance, is what are the important goals used to achieve a solution? To answer this question, we identify the goal associated with each episode in the protocol and then look at the frequency of occurrence of that goal in the protocol record of each subject. An example of the data from one such analysis is shown in Figs 3 and 4.

Inspection of Figs 3 and 4 shows that:

For a given problem, there is strong correspondence between the kinds of goals invoked by each expert (Fig. 3).

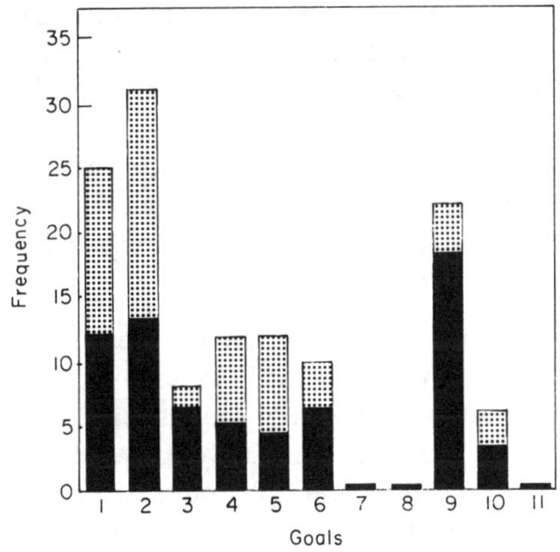

FIG. 3. Frequency of goals in Problem 2. ▦, E2 problem 2; ■, E1 problem 2.

Some goals are consistently invoked more than others across problems. These goals define a minimum expertise of the problem solver (Fig. 4).
For each problem PROPOSE DESIGN (goal #3) is invoked a relatively small number of times, suggesting that part of the expertise in designing experiments consists of making good initial guesses (Figs 3 and 4).
A large part of the problem-solving process seems to be devoted to EVALUATE DESIGN (goal #5). This suggests that part of the expertise consists of good evaluation capabilities. In other words, the expertise permits the artifact which embodies it to suggest a design at any point, and then evaluate this design against the constraints posed by the client (Fig. 4).
Although expertise in experimental design embodies good guessing abilities, the large effort concerning evaluation indicates that the design process is carried out in an evolutionary manner.

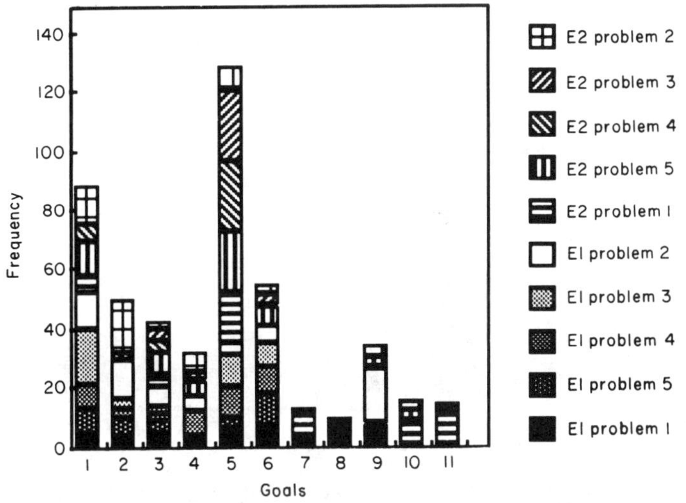

FIG. 4. Frequency of goals in five problems for E1 and E2.

A second question that can be addressed by examining the frequency of occurrence of goals in the protocol record is whether goals can serve as problem sequencers for developing a specification of expertise in this instance (i.e. in our case this consists of determining how the problem solver makes good first guesses about an appropriate design). To answer this question, we examine the frequency of all goals before the PROPOSE DESIGN goal occurs. The data in Fig. 5, show that understanding the product or process (goal #1) and relating factors to response (goal #2), play a major role in proposing a design.

Temporal analysis. The second kind of analysis is based upon the occurrence of stable states in the problem-solving record. Based upon the previous frequency analysis, we shall consider goals as problem sequencers for expertise in the domain of experimental design. To perform this analysis we plot goals against problem conditions (see Figs 6 and 7). The multiple numbers shown vertically along the abscissa in Figs 6 and 7 result from the program used to produce the figure—the two numbers in each column should be interpreted sequentially, i.e. is 11, is 13, etc. Each point on the graphs shown in Figs 6 and 7 corresponds to the activation of a particular goal during the sequence of problem solving. We call such a diagram a goal trace. Figure 6 gives the goal trace for problem 1 for both our experts. Figure 7 gives the goal trace for problem 2 for the same two experts.

We shall construct a specification of expertise based upon the temporal intervals and goals for PROPOSE DESIGN as a sequencer in problems 1 and 2. Table 1 summarizes the specific designs (solutions) proposed by each subject in problems 1 and 2 and in three additional problems. The data in Table 1 specify sequences of designs proposed by each subject, namely, factorial designs and screening designs are related to fractional factorial designs, which in turn are related to Latin square designs.

We now describe how a specification for expertise in experimental design can be developed by using the goal traces and problem sequencer shown in Figs 6 and 7 plust the data in Table 1. For the sake of simplicity, we will show only that part of

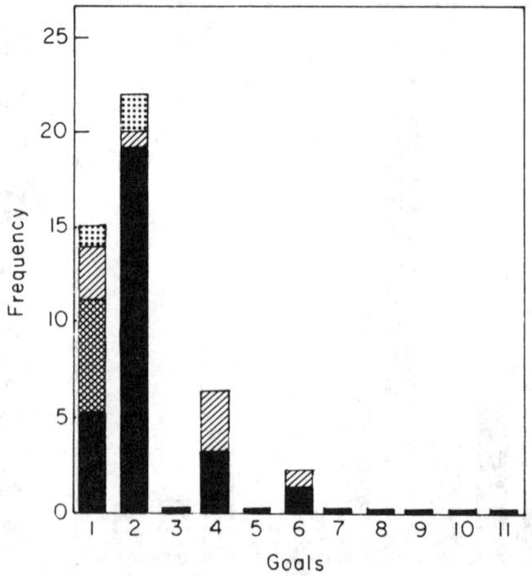

FIG. 5. Frequency of goals before a design is proposed. ▦, E1 before first propose: problem 1; ▨, E1 before first propose: problem 2; ▩, E2 before first propose: problem 1; ■, E2 before first propose.

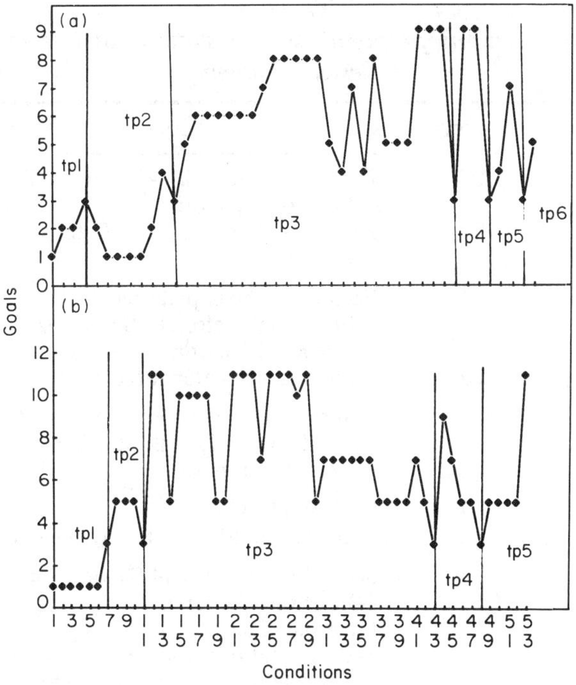

FIG. 6. Goal trace for Problem 1. (a), E1 problem 1; (b), problem 1. tp = temporal interval.

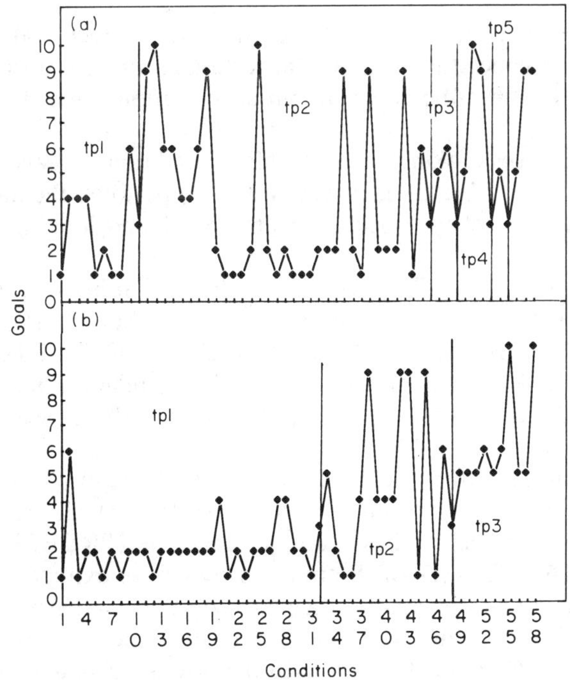

FIG. 7. Goal trace for Problem 2. (a), E1 problem 2; (b), E2 problem 2; tp = temporal interval.

TABLE 1
Sequences of designs proposed by two subjects in five experimental design problems

Problem	Subject	Sequence
1	E1	Factorial—Factorial—Factorial—Factorial —Fractional factorial
	E2	Factorial—Latin square—Latin square —Latin square
2	E1	Factorial—Incomplete blocked —Fractional factorial—Screening —fractional factorial
	E2	Screening—Fractional factorial
3	E1	Central composite—Fractional factorial
	E2	Central composite—B-Bunkhen —Central composite
4	E1	Sequential—C.M.
	E2	Factorial—C.M.—B. Bunkhen—Factorial
5	E1	Screening—Fractional factorial—P. Burman. —Fractional factorial
	E2	Fractional factorial—P. Burman —Fractional factorial—Fractional factorial —Latin square—Fractional factorial

the solution space consisting of full factorial designs, fractional factorial designs, screening designs, balanced incomplete block designs and Latin square designs. The set of such designs defines the circles in our specification: each circle corresponds to a type of design.

The first step involves looking at the problem conditions in each protocol for each expert to determine possible transitions (strategies) between the alternative designs. We illustrate this process by considering problem 1 and looking at the goal trace generated by E_1 (Fig. 6).

Each temporal interval in Fig. 6 defines a relationship between the designs on its boundaries. Consider temporal interval #5. Its boundaries are PROPOSE FACTORIAL DESIGN and PROPOSE FRACTIONAL FACTORIAL DESIGN (see also Table 1). We conclude that there exists a relationship between factorial and fractional factorial designs in the specification with respect to the sequencer PROPOSE DESIGN.

We mark an arrow from factorial to fractional factorial in the specification corresponding to this temporal interval. Next, we look at all the episodes that are activated in this interval. These are episode 4 (PROBE) and episode 6 (CALCULATE SAMPLE SIZE). From the protocol we extract the goals related to these episodes. We draw the cloud associated with the relationship between the two designs and include the goals. Next, we look at the conditions associated with each of these goals, and include the conditions in the box associated with the relationship. Finally, we identify the set of abilities that are required to achieve these goals. We include "the ability to calculate sample size" and the "ability to probe for information" in the triangle of abilities (see Fig. 8).

The above description identifies the process by which a relationship was determined from a temporal interval. The same process was repeated for each interval in the two problems. For all those intervals corresponding to a relationship between the same two parts of the solution (and having the same direction), the union of the corresponding set of goals, set of conditions, and set of abilities for each interval, defines the corresponding sets for the relationship included in the specification.

The first goal in each interval in Figs 6 and 7 has the special property that the problem condition associated with the left boundary of the first interval does not have any solution associated with it. Such intervals form special kinds of relationships in the specification, namely, the transition is made in the absence of any information. In the specification, the absence of information is represented by an == sign.

The final specification for our example is given in Fig. 9. This specification is a "road map" of the possible transitions between parts of the solution in the design problem. For example, to move from considering a factorial design to considering a latin square design, one needs the ability to "evaluate design" and the ability to "calculate sample size". The map also shows that the discrete nature of variables,

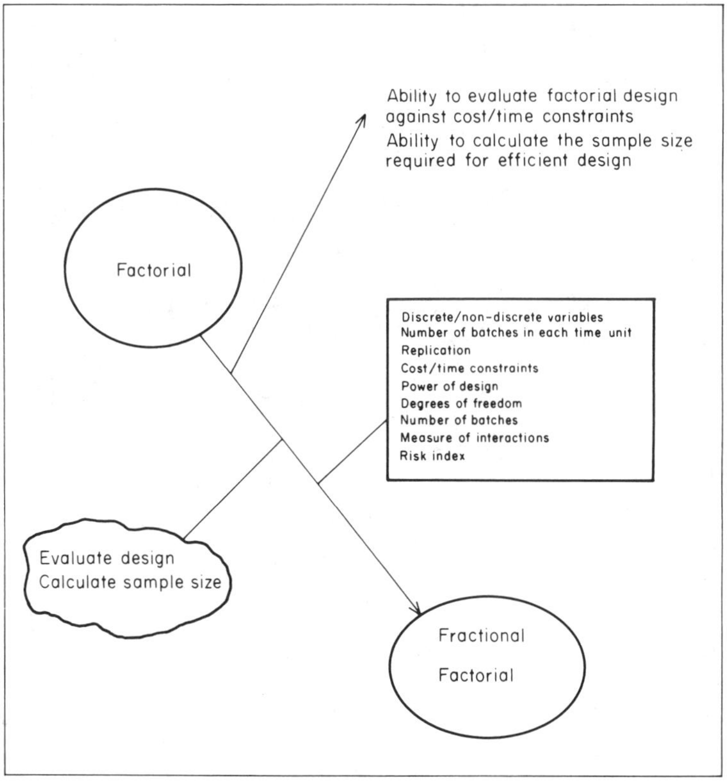

FIG. 8. Partial specification of expertise.

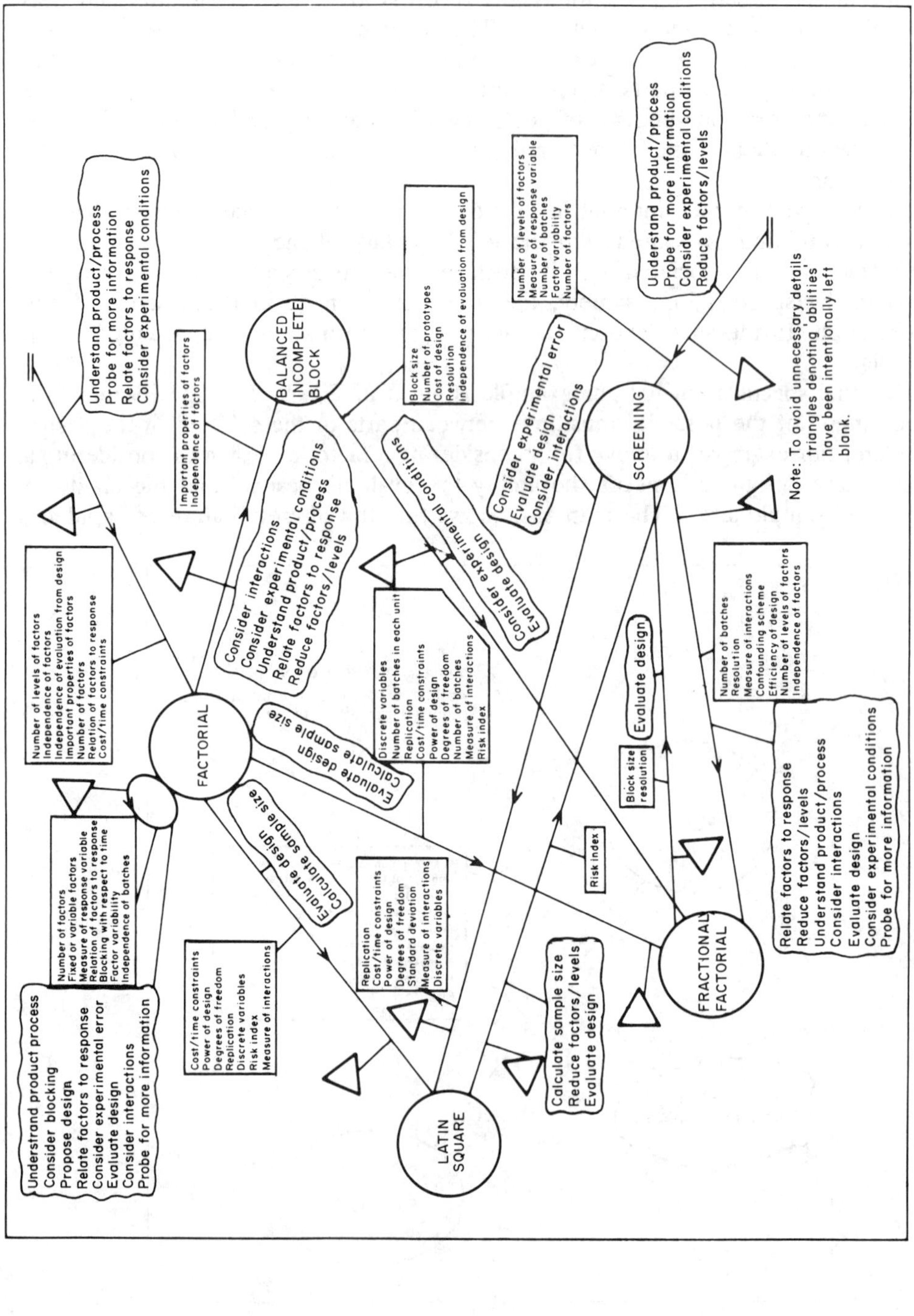

FIG. 9. Specification of expertise for experimental design.

number of runs, interactions to be estimated etc., are problem conditions that trigger this transition. We can observe from the specification that the union of all abilities associated with the relations stemming from the empty state is the minimum set of abilities required to propose an initial design. The conditions associated with this union identify the set of data cues that needs to be extracted from the environment before a consideration of any part of a solution can be made. Finally, the set of all conditions corresponds to the vocabulary the expert uses to formulate the problem.

Expertise in protocol analysis

Before concluding, we provide another example of the application of our methodology: a specification of the expertise required to perform the task of protocol analysis. In this work we chose as our expert an individual who had coded numerous protocols in several fields. In our example, we gave this individual a protocol from one of our experimental design problems and asked him to code for episodes and goals.

As the expert scanned the protocol, he read aloud the portion he was considering. He also marked the protocol each time he identified a goal or episode. The session was tape recorded and analysed as described below.

Analysis of data

The expert analysed the protocol using the caterories of text, goals and episodes. The "solutions" sought by the expert to the problem of coding the protocol were embodied in four kinds of mappings. These were the mappings between text and goals, between two or more separate goals, between goals and episodes, and between text and episodes. For example, mapping between texts and goals corresponds to saying that in lines (L1, L2) of protocol text, the expert is following a goal G. Each mapping corresponded to a solution or part of a solution. Each mapping had a number of goals associated with it. These goals are listed below.

(1) PROPOSE GOAL: propose a new goal based on reading of text;
(2) ACCEPT GOAL: accept a goal based on reading of text;
(3) COMPARE: compare current goal with all previous goals;
(4) COMPARE 1: compare current goal with immediately previous goal;
(5) ACCEPT GOAL EPISODE: accept current episode based on goal information;
(6) PROPOSE GOAL EPISODE: propose episode based on particular goal;
(7) SCAN: scan text for evidence of episode;
(8) ACCEPT TEXT EPISODE: accept episode based on reading of text;
(9) PROPOSE TEXT EPISODE: propose episode based on reading of text.

As in Case Study 1, we next performed both frequency and temporal analyses of these data.

Frequency analysis. A frequency plot of the above goals is presented in Fig. 10. As shown in Fig. 10, very little time was spent in considering goals of PROPOSE GOALS or PROPOSE EPISODES or SCAN. The most frequent goals were

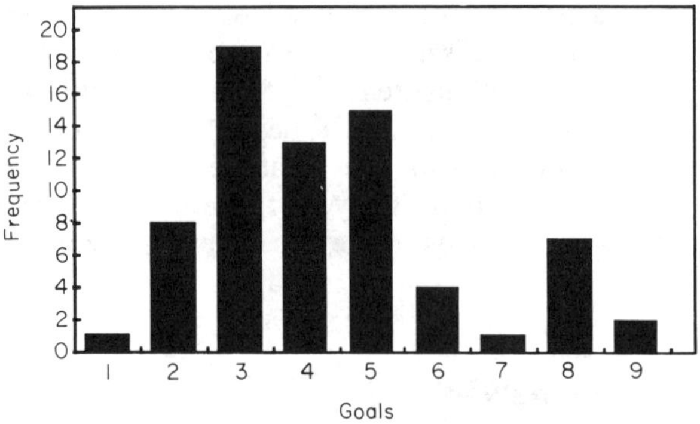

FIG. 10. Frequency of goals for protocol analysis.

COMPARE and COMPARE 1. Frequently, the expert compared a goal with the immediately previous goal. More often, he made a broader comparison between a goal and several previous goals. Sometimes, the expert considered ACCEPT GOAL or ACCEPT GOAL EPISODE directly after reading the text. However, he was much more likely to consider ACCEPT GOAL EPISODE after performing an additonal analysis of the protocol looking for goals.

Temporal analysis. The goal trace is illustrated in Fig. 11. As shown in Fig. 11, one often sees a series of COMPARE goals or a series of COMPARE 1 goals. However, it is unusual to see a COMPARE goal followed by a COMPARE 1 goal or vice versa. That is, if the expert is making comparisons of a goal with the previous goal, he will continue in that sequence. If, however, he is comparing a goal with one which is removed by one or more episodes, he will tend to continue to compare the goal with goals which are not adjacent. COMPARE frequently occurs after a goal has been accepted, signalling the start of a new cycle. The expert never moves from ACCEPT GOAL to the acceptance of another goal. Instead, he moves to either a COMPARE goal or a PROPOSE EPISODE goal before moving to another ACCEPT GOAL. By contrast, he often accepts several episodes in sequence. This suggests that once the expert has identified one episode, he can identify new, perhaps related episodes quite readily.

We first selected as problem sequencers those goals which occurred most

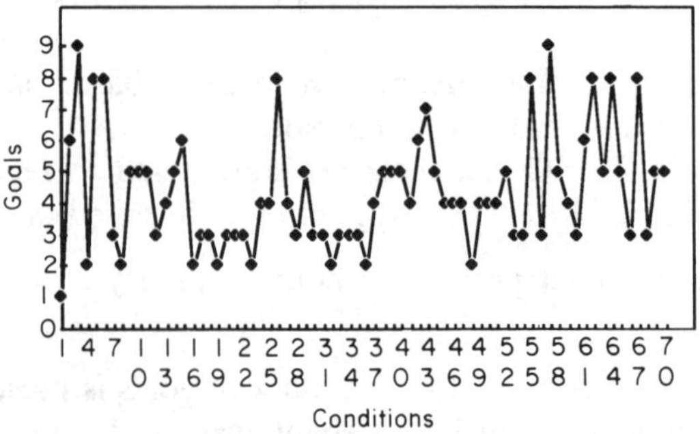

FIG. 11. Goal trace for protocol analysis.

frequently. Unlike the previous case study, however, when we divided the protocol record into temporal intervals based upon these sequencers we found that none of the goals individually provided a useful specification of expertise. This result was due to the fact that in this domain there is a strong correlation between the occurrence of goals and parts of solutions (so that the specification of expertise using goals as sequencers would consist of relatively independent solution parts). Therefore, we next selected a problem sequencer consisting of single problem solutions. In this case the specification of expertise consisted of a single solution and its transitions to and from other solutions. However, the transitions among the other solutions were omitted. Consequently, we finally examined pairs of solutions as a problem sequencer and found that all transitions between all solutions could be accounted for.

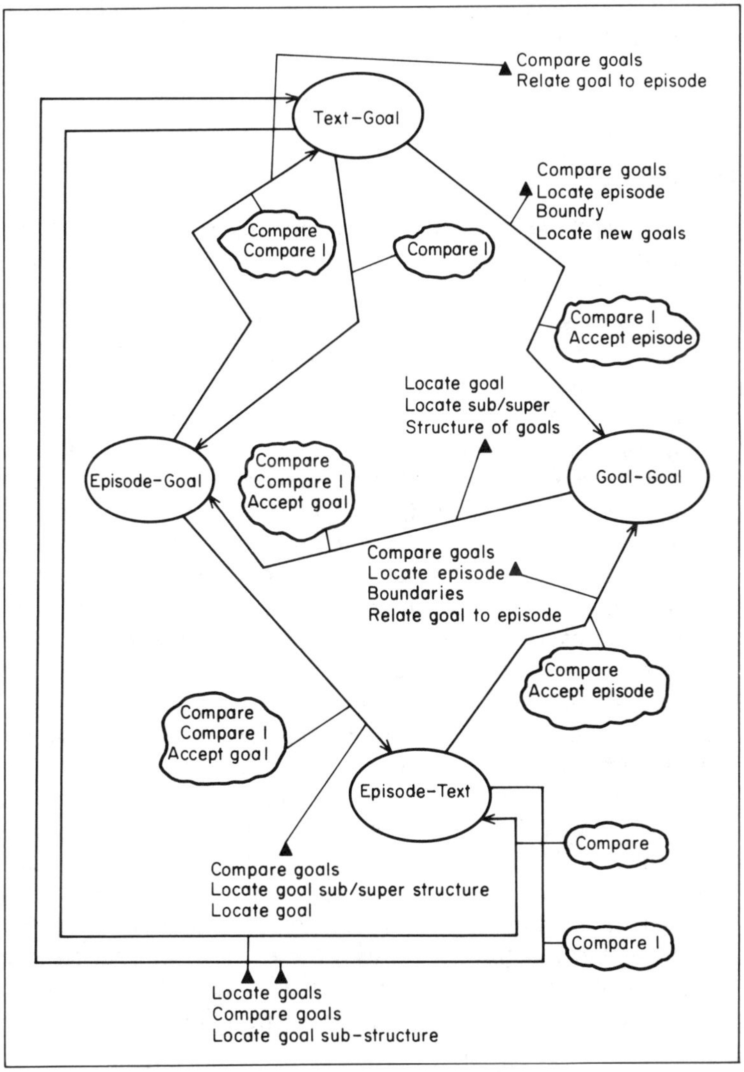

FIG. 12. Specification of expertise for protocol analysis.

For example, to develop that portion of the specification for the solution pair (text–goal, episode–goal) we located the first place on the goal trace in which a problem condition coincided with the text–goal solution category. This corresponded with the left boundary of the temporal interval. We then located the first place on the goal trace in which a problem condition coincided with the episode–goal solution. This formed the right boundary of the temporal interval. This process was continued for all pairs of text–goal, episode–goal, goal–goal, and episode–text solutions. The resulting temporal intervals were then analysed for goals, conditions, abilities, strategies (transitions), and solutions in the same manner as in Case Study 1. The resulting specification of expertise is illustrated in Fig. 12. It should be noted that only eight of the 12 possible arrows representing relationships are shown in the figure. This is because in order to base our specification on the major goals of problem solving we excluded an arrow when the interval between the two relevant solutions did not include a minimum of two instances of at least one goal.

CONCLUSIONS

In this paper we have presented a framework for representing expertise. When we talk about our representation as a specification of requirements for expertise, however, we do not mean that this framework can represent all kinds of expertise. We qualify our idea by saying that our representation formalism currently seems to describe expertise in certain types of design and diagnostic tasks (Chandrasekaran, 1983; Clancey, 1985). Since our specification has the "competency" to generate a very large number of possible solution sequences from a finite set of inputs for such tasks, we might consider it to be a "grammar" of expertise for this class of tasks. Validation of such a claim must await further research.

In concluding, we wish to make two additional points. First, the choice of a sequencer in our framework has implications for the specification of expertise. For instance, a given problem sequencer determines possible solutions and relationships among these solutions for a class of tasks. Different sequencers might lead to different representations of expertise. Our study of expertise in a number of different domains has helped us to identify a set of possible sequencers, but in general the choice of sequencer requires further research.

Our second point is that there is a relationship between the nature of the expertise practiced by human experts in a class of tasks, and the way that expertise is represented in our framework. For example, human experts often use a small number of cues to distinguish between competing sets of possible solutions (Johnson *et al.*, 1981; Johnson, Moen & Thompson, in press). The implication of this observation for the representation of expertise is that there should not be a large number of problem conditions in any relationships between possible solutions corresponding to such competing sets of hypotheses. This implication extends to the expertise for a class of tasks, and for specific domains in that class. The "carry-over" of findings from the empirical study of human experts to our proposed framework also requires further investigation.

Obviously, the work we have described is in progress, and while we are reasonably comfortable with the overall shape of our results, many of the details will undoubtedly change as our investigation continues and our theory improves. At this point, we claim only that our representation can serve as a specification for a computational model of expertise. We hope that it will also provide a guide for further thinking on the nature of expertise and its importance in understanding problem solving.

This work was supported in part by grants to the first author from the University of Minnesota Microelectronic and Information Sciences Center, and from IBM and the Control Data Corporation.

We wish to thank Keith Bellairs, James Moen and Elizabeth Stuck for comments on earlier drafts of this paper.

References

ANDERSON, J. R., GREENO, J. G., KLINE, P. J. & NEVES, D. M. (1981). Acquisition of problem solving skill. In ANDERSON, J. R. Ed., *Cognitive Skills and Their Acquisition.* Hillsdale, New Jersey: Lawrence Erlbaum Associates.

CHANDRASEKARAN, B. (1983). Towards a taxonomy of problem solving types. *AI Magazine,* **4**(1), 9–17.

CLANCEY, W. J. (1985). Heuristic classification. *Artificial Intelligence,* **27,** 289–350.

DEGROOT, A. (1965). *Thought and Choice in Chess.* New York: Basic Books, Inc.

JOHNSON, P. E., DURAN, A. S., HASSEBROCK, F., MOLLER, J., PRIETULA, M., FELTOVICH, P. J. & SWANSON, D. B. (1981). Expertise and error in diagnostic reasoning. *Cognitive Science,* **5,** 235–283.

JOHNSON, P., MOEN, J. & THOMPSON, W. (1987). Garden path errors in diagnostic reasoning. In BOLC, L. & COOMBS, M. J. Eds, *Computer Expert Systems.* Springer–Verlag. In press.

MARR, D. (1982). *Vision.* San Francisco, California: W. H. Freeman & Co.

NEWELL, A. (1981). The knowledge level. *AI Magazine,* **1,** 1–20.

NEWELL, A. & SIMON, H. A. (1972). *Human Problem Solving.* Englewood Cliffs, California: Prentice–Hall.

NEWELL, A. (1979). Reasoning, problem solving and decision processes: the problem space as a fundamental category. In NICKERSON, R. Ed., *Attention and Performance,* Vol. VIII. Lawrence Erlbaum & Associates.

RAMAMOORTHY, C. V., PRAKASH, A., TSAI, W. & USUDA, Y. (1984). Software engineering: problems and prospectives. *IEEE Computer,* **17**(10), 191–209.

SIMON, H. A. (1980). *Sciences of the Artificial* (2nd edn), Cambridge, MA: MIT Press, 1980.

The use of alternative knowledge-acquisition procedures in the development of a knowledge-based media planning system

ANDREW A. MITCHELL

Faculty of Management Studies, University of Toronto, Toronto, Ontario M56 1V4, Canada

The knowledge-acquisition procedures used in developing a knowledge-based media planning system are discussed. The approach used in developing the system and the resulting system will have a number of unique characteristics. First, in developing the system, we first constructed a system which we call a decision frame. This system structures the problem for the media planner and contains little expertise. We are currently adding expertise to the system so that our final system will be able to operate both as a decision frame and as an expert system. A number of different knowledge-acquisition procedures are currently being used to obtain the requisite knowledge from media planners. These include: (1), elicitation procedures; (2), problem-sorting techniques; (3), protocol analysis; and (4), having experts use the decision frame system to acquire knowledge and to determine the validity of the system.

Introduction

In this paper, we discuss the development of a knowledge-based media planning system and the procedures that we are using to acquire the declarative and procedural knowledge that will be used in the system. Our approach to developing the system differs from other approaches that have been reported in the literature such as rapid prototyping (e.g. Hayes-Roth, Waterman & Lenat, 1983). Initially, we obtained an understanding of how media planners develop media plans and then we developed a system, which we call a decision frame, which essentially structures the problem for the media planner in a way that is compatible with the media planning process. The decision frame system contains little knowledge, so most of the knowledge resides in the media planner. Next, we plan to obtain both declarative and procedural knowledge from expert media planners which will be added to the system. The development of the decision frame systems allows us to understand exactly what type of knowledge is needed at various points in the decision making process. Consequently, we are able to use very specific knowledge-acquisition procedures to obtain the requisite knowledge.

Our system is also designed to be used by different media planners. Consequently, we need to obtain information from multiple experts (cf. Mittal & Dym, 1985) and our knowledge-acquisition procedures must be designed so we can identify differences and similarities between media planners in the media planning process.

Since the knowledge-acquisition procedures that we are using are specific to the model that we are developing, we first need to discuss the media planning problem and its characteristics. We then discuss how media planners develop a media plan

KNOWLEDGE-BASED SYSTEMS Vol. 1
ISBN 0-12-273251-0

and present an overview of the system that we are developing. Finally, we discuss the knowledge-acquisition procedures that we are currently using to develop the system.

THE MEDIA PLANNING PROBLEM

The media planning problem involves the allocation of an advertising budget for a specific product or service (e.g. Tide detergent) to a number of media options (e.g. Time magazine) over a specific planning horizon (e.g. 1 year). The resulting media plan differs somewhat by the media used in the plan. For instance, for magazines, the media plan will give the size (e.g. full page) and type of advertisement (e.g. four color) to run in specific issues of specific magazines during the planning horizon. For network television, the media plan will give the amount of money to be spent weekly in different dayparts (e.g. daytime television) over the planning horizon. These dollar amounts are translated into Gross Rating Points (GRPs), which is a measure of the amount of advertising directed at a particular audience. An advertising campaign that achieves 100 gross rating points means that the number of advertising exposures is equal to the size of the targeted audience.

An example of a media plan for a liquid dishwashing detergent is shown in Fig. 1. In this media plan, advertisements will be placed in network television, spot or local television markets and Sunday Supplements. Two different dayparts are used in network television, daytime and early news. Thirty second advertisements will be used for the television advertisements and $\frac{1}{2}$-page, 4-color advertisements will be used in the Sunday supplements.

There are two primary players that interact in the development of a plan. The first is the advertiser (e.g. Proctor & Gamble) who has a product or service to sell. The second is the advertising agency (e.g. Leo Burnett), which is an independent company hired by the advertiser to develop and place advertisements in the media. Within the advertiser there are a group of individuals in charge of a specific brand

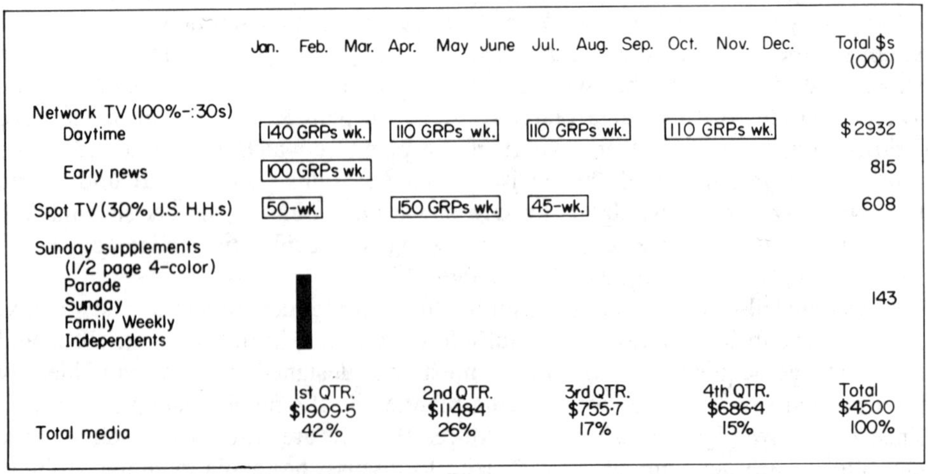

FIG. 1. Media plan for a liquid dishwashing detergent.

TABLE 1
Information used to develop a media plan

Information from advertiser
Marketing strategy
Target market
Advertising budget
Sales and market share data by market
Consumer information
Information from advertising agency
Competitors' media strategies and advertising expenditures
Audience data for media options
Cost data for media options
Creative strategy for brand

(e.g. Tide detergent). This group develops a marketing plan for the brand which includes sales objectives, marketing expenditures (e.g. advertising expenditures) and a marketing strategy for achieving the sales objectives.

Within the advertising agency there is a group of individuals that develop and place the advertisements for that brand. The media planner is one of these individuals. One of his or her tasks is to develop a media plan that will effectively use the marketing expenditures allocated in the marketing plan to achieve the sales objectives, in a manner that is consistent with the marketing strategy for the brand.

A considerable amount of information is used in developing a media plan (Table 1). The marketing plan generally contains the tentative advertising budget and the target market for the brand (e.g. women, 18–35 years of age). In addition, the advertiser also provides the sales and market share data for the brand and its competitors across markets and additional information about users of the brand and its competitors, and their purchase habits (e.g. how often they purchase brands from the product category).

The media planner in the advertising agency acquires data on competitors' media strategies and advertising expenditures, the costs of running advertisements in the various media options, audience data on the various media options and the creative strategy for the product or service. All this information is then synthesized by the media planner in developing a media plan.

Based on this brief description, it can be seen that the media planning problem has a number of characteristics. First, while the solution or problem space is well defined, the exact procedures for obtaining a solution and the criteria for testing a solution are somewhat vague. These characteristics cause it to be a semi-structured or ill-structured problem (cf. Keene & Scott-Morton, 1978; Simon, 1973).

Second, large amounts of data must be synthesized by the media planner in developing the media plan. Media planners must use information about the audiences and costs of different media options, duplication of audience between media options, predicted changes in viewing and readership habits over the time horizon, and competitors' media expenditures and strategies.

Third, the solution space for a particular media planning problem is very large. In fact, if one considers all the possible combinations of media options (e.g. all the magazines, all the possible dayparts in radio and television, etc.) over the planning horizon, the number of possible solutions approaches infinity. Since the number of possible media problems is also very large, solutions probably must be constructed, although classification procedures may be part of the decision process (cf. Clancey, 1985).

Finally, it is difficult to assess the quality of a particular solution. The ultimate criteria for evaluating a media plan is the resulting sales or market share of the advertised brand, however, the media plan used to advertise the brand is only one of a number of variables that affect sales. Therefore, it is difficult to attribute any change in sales specifically to the media plan. This means that other criteria must be used to evaluate a media plan and that it will be difficult to relate these criteria directly to sales. Consequently, the relationship between specific elements of the media plan and their affect on the external environment are not understood very well (cf. Einhorn & Hogarth, 1981). This should be contrasted against other planning environments where these relationships are well understood, such as job shop scheduling (e.g. Smith, Fox & Ow, 1986), and seems to typify many marketing problems (e.g. Chakravarti, Mitchell and Staelin, 1981).

WHAT MEDIA PLANNERS SEEM TO DO

Given the characteristics of the media plan outlined previously, it is not too surprising to find that media planners use goals and constraints to narrow the solution space and decompose the problem into a series of subproblems (e.g. Newell & Simon, 1972). In this section we discuss the different goals and constraints that media planners use, how media planners decompose the problem and, finally, the different stages of the media planning process.

Goals and constraints

There are a number of different types of goals and constraints that are generally used in developing a media plan. These are: (1), reach and frequency goals; (2), scheduling goals and constraints; (3), budget; and (4), problem specific constraints.

Reach and frequency are measures of the impact of the advertising campaign based on the number of exposures the target market will receive in a given time period. Reach is the percentage of the target audience that have the potential to see at least one advertisement from the advertising campaign, while frequency is the potential number of advertisements that a member of the target audience will have the opportunity to see. Note that there is a tradeoff between reach and frequency. For instance, in order to increase reach with a specific budget, you have to sacrifice frequency.

Reach and frequency are generally used as goals for a particular campaign, however, they also may be used as constraints. For instance, a media planner may set minimum reach and frequency levels for a particular time period.

Scheduling goals concern the placement of advertisements over the planning horizon. There are a number of possible scheduling goals and constraints. A media planner, for instance, may try to achieve specific reach and frequency goals or may try to match a competitor's media expenditures during different time periods in the planning horizon. Examples of scheduling constraints are: do not advertise between

Thanksgiving and New Years or do not advertise on television during the summer because the audiences are small.

The budget is almost always treated as a constraint. Generally, the media planner will try to achieve certain reach/frequency and scheduling goals given the budget constraint.

Finally, there may be problem specific constraints. An example of this type of constraint is the need to run print ads in April to support a coupon promotion.

From this discussion, it can be seen that media planners can structure the problem in a number of different ways through the use of different goals and constraints. One media planner, for instance, might set reach and frequency goals and then, given these goals, advertise in as many time periods as the budget will allow. Another media planner may set scheduling goals by deciding how many time periods to advertise and then spend as much money in each time period as the budget will allow.

The use of these goals and constraints significantly reduces the size of the solution space and, in some cases, no media plan will satisfy all the goals and constraints. When this occurs, media planners must consider tradeoffs between the different goals and constraints. In other situations, where a number of media plans may satisfy the goals and constraints, other criteria must be applied.

Problem decomposition

Media planners also decompose the problem into a series of subproblems. Although there are some weak interactions between subproblems, they usually can be solved independently (cf. Simon, 1969). These interactions generally can be accommodated by making adjustments to the final plan after all the subproblems have been solved. The first subproblem involves an allocation of the budget between media. For instance, it may be decided to put 70% of the budget into television and the remainder into magazines. The second subproblem is the selection of media options within a media. This is usually accomplished by selecting the most efficient media options for reaching a particular target market. For example, daytime television is the most efficient daypart for reaching women and weekend sports is the most efficient daypart for reaching men. The final subproblem is the allocation of the budget over the planning horizon. To solve this subproblem the scheduling goals and constraints are used.

Steps in the media planning process

In developing a media plan, a media planner goes through a number of steps (Fig. 2). The first is to gather and synthesize all the information that was discussed previously. This information defines the dimensions of the problem. Next the media planner sets goals and formalizes the constraints. These goals and constraints are then used to develop one or more rough media plans. These media plans are then checked to see if they achieve the goals and satisfy the constraints. In order to determine if they achieve the reach and frequency goals, a number of models have been developed to estimate the reach and frequency of a particular media plan (see Rust, 1986). If the goals are *not* achieved or the constraints are *not* satisfied, the plans are adjusted and then the goals and constraints are checked again. If it is not

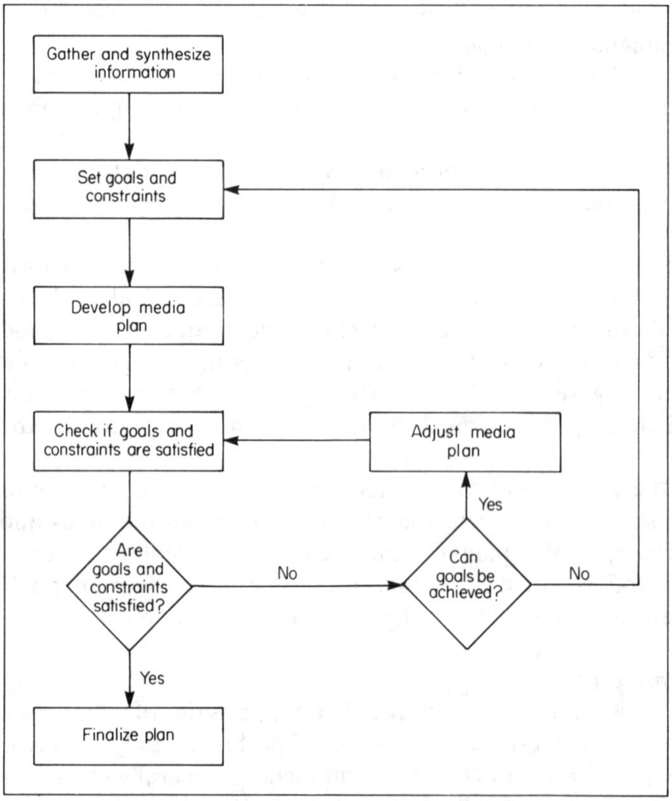

FIG. 2. Stages of the media planning process.

possible to develop a media plan that will satisfy the goals and constraints, they are adjusted and a new media plan is developed.

In developing a media plan, expert media planners do not seem constantly to refer to data. Instead, they rely primarily on a number of heuristics and beliefs that they have acquired. Here, we use the term heuristic to refer to specific procedures or rules, and beliefs as specific statements of declarative knowledge.

Some examples of heuristics that media planners use are: (1), to increase reach, add another daypart or media option; and (2), to increase frequency, increase expenditures in current daypart or media option.

Some examples of beliefs are: (1), pulse media schedules are generally more effective than continuous schedules; (2), at least three advertising exposures are needed within a purchase cycle; and (3), a budget of at least $3 million is needed for network television. These heuristics and beliefs seem to be based on both experience and empirical research (e.g. Krugman, 1972; Naples, 1979).

System characteristics

The system that we are developing is a frame and rule based system that is programmed in LOOPS (e.g. Stefik & Bobrow, 1986). The frames contain information about different brands and about different media options. These latter frames are organized hierarchically. In addition, there will be a frame for each media planner which will contain his or her specific heuristics and beliefs.

The system is organized into modules which are designed to solve the different subproblems. For instance, for network television, there are modules for selecting dayparts, allocating the budget by time period, allocating the budget within time periods and for integrating the different budget allocations for each time period into an integrated media plan for the planning horizon. This latter module also checks the resulting plan against the goals and constraints and adjusts the plan if any of the goals and constraints are not satisfied.

The resulting system is similar to a number of other planning systems that have been developed, in that constraints are used to reduce the size of the solution space (e.g. Stefik, 1981); however, the constraints imposed are generally not enough to achieve a single solution. The resulting system is not a pure top down system like NOAH (Sacerdoti, 1977) nor a highly opportunistic system as proposed by Hayes-Roth & Hayes-Roth (1979).

When it is completed, the system will operate at a number of different levels to accommodate the expertise of the user. At one level the system will operate as a decision frame. It will structure the problem for the media planner, provide the required information and make the necessary calculations at each stage of the decision process. For instance, the system will calculate cost per thousand values for the different media options to help the media planner select between these options. The system will also calculate tradeoffs with respect to different goals and constraints to help the media planner select alternative allocation plans and calculate the marginal cost of various constraints that the media planner imposes. At this level, the system is primarily a decision support system (e.g. Keene & Scott–Morton, 1978) and our approach to developing the system is similar to the approach discussed by Hollnagel & Woods (1983) and Woods (1986).

For media planners with little expertise, the system will operate as a pure expert system by using the dimensions of a particular problem to generate goals and constraints and then develop a media plan. At intermediate levels, the media planner has the option of either making the particular decisions or letting the system make the decision.

Currently, we are completing the programming of the decision frame for network television. This portion of the system uses as input the goals and constraints provided by the user and provides an interactive menu-driven framework for structuring the problem. As discussed above, this portion of the system contains four modules. This modular framework allows the user to move back and forth between different subproblems in developing a media plan, so that potential interactions between subproblems can be taken into account. For instance, while allocating the budget within a time period, the planner may decide to change dayparts. By calling up the daypart selection module, these changes can be made. Then the media planner can switch back to the budget allocation within a time period module.

We are now acquiring the knowledge necessary to make a number of extensions to this basic system. First, we are acquiring the knowledge necessary to turn this portion of the system into an expert system. The decision frame system has indicated what decisions need to be made and we are now acquiring the knowledge from media planners so the system can make these decisions.

Second, we are acquiring the knowledge necessary to add additional media to the system. Initially, we plan to add spot television and magazines. Finally, we are acquiring the knowledge necessary to add goal and constraint generation. This module will take the basic characteristics of the problem and generate the goals and constraints.

Knowledge-acquisition procedures

In developing our knowledge-based media planning system, there are three distinct stages in the information acquisition process. During the first stage, our goal was to obtain an understanding of the process that media planners went through in developing media plans and the critical factors that affected the resulting media plan. To achieve this goal, we reviewed a number of text books on media planning (e.g. Barban, Cristol & Kopec, 1985; Surmanek, 1985), the manuals used by a number of advertising agencies in training programs, and held discussions with a number of media planners. These sources provided an understanding of the critical variables that affect the resulting media plan and provided a broad understanding of the media planning process. This information, however, was not detailed enough to allow us to write a computer program for the decision frame system. These sources did indicate, however, that the general procedures used in developing a media plan (e.g. setting reach and frequency goals) seemed to be the same across media planners. They also indicated that there were some differences at different stages of the process, but it appeared that these differences could be accommodated in our system. These differences seemed to be primarily in how media planners use audience data in making comparisons between media options. Some planners, for instance, weight audience data by estimates of the amount of attention devoted to watching television for different dayparts.

In order to obtain a more detailed understanding of the media planning process, we conducted a series of extensive interviews with a media planner at a major advertising agency in the U.S.A. These interviews varied greatly in structure. Sometimes, we would ask general questions about the media planning process and ask for diagrams of different stages of the process. Other times, we would give him specific media problems and ask him to think out loud while developing a media plan. During this process, he would sometimes ask us for additional information about the problem. These types of questions were especially helpful to us in understanding what information was used at different stages of the media planning process.

The thinking out loud protocols were the most useful to us in developing our decision frame system. The actual writing of a program for this system identified the gaps in our knowledge of the media planning process and allowed us to go back to the media planner with specific questions. As mentioned previously, the decision frame system for network television has now been designed and we are currently completing the programming of it.

The second stage of the knowledge-acquisition process has three goals. The first goal is to acquire the knowledge that is necessary to turn our decision frame system into a true expert system. The second goal is to obtain an understanding of how media planners use the different dimensions of the problem (e.g. the size of the advertising budget) to form goals and constraints and, eventually, the media plan. The third goal is to understand how budget allocations are made between media, how media plans are developed for other media (e.g. magazines), and to acquire the knowledge necessary to make these addition to the system. In order to acquire this information, we plan to interview a number of additional media planners.

For the third stage of the knowledge-acquisition process, we plan to develop standardized materials that we can give to a relatively large number of media planners. The purpose of these materials is to determine if the general structure of our system is consistent with how all media planners develop media plans, whether the declarative and procedural knowledge built into our system is correct and to identify differences between media planners. In other words, we want to determine if the flexibility we have built into our system is consistent with the differences between media planners and whether the rigidities in the system are consistent with the similarities.

Since, we are currently at the second stage of the knowledge-acquisition process, we will focus on the procedures that we are using at this stage.

ALTERNATIVE PROCEDURES

In developing our procedures for knowledge acquisition, we have relied heavily on current models of memory. In these models, it is generally recognized that there are two types of knowledge—declarative and procedural knowledge (e.g. Anderson, 1983*a*; Tulving, 1983). Declarative knowledge is our knowledge about the external world while procedural knowledge is our knowledge about how to perform certain skills such as solving geometry problems or driving an automobile. Spreading activation network models are the most commonly accepted models for representing declarative knowledge and production systems are generally used for representing procedural knowledge (e.g. Anderson, 1983*a*). Consequently, these models underlie our knowledge-acquisition procedures.

It is also currently believed that in problem-solving, individuals first categorize problems and the solution procedures used for solving a problem depend on how the problem is categorized (e.g. Chi, Feltovich & Glaser, 1981; Lewis and Anderson, 1985). This finding also underlies our knowledge-acquisition procedures.

Four different procedures are being used to acquire knowledge from media planners to achieve the three goals at the second stage. These are elicitation procedures, problem-sorting procedures, protocol analysis, and problem-solving with decision frame systems. These procedures and their application to knowledge acquisition at the second stage will now be discussed.

Elicitation procedures

These procedures are used to obtain an understanding of the declarative knowledge that an individual has within a domain (e.g. Mitchell & Chi, 1986). For these procedures, subjects are given a number of important concepts from the domain and are asked to mention everything that comes to mind when they think of these concepts. According to a spreading activation model of memory, thinking about a particular concept activates that node in memory. Activation then spreads to all the nodes linked to the activated node and, if the amount of activation at a linked node exceeds a threshold level, the contents of that node is recalled (e.g. Anderson, 1983*b*).

During the second stage of the knowledge-acquisition process, we are using this procedure to determine what associations media planners have linked to the different media options, target audiences, product categories, brands and companies that they use in developing media plans. The purpose behind the use of this procedure is to identify any idiosyncratic knowledge that may pertain to specific media planning problems. For instance, the media planner may know that readers of a specific magazine have a particular psychographic profile that matches heavy users of a particular product category.

Problem sorting

The second procedure is a problem-sorting procedure. For this procedure, media planners are given a series of media planning problems. The problems differ along a number of dimensions such as target audience and size of the advertising budget. The media planners are then asked to group these problems in terms of the similarity of the resulting media plans. The purpose of this procedure is to determine if media planners first categorize media problems and, if they do, the different categories that media planners may have of media problems. If it turns out that media planners first categorize media problems, this may greatly facilitate the ease with which we can obtain an understanding of the relationships between the dimensions of the problem, the goals that are formed and the resulting media plan. It also means that heuristic classification procedures could be used for part of the problem-solving process (cf. Clancey, 1985).

Protocol analysis

The third procedure is the analysis of think out loud protocols that are obtained while media planners develop a media plan. Ericsson & Simon (1984) have discussed a number of conditions that must be met for this type of data to be valid. One of the most important of these conditions is that individuals must use verbal thought processes to solve the problem. Based on our initial experiences with media planners, this condition seems to be met.

We are currently using protocol analysis to understand four different types of problem-solving which occur during the development of media plans. These problems occur at the different stages of the media planning process, which were discussed earlier (see Fig. 2). The first type of problem-solving is goal and constraint setting. In order to examine this, media planners are given a number of different media problems that differ along a number of dimensions, such as size of budget and target audience. The media planner is then asked to set reach/frequency and scheduling goals and to think out loud while they are doing this task.

The second type of problem-solving is the use of these goals to develop an actual media plan. Here, the media planner is given a number of media problems that differ with respect to reach/frequency and scheduling goals and he or she is asked to think out loud while they develop a media plan. The purpose of obtaining protocols for these first two types of problems is to understand how media planners map the dimensions of the media problem into goals and constraints and, then, how they

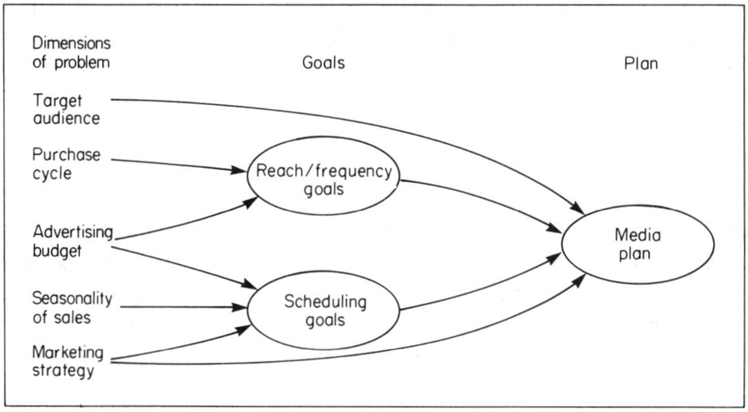

FIG. 3. Mapping of the dimensions of the problem into goals and constraints and the media plan.

map these goals and constraints into a media plan. These mappings are displayed in Fig. 3. As can be seen from the diagram, different dimensions of the problem map into different goals. What somewhat complicates the problem is that some dimensions of the problem directly affect the media plan without going through reach/frequency and scheduling goals.

The last two types of problem-solving occur at the final stages of the media planning process. These are how media planners adjust the media plan when specific goals and constraints are not satisfied and how media planners evaluate media plans. In the first case, media planners are given a media plan, and the goals and constraints that the media planner was trying to satisfy in developing the media plan. In addition, the level of the goals achieved with the plan and whether or not the constraints were satisfied are also given. Different problems contain evidence that different goals were not achieved or different constraints were not satisfied. Media planners will then be asked to think out loud while they determine what they would do in these situations. In the second case, media planners are given a series of media problems and a number of media plans that were developed for the problems. They are then asked to evaluate them. Again, the media planners are asked to think out loud while giving their evaluations.

Problem-solving with the decision frame system

The fourth procedure is to have experts solve media problems using our decision frame system. Here media planners will be given a number of media problems that vary along a number of dimensions and they will be asked to use the system to solve the problem. The system is designed to record each decision that is made and the sequence in which the decisions were made. By using some of the same media problems that media planners were asked to solve while think out loud protocols were taken, we will be able to check whether the structure of our decision frame system has an effect on the problem-solving process.

Finally, after we have additional knowledge built into the system, we will be able to check the media plans developed by the system against the media plans developed from both the problem-solving tasks and the media plans developed by having 10 media planners use the system to develop a media plan. If the solutions are similar, this will provide evidence as to the validity of the knowledge contained in the system.

Conclusion

Based on our experiences in developing our knowledge-based media planning system, we have come to a number of conclusions concerning the use of knowledge-acquisition procedures. First, we believe that knowledge-acquisition procedures should be based on the system being developed and the particular stage of system development. Our system is a frame and rule-based system, so we have used procedures that attempt to uncover the declarative and procedural knowledge of media planners.

Second, we have found that the development of an initial decision frame system to be very useful. The development of this system gives us a better understanding of how media planners develop media plans and the specific types of knowledge that are required to add expertise to systems.

Third, we have found ourselves continually moving back and forth between the decision frame system and the development of knowledge-acquisition procedures. In developing this system, we would identify the gaps in our knowledge of the decision-making process and then develop procedures for understanding these gaps.

Fourth, there seems to be considerable subjectivity in interpreting knowledge-acquisition data. We believe that much of this data can only be understood in the context of the decision frame system that we are developing. Finally, we believe that a number of different knowledge-acquisition procedures are required to develop any system. This is evident from our discussion of the procedures that we are using at the second stage of knowledge acquisition.

This research is partially funded by the Marketing Science Institute and Xerox Canada. The author would like to thank Arvind Sathi of the Carnegie Group, Inc. for his many helpful comments during the early stages of this project and to Christopher August and Nazila Naderi, who made substantial contributions to the development and programming of the current system.

References

Anderson, J. R. (1983*a*). *The Architecture of Cognition.* Cambridge, Massachusetts: Harvard University Press.

Anderson, J. R. (1983*b*). A spreading activation theory of memory. *Journal of Verbal Learning and Verbal Behavior,* **22,** 261–295.

Barban, A. M., Cristol, S. M. & Kopec, F. J. (1986). *Essentials of Media Planning.* Lincolnwood, Illinois: NTC Business Books.

Chakravarti, D., Mitchell, A. A. & Staelin, R. (1981). Judgment based marketing decision models: problems and possible solutions. *Journal of Marketing,* **45,** 13–23.

Chi, M. T. H., Feltovich, P. J. & Glaser, R. (1981). Categorization and representation of physics problems by experts and novices. *Cognitive Science,* **5,** 121–152.

Clancey, W. J. (1985). Heuristic Classification. *Artificial Intelligence,* **27,** 289–350.

Dym, L. C. & Mittal, S. (1985). Knowledge Acquisition from Multiple Experts. *AI Magazine,* **7,** 32–37.

EINHORN, H. J. & HOGARTH, R. M. (1981). Behavioral decision theory: processes of judgment and choice. *Annual Review of Psychology,* **31,** 53–88.
ERICSSON, K. A. & SIMON, H. A. (1984). *Protocol Analysis: Verbal Reports as Data.* Cambridge, Massachusetts: MIT Press.
HAYES-ROTH, F., WATERMAN, D. A. & LENAT, D. B. (1983). *Building Expert Systems.* Reading, Massachusetts: Addison–Wesley.
HAYES-ROTH, B. & HAYES-ROTH, F. (1979). A cognitive model of planning. *Cognitive Science,* **3,** 275–310.
HOLLNAGEL, E. & WOODS, D. D. (1983). Cognitive systems engineering: new wine in old bottles. *International Journal of Man–Machine Studies,* **18,** 583–600.
KEEN, P. G. W. & SCOTT-MORTON, M. S. (1978). *Decision Support Systems.* Reading, Massachusetts: Addison–Wesley.
KRUGMAN, H. E. (1972). Why three exposures may be enough. *Journal of Advertising Research,* **12,** 11–14.
LEWIS, M. W. & ANDERSON, J. R. (1985). Discrimination of operator schemata in problem solving: learning from examples. *Cognitive Psychology,* **17,** 26–65.
MITCHELL, A. A. & CHI, M. T. (1986). Measuring knowledge within a domain. In NAGY, P. Ed., *Representation of Cognitive Structures.* Toronto, Ontario: Ontario Institute for Studies in Education, pp. 85–109.
NAPLES, M. J. (1979). *Effective Frequency: The Relationship Between Frequency and Advertising Effectiveness.* New York: Association of National Advertisers.
NEWELL, A. & SIMON, H. A. (1972). *Human Problem Solving.* New York: Prentice–Hall.
RUST, R. T. (1986). *Advertising Media Models.* Lexington Massachusetts: Lexington Books.
SACERDOTI, E. D. (1977). *A Structure for Plans and Behavior,* Amsterdam: Elsevier-North Holland.
SIMON, H. A. (1973). The structure of ill structured problems. *Artificial Intelligence,* **4,** 181–201.
SIMON, H. A. (1969). *Sciences of the Artificial.* Cambridge, Massachusetts: MIT Press.
SMITH, S. F., FOX, M. S. & OW, P. (1986). Constructing and maintaining detailed production plans: investigations into the development of knowledge-based factory scheduling systems. *AI Magazine,* **7,** 45–61.
STEFIK, M. (1981). Planning with constraints (MOLGEN Part 1). *Artificial Intelligence,* **16,** 111–140.
STEFIK, M. & BOBROW, D. (1986). Object-oriented programming: themes and variations. *AI Magazine,* **6,** 40–62.
SURMANEK, J. (1985). *Media Planning.* Lincolnwood, Illinois: Crain Books.
TULVING, E. (1983). *Elements of Episodic Memory.* New York: Oxford University Press.
WOODS, D. D. (1986). Cognitive technologies: the design of joint human–machine cognitive systems. *AI Magazine,* **6,** 86–92.

Learning Techniques for Knowledge Acquisition

A formal approach to learning from examples

JAMES P. DELGRANDE

School of Computing Science, Simon Fraser University, Burnaby, B.C., Canada V5A 1S6

A formal, foundational approach to learning from examples is presented. In the approach, it is assumed that a domain of application is describable as a set of facts, or ground atomic formulae. The task of a learning system is to form and modify hypothesised relations among the relations in the domain, based on a known finite subset of the ground atomic formulae. The subset of known ground atomic formulae is also assumed to grow monotonically, and so the set of hypotheses will require occasional revision.

Formal systems are derived by means of which the set of potential hypotheses that can be formed is precisely specified. A procedure is also derived for restoring the consistency of a set of hypotheses after conflicting evidence is encountered. The framework is intended both as a basis for the development of autonomous systems that learn from examples, and as a neutral point from which such systems may be viewed and compared.

1. Introduction

An important area of machine learning that has received widespread attention is that of *learning from examples*. In this approach a learning system is presented with a stream of facts describing some domain of application. The task of the learning system is to induce general statements characterizing this domain. As further facts are encountered, new hypothesised statements may be formed while other hypotheses may be falsified. Thus the consistency of the set of hypothesised statements must continually be maintained as new information is discovered. In some learning systems the facts are arranged into complex entities representing examples and counterexamples of a particular concept. The object in this case is to form a description of the concept given in the examples.

This paper presents a formal, foundational approach to learning from examples. The basic idea is to ignore, insofar as possible, pragmatic concerns and instead to investigate the underlying formal aspects of such learning. To this end the starting point is the same as that of virtually all knowledge representation schemes: that the world consists of, or may be described by, a collection of individuals and a collection of relations on these individuals. The task of a learning system then is to form and modify hypothesised relations concerning the relations in the domain, based on a partial knowledge of the world. This partial knowledge of the world is expressed solely in terms of relations among individuals or, more precisely, as a set of ground atomic formulae.

These notions are expanded and amplified in the rest of this section. In the second section, a language for expressing conjectures, together with naive criteria for forming conjectures, is introduced. In the third section, formal systems are developed for guiding the formation of conjectures, and in the subsequent section a procedure is described for restoring the consistency of a set of conjectures, based on

KNOWLEDGE-BASED SYSTEMS Vol. 1
ISBN 0-12-273251-0

the formal systems. An extension to the approach is briefly described in the fifth section, and in the sixth section the approach is compared with representative AI learning systems. Further details, proofs of theorems, etc. are given in Delgrande (1985).

1.1. APPROACH

The world is assumed to consist of a collection of individuals and relations on these individuals. This, of course, is the starting point for the standard Tarskian definition of an interpretation for (function-free) first-order logic. At any point in time, some portion of the domain, described by a finite set of ground atomic formulae (or, informally, "facts"), is assumed to be known by the learning system. As time progresses the learning system will presumably encounter new information, and so this set of known ground atomic formulae will monotonically increase. Moreover, as time progresses, it is quite possible that not only will new individuals be encountered, but new predicates (relations in the domain) may also be encountered and need to be incorporated into the hypothesis set.

Given a set of ground formulae, the aim of the learning system is to propose a structure for this set, and hence hypothesize relations among the known relations. Thus, for example, if all ravens we had encountered were black, we might form the hypothesis "all ravens are black". However, since the set of known ground instances increases monotonically, new hypotheses may become tenable while others will be falsified. So the question arises as to how a set of hypotheses may be modified as falsifying instances are encountered.

The problem then consists of forming hypotheses based on a stream of ground atomic formulae, where the hypotheses are phrased in terms of set relations among (the extensions of) predicates. However the emphasis in this paper is on what hypotheses may *potentially* be formed, and not on which hypotheses may *justifiably* be formed. Thus for example if we have a set of black ravens and know of no non-black ravens, we could hypothesise that ravens are black. However the approach at hand gives no indication as to when such a hypothesis should be formed or what constitutes adequate evidence for such an assertion. So the problem is to determine formal criteria which prescribe the set of potential conjectures, rather than to determine pragmatic criteria whereby an acceptable set of conjectures may be formed. A similar distinction can be made in a deductive system, where an underlying logic specifies what *could* be derived, but now what *should* be derived.

The overall approach is as follows. Hypotheses are proposed and modified on the basis of a finite, monotonically increasing set of ground instances. The hypotheses are expressed in a language, HL, that is a simple variant of the language of elementary algebra. The criteria for proposing a hypothesis are straightforward: there is a reason to do so (i.e. some minimal notion of evidence is satisfied) and the hypothesis is not known to be falsified. These criteria though are far too simplistic, and in general the resultant set of hypotheses will be inconsistent.

However, with each term in a sentence of HL we can associate two subsets of the ground instances, consisting of those known to satisfy the term and those known to not. For example, to the term "black raven" we can associate the set of individuals known to be black and a raven, and the set known to be either non-black or non-raven. Formal systems are developed to characterize relations between terms in HL by means of these sets. From this, ground instances whose truth values are unknown can be iteratively located so that determining their truth value leads to a convergence of the hypothesis set to consistency. These "knowable but unknown ground instances" are "constructed" from the set of individuals and predicate names encountered in the set of known ground atomic formulae and, informally, correspond to unknown but potentially knowable "facts" in the domain. (This capability of testing individuals for membership in a relation, where both individuals and relations have been encountered in the set of known ground atomic formulae, will prove essential for restoring consistency in a set of hypotheses).

This approach provides a sharp separation between the inductive and deductive aspects of the problem of learning from examples. Induction, as such, plays a relatively minor role: it is used to suggest an initial (and usually inconsistent) set of hypotheses, which then are modified using strictly deductive techniques. The set of hypotheses that may be formed is shown to be perhaps surprisingly general and in fact (with respect to expressiveness) subsumes a number of systems for learning from examples and by discovery.

To recapitulate, I assume only that:

(1) the domain is describable as a set of ground atomic formulae;
(2) some finite subset of the ground atomic formulae is known;
(3) the set of known ground atomic formulae is correct and error-free;
(4) the set of known ground atomic formulae may grow monotonically with time;

and

(5) known individuals or tuples of known individuals may be tested for membership in a known relation.

The first assumption may seem somewhat restrictive; certainly any general system that may induce statements from examples should also be able to deal with general statements concerning the domain, given to it *a priori* (presumably by "being told"). However, if more general statements are permitted—for example existentials, universal generalizations, or disjunctions—their presence would beg the question of their origin or would presuppose an agent who had determined such relations. Such an agent then would constitute an entity that had to some extent already performed some knowledge acquisition. This in turn would run counter to our aim of delimiting the fundamental properties of learning from examples. However, in Delgrande (1985) consequences of relaxing this assumption and allowing arbitrary statements to be given to the learning system are investigated.

Interestingly also, there are significant applications which make similar assumptions. Relational databases clearly consist of ground formulae, and sentences, of, for example, relational algebra are equivalent to such sets. In addition, many semantic networks expressly limit or omit disjunctive and existential sentences, offsetting the loss of expressibility with a gain in computational efficiency. Thus the approach at hand may be directly applicable to such schemes.

The third assumption also deserves further comment. Consider where we have some conjecture (say, "ravens are black") and an exceptional individual (say, we encounter an albino). If we do not want to totally abandon our original hypothesis, then there seems to be two ways we can discharge the exception. First, we could amend the conjecture to something like "normally ravens are black", and perhaps also introduce "normally albino ravens are white". Formal aspects of this approach are investigated in Delgrande (1986)—in any case the exception is "excused". Second we could determine, or simply declare, that the observation is erroneous: either the individual is not a raven or it really is black. However, *this* procedure of determining that an observation is incorrect is a pragmatic concern, and is quite distinct from our concern of what hypotheses "follow" potentially from a given set of observations.

1.2. RELATED WORK

Much work has, of course, been carried out in AI addressing the problem of learning from a stream of examples. In the early, influential work of Patrick Winston (Winston, 1975), descriptions of concepts are formed from a set of examples of a concept and "near misses". The work presented in Brown (1973); Buntine (1986); Hayes-Roth (1978); Mitchell (1977); Shapiro (1981); Solway and Riseman, (1977); and Vere (1978) also falls into this category. The more recent work of Michalski, presented in Michalski (1983), is a particularly detailed approach to learning from examples. An extensive survey of AI learning systems is given in Dietterich, London, Clarkson and Dromey, (1982), while Smith, Mitchell, Chestek and Buchanan, (1977) describes a proposed "model" for learning from examples and Dietterich and Michalski (1983) compare four particular generalization programs.

Much of this work is concerned with proposing and refining a description of a concept, and many of the approaches detail particular rules or strategies for forming a general concept from a set of instances. In contrast, the work at hand deals with characterizing the hypotheses formable under a set of (arguably) minimal assumptions and hence is more concerned with exploring intrinsic properties and limitations of such approaches. In the sixth section I return to this distinction and compare the work at hand with three systems for learning from examples.

Most formal approaches to learning from examples have been concerned with inducing instances of a given type of formal language. The area of *learning theory* studies systems that implement functions from evidential states to languages. Learning theory was given a rigorous foundation with the work of Gold (Gold, 1967). A survey of such approaches is presented in Angluin & Smith (1982), while Osherson, Stob and Weinstein, (1983) gives recent results in this area. The key difference between such approaches and the present work, clearly, is that no underlying formal grammar is presupposed, beyond that for elementary set theory.

2. Introducing conjectures: initial considerations

As mentioned, the domain of application is assumed to be described by a presumably infinite set of ground atomic formulae, formed from presumably infinite sets of individuals and predicates. However, given a particular predicate, all that can be known of it is a subset of those individuals (or tuples) which satisfy it and a subset of those individuals which do not. I will speak of an individual as being *known,* if it is known to be or not be part of the extension of a known predicate. A predicate will be referred to as *known* if its truth value on a given individual (tuple) can be determined. Thus perhaps, for a known predicate, a *verification procedure* is assumed to be known for determining if a given individual is part of the extension. Informally, a known individual or predicate is one "encountered" by a learning system. The sets of tuples known to belong to the extension of a predicate and known to not belong to the extension are referred to as the *known extension* and the *known antiextension* respectively. So for a known n-place predicate P and known individuals $a_1, \ldots, a_n$ there are three possibilities:

(1) $P(a_1, \ldots, a_n)$ is known to be true.
(2) $\neg P(a_1, \ldots, a_n)$ is known to be true.
(3) Neither $P(a_1, \ldots, a_n)$ nor $\neg P(a_1, \ldots, a_n)$ are known to be true.

Definition: For each known predicate symbol P define sets P_+ and P_- by:

$$P_+ = \{\langle a_1, \ldots, a_n\rangle \mid P(a_1, \ldots, a_n) \text{ is known to be true}\}.$$

$$P_- = \{\langle a_1, \ldots, a_n\rangle \mid \neg P(a_1, \ldots, a_n) \text{ is known to be true}\}.$$

Conjectures are expressed in a language HL. This language is analogous to that of elementary set theory, except that operators and relations are subscripted with the character "h". I will use the symbol "$|_h$" for the (hypothesised) disjointness relation, and "$\propto_h$", "$\circ_h$", and "$\uparrow_h$" for the converse, composition, and image operations. A "ply" operator $\supset_h$ is also introduced. When we come to consider the algebra of terms of HL, we will also want to consider the corresponding propositional logic. The ply operator will serve as the analogue in the algebra of the material conditional in the logic. This operator is discussed further in the third section.

Definition: If P is the set of known predicate names, then the *terms* of HL are exactly those given by:

(1) If $\alpha \in P$ then α is a term of HL.
(2) If α, β are 2-place terms and γ a 1-place term of HL not containing $\supset_h$, then $\propto_h \alpha$, $\alpha \circ_h \beta$, and $\alpha \uparrow_h \gamma$ are terms of HL.
(3) If α and β are terms of HL, then so are $\alpha \cap_h \beta$, $\alpha \cup_h \beta$, $\neg_h \alpha$, and $\alpha \supset_h \beta$.

Definition: The *sentences* of HL are exactly given by:
If α, β are terms of HL, then $\alpha =_h \beta$, $\alpha \subset_h \beta$, $\alpha \subseteq_h \beta$, $\alpha \,|_h\, \beta$, $\alpha \neq \beta$, $\alpha \not\subset \beta$, $\alpha \not\subseteq \beta$, $\alpha \nmid \beta \in$ HL. So, for example,

$$\textit{Raven} \cup_h \textit{Penguin} \cup_h \textit{Robin} \subseteq_h \textit{Bird}$$

$$\textit{Uncle} =_h (\textit{Brother} \circ_h \textit{Parent}) \cup_h (\textit{Husband} \circ_h \textit{Sister} \circ_h \textit{Parent})$$

have the respective readings "the set of ravens, penguins, and robins is hypothesized to be contained in the set of birds" and "(the binary relation) uncle is hypothesized to be equivalent to the union of the composition of brother with parent, and the composition of husband with sister with parent".

The known extension and antiextension corresponding to terms in HL can easily be determined. Thus, for example, the *hypothetical intersection* of P and Q is known to contain just those elements that both P and Q are true of, and is known to not contain just those elements that either P or Q is known to not be true of. For the hypothesised operations we obtain:

Proposition:

Complement: $\neg_h P$ is (P_-, P_+)

Union: $P \cup_h Q$ is $(P_+ \cup Q_+, P_- \cap Q_-)$

Intersection: $P \cap_h Q$ is $(P_+ \cap Q_+, P_- \cup Q_-)$

Ply: $P \supset_h Q$ is $((\neg P_+ \cup Q_+) \cap (\neg Q_- \cup P_-), Q_- \cap \neg P_-)$

Converse: $\propto_h P$ is $(\{\langle y, x\rangle \mid \langle x, y\rangle \in P_+\}, \{\langle y, x\rangle \mid \langle x, y\rangle \in P_-\})$

Image: $P \uparrow_h Q$ is $(\{y \mid (\exists x)(\langle x, y\rangle \in P_+ \wedge x \in Q_+)\}, \varnothing)$

for binary relation P and one-place predicate Q

Composition: $P \circ_h Q$ is $(\{\langle x, z\rangle \mid (\exists y)(\langle x, y\rangle \in P_+ \wedge \langle y, z\rangle \in Q_+)\}, \varnothing)$.

The expression for the ply is chosen so that it corresponds to the material conditional in the logic (following). These operations generalise easily to ternary and higher-order predicates. Other operations such as domain and range may be defined in terms of these.

It will prove essential that occasionally we be able to determine the value of $P(a_1, \ldots, a_n)$, provided that P and $a_1, \ldots, a_n$ are known. However, in general we would not want to determine the truth values of all known predicates applied to all combinations of known individuals. The reason for this is combinatorial: given p known n-place predicates and m individuals, there are pm^n knowable ground instances. In the approach to be described, at most $p(p-1)$ of these combinations need to be known for hypothesising relations.

Naively, two terms of HL may be conjectured to be equal when there is some reason to do so (i.e. the intersection of their known extensions is non-empty) and there are no known counter-examples. While there are conditions other than these that can be used for forming conjectures, any set of alternative conditions arguably must include at least these, and so these conditions represent a set of minimal criteria for hypothesis formation. (See Delgrande, 1985) for an examination of other such conditions.) In a similar manner, conditions for containment and disjointness may also be specified. We obtain:

Definition:

$\alpha =_h \beta$ when $\alpha_+ \cap \beta_+ \neq \varnothing$ and $\alpha_+ \cap \beta_- = \varnothing$ and $\alpha_- \cap \beta_+ = \varnothing$.
$\alpha \subset_h \beta$ when $\alpha_+ \cap \beta_+ \neq \varnothing$ and $\alpha_+ \cap \beta_- = \varnothing$ and $\alpha_- \cap \beta_+ \neq \varnothing$.
$\alpha \subseteq_h \beta$ when $\alpha_+ \cap \beta_+ \neq \varnothing$ and $\alpha_+ \cap \beta_- = \varnothing$.
$\alpha \mid_h \beta$ when $\alpha_+ \cap \beta_+ = \varnothing$ and $\alpha_+ \cap \beta_- \neq \varnothing$ and $\alpha_- \cap \beta_+ \neq \varnothing$.

The problem with this approach to forming conjectures, of course, is that it is hopelessly simplistic. For example, assume that M means that one can supervise M.Sc. students, while P and HP means that one can supervise Ph.D. students or has a Ph.D. (respectively). If all that is known is that:

$M(John)$, $P(John)$, $HP(John)$, together with
$M(Mary)$, $\neg P(Mary)$

we would have:

$$P \subset_h M,\ P =_h HP, \text{ along with}$$
$$HP =_h M.$$

This clearly is inconsistent. A potential solution to this difficulty is to determine the truth values of select ground instances, where both the predicate and the individual are known, but where the truth value of the ground instance is not known. In the above example, if $HP(Mary)$ was determined to be true, then $P \subset_h HP$ could be formed; if $HP(Mary)$ was determined to be false, then $HP =_h M$ could be weakened to $HP \subset_h M$.

So two questions arise. The first concerns how such "select" ground instances can be determined for the restoration of consistency. The second concerns specifying or characterizing the conjectures to which this procedure may be applied. Both questions are answered by examining the algebra of the known extensions and antiextensions of terms of HL and, from this, the corresponding propositional logic.

3. An algebra and a logic for forming conjectures

Two terms of HL are defined to be (strictly) equal when their known extensions and antiextensions coincide. Containment ($\leq$) is introduced by the usual definition. Hence:

Definition: For α, β terms of HL.

$$\alpha = \beta \text{ iff } \alpha_+ = \beta_+ \text{ and } \alpha_- = \beta_-.$$
$$\alpha \leq \beta \text{ iff } \alpha \cap_h \beta = \alpha.$$
$$\alpha < \beta \text{ iff } \alpha \leq \beta \text{ but } \alpha \neq \beta.$$

The resultant algebra HLA is given by $\text{HLA} = [\text{H}; \neg, \cap_h, \cup_h, \supset_h]$, where the carrier H is given by:

$$\text{H} = \{\langle a, b\rangle \mid a, b \subseteq \text{I and } a \cap b = \varnothing\}$$

for a set of known individuals I. The pair of elements in a member of H corresponds to a possible known extension/antiextension pair. Upper and lower bounds of H are defined by:

$$1 =_{df} (\text{I}, \varnothing) \qquad 0 =_{df} (\varnothing, \text{I}).$$

We obtain the following postulates:

Postulates:

P1 $\alpha \cap_h \beta = \beta \cap_h \alpha$

$\alpha \cup_h \beta = \beta \cup_h \alpha$

P2 $\alpha \cap_h (\beta \cap_h \gamma) = (\alpha \cap_h \beta) \cap_h \gamma$

$\alpha \cup_h (\beta \cup_h \gamma) = (\alpha \cup_h \beta) \cup_h \gamma$

P3 $\alpha \cap_h (\alpha \cup_h \beta) = \alpha$

$\alpha \cup_h (\alpha \cap_h \beta) = \alpha$

P4 $\alpha \cap_h (\beta \cup_h \gamma) = (\alpha \cap_h \beta) \cup_h (\alpha \cap_h \gamma)$
$\alpha \cup_h (\beta \cap_h \gamma) = (\alpha \cup_h \beta) \cap_h (\alpha \cup_h \gamma)$

P5 $\alpha \cap_h \alpha = \alpha$
$\alpha \cup_h \alpha = \alpha$

P6 $\alpha \cap_h (\beta \cup_h (\alpha \cap_h \gamma)) = (\alpha \cap_h \beta) \cup_h (\alpha \cap_h \gamma)$
$\alpha \cup_h (\beta \cap_h (\alpha \cup_h \gamma)) = (\alpha \cup_h \beta) \cap_h (\alpha \cup_h \gamma)$

P7 $\alpha \cap_h 0 = 0 \quad \alpha \cup_h 0 = \alpha$
$\alpha \cap_h 1 = \alpha \quad \alpha \cup_h 1 = 1$

P8 $\alpha = \neg_h \neg_h \alpha$

P9 $\neg_h(\alpha \cup_h \beta) = \neg_h \alpha \cap_h \neg_h \beta$
$\neg_h(\alpha \cap_h \beta) = \neg_h \alpha \cup_h \neg_h \beta$

P10 $\alpha \cap_h \neg_h \alpha \leq \beta \cup_h \neg_h \beta$

P11 $\alpha \cap_h (\alpha \supset_h \beta) \leq \beta$

R1 If $\alpha \cap_h \gamma \leq \beta$ then $\gamma \leq (\alpha \supset_h \beta)$

P1–P10 then characterise $\cap_h$, $\cup_h$, and $\neg_h$. These postulates very nearly, but do not quite, characterize Boolean algebras. Instead of a postulate for a universal complement,

$$\alpha \cap_h \neg_h \alpha = 0, \qquad \alpha \cup_h \neg_h \alpha = 1$$

we obtain the weaker "Kleene" postulate P10. However we retain postulates governing universal bounds (P7) and involution (P8) as well as De Morgan's laws (P9). The weakened complement arises from the fact that the known extension and antiextension of a predicate typically do not together constitute the set of known individuals I. This algebra has been investigated under the names of *normal involution lattices* (Kalman, 1958) and *Kleene algebras* (Kleene, 1952).

We also want to derive the propositional logic corresponding to HLA. The operations of hypothesised intersection, union, and complement will clearly be analogous to the logical operations of conjunction, disjunction, and negation. Corresponding to the material conditional in the logic we need to introduce a fourth operation, called the *ply* (Curry, 1963). Postulates P11 and R1 are used to characterize this operation. P11 corresponds to *modus ponens,* while R1 says that the ply is maximal among solutions to P11. Note though that $\supset_h$ is of limited usefulness in forming conjectures: while it does in fact correspond to the material conditional in the logic (following), logical entailment corresponds to inclusion ($\leq$) in the algebra. Thus the ply serves basically to facilitate development of the formal results.

Given these postulates, the corresponding propositional logic HLL is derived. This logic, which seems to have not appeared in the literature, is given below.

Axiom schemata

A1 $\alpha \supset (\beta \supset \alpha)$

A2 $(\alpha \supset (\beta \supset \gamma)) \supset ((\alpha \supset \beta) \supset (\alpha \supset \gamma))$

A3 $\alpha \wedge \beta \supset \alpha$

A4 $\alpha \wedge \beta \supset \beta$

A5 $\alpha \supset (\beta \supset (\alpha \wedge \beta))$

A6 $\alpha \supset (\alpha \vee \beta)$

A7 $\beta \supset (\alpha \vee \beta)$

A8 $(\alpha \supset \gamma) \supset ((\beta \supset \gamma) \supset (\alpha \vee \beta \supset \gamma))$

A9 $\alpha \equiv \neg\neg\alpha$

A10 $(\alpha \supset \neg\alpha) \vee \neg(\alpha \supset \neg\alpha)$

Rules of inference

MP From $\vdash \alpha$ and $\vdash \alpha \supset \beta$ infer $\vdash \beta$

HN $\vdash \alpha \supset \beta$ iff $\vdash (\alpha \supset \neg\alpha) \vee \beta$ and $\vdash (\neg\beta \supset \beta) \vee \neg\alpha$

A semantic account for the formulae of HLL follows easily from the algebra HLA.

Definition: $\models \alpha'$ in HLL iff $\alpha = 1$ in HLA.

where α' is the formula obtained from α by replacing intersection with conjunction, etc. in the obvious way. The material conditional is linked to containment, via entailment, by the following.

Proposition: $\models \alpha' \supset \beta'$ in HLL iff $\alpha \leq \beta$ in HLA.

We obtain:

Theorem: HLL is sound and complete with respect to HLA.

Corollary: HLL is decidable.

Not surprisingly, negation in HLL is weaker than in PC: we lose *reductio ad absurdum* as a method of proof; also we lose the law of the excluded middle. PC is obtained if A10 and HN are replaced by $(\alpha \supset \beta) \supset ((\alpha \supset \neg\beta) \supset \neg\alpha)$. We also obtain:

Theorem:

(1) $\vdash \alpha \supset \beta$ iff $\vdash \neg\beta \supset \neg\alpha$;
(2) If $\vdash \neg\alpha \vee \beta$ then $\vdash \alpha \supset \beta$;
(3) If $\vdash \beta$ then $\vdash \neg\beta \supset \alpha$;
(4) If $\vdash \beta$ and $\vdash \alpha \supset \neg\beta$ then $\vdash \neg\alpha$;
(5) If $\vdash \neg(\alpha \supset \beta)$ then $\vdash \neg\beta$.

Note however that none of the formulae obtained by replacing the meta-theoretic "if · · · then · · · " in the above, with the material conditional, is a theorem of HLL.

The logic resembles the system of first degree entailment, E_{fde}, of Anderson and Belnap (1975). The principal difference is that axiom A1 and the theorem $\alpha \supset (\beta \supset \beta)$ of HLL is rejected by Anderson and Belnap, while their axiom $(\alpha \supset \beta) \supset (\neg\beta \supset \neg\alpha)$ is not a theorem of HLL.

4. Restoring the consistency of conjectures

For restoring the consistency of a set of conjectures, we subsequently obtain:

Theorem: If $\alpha_1, \ldots, \alpha_n, \alpha \in$ HL and $\alpha_1, \ldots, \alpha_n$ have been hypothesised according to our naive criteria, where α is derivable from $\alpha_1, \ldots, \alpha_n$ in HLA, then ground instances can be determined from the set of confirming instances for $\alpha_1, \ldots, \alpha_n$ that will either:

(1) refute one of $\alpha_1, \ldots, \alpha_n$;
(2) allow α to be hypothesized.

Outline of proof: Equality is characterized in HLA in the standard fashion as a reflexive predicate which obeys the principle of substitution of equals into formulae. An equivalent, but much more basic, characterization is provided and shown to be equivalent to the original. This alternative set of rules then may be used for forming proofs in HLA. However, this set of rules has the important property that if there is evidence (confirming ground instances) for the rule premiss(es), then there is (immediately and simply) evidence for the conclusion, or else there is an instance that falsifies the rule premiss. Since a proof is a sequence of steps from original premisses to desired conclusion according to the rules of inference, the theorem follows immediately. Note that it does not matter *which* proof of α from $\alpha_1, \ldots, \alpha_n$ is selected. This guarantees that if our naive criteria allowed us to hypothesize that:

$$A =_h B, \qquad B =_h C, \qquad C =_h D,$$

but not

$$A =_h D,$$

that we could identify an instance g with the following characteristics. The predicate symbol of g is among the set of known predicates, and the constants (individuals) of g are among the known individuals. The truth value of g is unknown, but determining the truth value of g will refute one of the premisses, or, via our naive criteria, allow the conclusion $A =_h D$ to be hypothesized.

The proof of this theorem is constructive, and leads immediately to a procedure which will locate evidence (i.e. ground instances) for a conjecture α that follows from a set of premisses $\alpha_1, \ldots, \alpha_n$, where each premiss has supporting evidence, or else will refute one of $\alpha_1, \ldots, \alpha_n$. This evidence is located from the instances supporting $\alpha_1, \ldots, \alpha_n$; the procedure is linear in the length of the proof of α. A corollary to the theorem allows for a generalization of the notion of evidence to include any metric that is applied uniformly to the conjectures. Consistency can thus be restored in a set of conjectures by repeatedly applying this procedure. Moreover, if we begin solely with a set of ground instances we will, by repeated application of our naive criteria for forming conjectures together with this procedure, arrive at a set of consistent conjectures.

This resolves the first question posed above concerning how consistency can be maintained. The second is answered via the logic HLL and a secondary result concerning the algebra: the conjectures to which the procedure may be applied correspond precisely to the sentences of elementary set theory, except that:

(1) we do not have a universal complement;
(2) the ply operator does not appear within the scope of a hypothetical converse, image, or composition operator.

The first condition is unavoidable in this approach (or any approach based on the five assumptions listed in the introduction). For the complement, though, we do retain involution, De Morgan's laws, laws concerning universal bounds, and the "Kleene" postulate. The second condition, which was given in the definition of HL, presents no real obstacle. Since containment in the algebra provides an analogue for entailment in the logic, the ply is of limited use in forming conjectures.

5. Further issues

Up to this point I have been considering the question of what conjectures may (potentially) be formed on the basis of a stream of ground instances, and how the consistency of a set of conjectures may be restored in the face of conflicting ground instances. In this section I further consider formal properties of the systems involved, along with the overall expressiveness of these systems. The approach is to consider elements of the algebra HLA (i.e. known extension and antiextension pairs) and compare them with standard (Boolean) sets.

In HLA we have entities such as (Red_+, Red_-) where Red_+ is the set of things known to be (say) red and Red_- is the set of things known not to be red. These entities represent what is known of predicates (relations) in the domain of application. In contrast, in standard naive set theory we simply have collections of entities together with the familiar notions of set intersection, union etc. If we use capitalized strings for the former entities and strings beginning with a lower case letter for the latter, we might say something like:

$$Red = (Red_+, Red_-) = (\{r_1, \ldots, r_n\}, \{t_1, \ldots, t_m\})$$

and

$$s = \{s_1, \ldots, s_k\}.$$

So entities such as *Red* must be distinguished from the more familiar sets such as s. The latter are wholly and entirely known, in that their constituents are completely known. Thus $s = \{s_1, \ldots, s_n\}$ exactly determines s; for any individual a, it is known whether $a \in s$ or $a \notin s$. *Red* on the other hand is used to represent what is known about a relation in the domain of application—i.e. the individuals known to be in the extension and the individuals known to not be in the extension. However there are further differences between these two types of entities.

First, if $s = \{s_1, \ldots, s_k\}$ then any occurrence of s can always be replaced by $\{s_1, \ldots, s_n\}$. Thus it seems reasonable to assert $Stack(s)$ or $Stack(\{s_1, \ldots, s_n\})$ with equal facility. This though is not the case with *Red*. While it seems reasonable to say, for example, *colour_type*(*Red*), we certainly do not want to say *colour_type*((Red_+, Red_-)). Even if we knew that $\{r_1, \ldots\}$ was the extension of *Red*, we would not want to say *colour_type*($\{r_1, \ldots\}$). The difference lies in the intensional nature of predicates such as *Red,* an aspect not shared by sets such as *s.* That is, *Red,* and in fact all the elements of P and all the terms of HL, are assumed to represent what is known about *properties*; the conjectures of HL then may be looked on as representing hypothesised interrelations among properties, based on what is known of their extensions. (This means also that, among other things, we cannot distinguish predicates with differing intensions but with the same extensions—for example, vertebrates with hearts from vertebrates with kidneys).

But sets such as *s,* which are formed from members of I and P, are defined by their extensions. Once again, if $s = \{s_1, \ldots, s_n\}$ there is nothing else to learn about the extension of *s.* These sets that can always be replaced by their extension I will refer to as *reducible.* An equally appropriate term is *knowable.* The (hypothetical) sets, such as *Red,* which cannot be replaced by their extension, I will refer to as *irreducible.* An equally appropriate term here is *unknowable.*

So there are two questions of interest:

(1) how can we formally characterize the irreducible sets?
(2) how do the reducible and irreducible sets interrelate?

These questions are addressed by specifying, via a set of axioms, the set of allowable reducible and irreducible sets. For both reducible and irreducible sets, the axioms developed will parallel those in the system of Zermelo–Fraenkel (ZF) (Fraenkel, Bar-Hill & Levy, 1973). Note however that there is no commitment made as to which set theory applies in the real world. That is, there are sets (maybe classes) in the real world that the irreducible sets correspond to; however there is no need to decide which theory governs those (unknowable) sets.

Omitting the development, we obtain:

Notation:

The letters $a, b, c, \ldots$ will stand for reducible sets.
The letters $A, B, C, \ldots$ will stand for irreducible sets.
The letters $\ldots, x, y, z$ will stand for either reducible or irreducible sets.

Set axioms:
Existence:

(i) if $a \in \mathrm{I} \cup \mathrm{P}$, $\{a\}$ is a set;
(ii) if $a, b \subseteq \mathrm{I} \cup \mathrm{P}$ and $a \cap b \neq \varnothing$ then (a, b) is an irreducible set.

Extensionality:

(i) $(a)(b)(x)(x \in a \equiv x \in b) \supset a = b$;
(ii) $(A)(B)(x)((x \in A_+ \equiv x \in B_+) \wedge (x \in A_- \equiv x \in B_-)) \supset A = B$.

Pairing:

$$(x)(y)(\exists a)(z)(z \in a \equiv (z = x \vee z = y)).$$

Sum:

(i) $(a)(\exists b)(x)(x \in b \equiv (\exists c)(x \in c \wedge c \in a))$;

(ii) $(a)((\exists D)(D \in a) \supset$
$(\exists B)(x)[x \in B_+ \equiv ((\exists c)(x \in c \wedge c \in a) \vee (\exists C)(x \in C_+ \wedge C \in a))]$
$\wedge [x \in B_- \equiv \neg(\exists c)((x \in c \wedge c \in a) \wedge (C)(x \in C_- \wedge C \in a))])$;
(iii) $(A)(\exists C)(x)([x \in C_+ \equiv (\exists B(x \in B_+ \wedge B \in A_+) \vee (\exists b)(x \in b \wedge b \in A_+))]$
$\wedge C_- = \varnothing)$.

Power set:

(i) $(a)(\exists b)(c)(c \in b \equiv c \subseteq a)$;
(ii) $(A)(\exists B)(c)((c \in B_+ \equiv c \subseteq A_+) \wedge (c \in B_- \equiv (c \subseteq (A_+ \cup A_-) \wedge c \nsubseteq A_+)))$.

Separation:

(i) $(a)(\exists b)(x)(x \in b \equiv x \in a \wedge \sigma(x))$ for b not free in σ, and σ reducible or irreducible;
(ii) $(A)(\exists B)(x)((x \in B_+ \equiv (x \in A_+ \wedge \sigma(x))) \wedge (x \in B_- \equiv (x \in A_- \vee \neg\sigma(x))))$ for B not free in σ, and σ reducible or irreducible.

Regularity:

$(a)(a \neq \varnothing \supset (\exists x)(x \in a \wedge (y)(y \in x \supset y \notin a)))$
where for x irreducible, $y \in x$ means $y \in x_+$.

This axiomatization then answers the preceding two questions regarding reducible and irreducible sets; as well it provides us with the following benefits. First, the hypothetical operators can be justified and defined in terms of the axioms for irreducible sets. For example, for intersection we have:

Theorem: $(A)(B)(\exists C)(x)((x \in C_+ \equiv (x \in A_+ \wedge x \in B_+)) \wedge (x \in C_- \equiv (x \in A_- \vee x \in B_-)))$ and C is unique.

So for any two irreducible sets there is a third unique set whose known extension consists of elements common to the known extensions of the first two sets and whose known antiextension consists of elements in either of the known antiextensions of the two sets. This in turn justifies the definition:

Definition: $A \cap_h B = C$ iff $(x)((x \in C_+ \equiv (x \in A_+ \wedge x \in B_+)) \wedge (x \in C_- \equiv (x \in A_- \vee x \in B_-)))$.

Thus, $A \cap_h B = (A_+ \cap B_+, A_- \cup B_-)$.

Moreover results of applying an operator to a reducible and an irreducible set can be justified and defined. For example, for intersection we obtain:

Definition: $A \cap_h b = c$ iff $(x)(x \in c \equiv (x \in b \wedge x \in A_+))$.

Thus $A \cap_h b = A_+ \cap b$. Thus the result of intersecting a known set of blocks with the (irreducible) predicate *Red* is the subset of the blocks known to be red. The other hypothetical operations can be similarly defined.

Given the extended system, the concept of transitive closure is introduced in Delgrande (1985). Given transitive closure, we can introduce a rudimentary means of relating predicates which apply to sets of objects and to predicates that apply to pairs (or some fixed number) of objects. Thus we can express the hypotheses that a stack of objects is a set of objects that satisfies the transitive closure of the *On* relation and, conversely, any set of objects so bounded is hypothesized to be a stack. As an example of the expressiveness that we obtain, consider the oft-cited example of an arch. In first order logic we may write:

$$(y_1)(y_2)(y_3)[Arch\_reln(y_1, y_2, y_3)$$
$$\equiv On(y_3, y_1) \wedge On(y_3, y_2) \wedge \neg Touching(y_1, y_2) \wedge Pillar(y_1) \wedge Pillar(y_2) \wedge Block(y_3)]$$

to express the fact that an arch relation obtains between y_1, y_2, and y_3. In HL we can express the hypotheses:

$$\mathbf{D}_{(3,1)h}(Arch\_reln) \subseteq_h On,$$
$$\mathbf{D}_{(3,2)h}(Arch\_reln) \subseteq_h On,$$
$$\mathbf{D}_{(1,2)h}(Arch\_reln) \,|_h\, Touching,$$
$$\mathbf{D}_{(1)h}(Arch\_reln) \subseteq_h Pillar,$$
$$\mathbf{D}_{(2)h}(Arch\_reln) \subseteq_h Pillar,$$
$$\mathbf{D}_{(3)h}(Arch\_reln) \subseteq Block.$$

$\mathbf{D}_{(i)h}$ and $\mathbf{D}_{(i,j)h}$ are extensions of the image operator and select the ith or the ith and jth arguments of a relation. The first relation expresses the fact that the third and first arguments of *Arch_reln* are bounded by the *On* relation. Thus it is equivalent to;

$$(y_1)(y_2)(y_3)[Arch\_reln(y_1, y_2, y_3) \supset On(y_3, y_1)].$$

A pillar could be defined by:

$$Pillar =_h Stack \cup_h Block,$$

which would tie our definition back to that of *Stack* and the transitive closure of *On*.

A third capability that the extended system provides is a means of forming conjectures about sets of irreducible predicates, and thus meta-conjectures about the domain. For any set of known individuals or predicate names $A = \{A_1, \ldots, A_n\}$, we can form reducible sets $A_r \subseteq A$ or irreducible sets $A_i = (A_{i+}, A_{i-})$ where $A_{i+}, A_{i-} \subseteq A$ and $A_{i+} \cap A_{i-} = \varnothing$ (or, of course, sets of such sets, etc.). Thus, one may decide to specify:

$$colour\_type = \{Red, \ldots, Violet\}$$

where *colour_type* is introduced as an "internal" name, or by definition, based on the hypothesis:

$$Colour =_h Red \cup_h \cdots \cup_h Violet.$$

Similarly one may specify:

$$Bear\_type = (\{Black\_bear, Polar\_bear, \ldots\}, \{Red, Stack, \ldots\})$$

based on the hypothesis:

$$Black\_bear \cup_h Polar\_bear \cup_h \cdots \subseteq_h Bear.$$

Bear_type may be taken as irreducible, perhaps as a result of knowledge of individuals, for example the fact that some individual is known to be a bear, but is known to belong to none of the known subgroups. Likewise, *colour_type* may perhaps be taken as reducible as a result of pragmatic considerations. This time, for example, there is no known individual known to be coloured that is not one of the known colours.

So the set axioms, as well as providing us with a more primitive basis for forming conjectures, also expand the expressiveness of the system. In particular we can relate some predicates on sets of objects to other predicates or individuals. Also we can form conjectures about other conjectures, or meta-conjectures. However, while the set of allowable individuals has been vastly expanded, the form of the conjectures is unchanged. The formal results of the last section still apply and, since in the intended model the property of being an individual is decidable, the overall system remains decidable.

6. Comparison with learning systems

It is worth pausing at this point to compare the approach with other related work on learning from examples in AI. Three systems are particularly relevant and serve to place the present work within the field. The early work of John Seely Brown (1973) on automatic theory formation is a direct precursor to mine. Patrick Winston's dissertation (Winston, 1975) on learning structural descriptions from examples is a well-known early AI learning system and serves as a good representative of a general approach to learning from examples. Ehud Shapiro's work (Shapiro, 1981) on the inductive inference of theories from facts is similar to mine in broad outline, except that he makes substantial assumptions concerning how the domain of application is described.

The task of Brown's system is to propose definitions for a set of binary relations based on knowledge of the extensions of the relations. The system begins with a set of binary relations $R = \{R_1, \ldots, R_n\}$ and a database containing all the tuples for which $R_i(x, y)$ is true for each $R_i \in R$. The database is assumed complete and error-free. A process is given for proposing definitions of the relations. However the body of a definition is restricted to be either the composition of relations or disjunctions of such compositions. This format though is adequate for characterising a range of domains, including that of kinship relations.

The system is restrictive in that it demands complete and static data, and deals only with composition and disjunction. No analysis is carried out with regard to what may be conjectured, nor is an algorithmic analysis of the system given. Nonetheless the system illustrates some important points. Foremost is the overall difficulty in constructing *any* general system that will induce definitions from primitive instances. Given Brown's approach, and a domain as elementary as binary kinship relations, a vast number of conjectures was produced. However, as he remarks, it is not at all obvious which definitions should be selected as axioms and which should be left as redundant.

Brown's system is heuristic and was intended for direct implementation. Thus it dealt with matters such as efficiently searching for possible definitions, proposing definitions in a "simplest first" manner, etc. In contrast I have not addressed implementation issues, but rather have attempted to address general problems of hypothesis formation, and thus issues dealing with characterizing a set of conjectures and maintaining the consistency of a set of conjectures.

Winston's program learns definitions of concepts in the blocks world domain by means of "successive refinements" based on carefully chosen examples. The program is given a training sequence of examples of the concept and of "near misses", where a near miss is an example that is quite similar to an instance of the concept but differs in a small number of significant details. Relevant features that the concept must have are extracted from the (positive) examples, while negative information is extracted from the near misses. The notion of near miss is centrally important, and serves to focus and accelerate the learning process.

Winston is largely interested in the pragmatic aspects of a learning system, and concentrates on techniques that will speed the learning process or will assist in an implementation. In some sense then his approach is complementary to the one taken here (which avoids the more pragmatic issues). In addition, the set of sentences of HL subsumes the conjectures that may be formed in Winston's system; thus anything that can be formed in his system can also be conjectured in HL. (Winston actually gives a semantic net representation for his concepts. This representation however is clearly equivalent to a set of binary relations, and is useful mainly as a notational or implementational device.) However, in HL the set of formable conjectures and the means of restoring consistency are precisely laid out whereas Winston does not address these issues.

Again, however, these are not Winston's concerns; rather he deals with problems of implementation. Thus the problems of dealing with a vast number of possibilities for conjectures are avoided by means of a carefully selected sequence of examples and near misses. By restricting the language of discourse, the number of possibilities is further reduced. The result is that only a single (tentative) definition is retained. If worst comes to worst, and the definition becomes inconsistent, the system is able to resort to a backtracking strategy.

Shapiro's work is, superficially, the most similar to my own. Shapiro assumes that a domain is described by a stream of ground instances; based on the ground instances, a set of conjectured axioms for the domain is proposed and refined. The key difference is that in his approach far more is assumed about the way the domain is to be described. First, the domain is assumed to be describable by a set of rules in the form of restricted Horn clauses. Also the names of the classes for which defining axioms are to be induced are known beforehand and, for purposes of efficiency, each class is dealt with individually before the next class is considered.

A general, incremental algorithm for proposing a set of rules which imply the known ground instances is developed and described. The algorithm functions by successively reading ground instances. If the set of hypotheses is too strong (i.e. a contradiction is encountered), a *contradiction backtracking algorithm* is applied to locate the refuted hypothesis. The refuted hypothesis is then removed. If the conjectures are too weak, and do not imply the new ground instance, then a new conjecture is added, or a refinement of a previously refuted hypothesis is added. The algorithm has tuneable parameters that determine the complexity of the structure of a hypothesis and the complexity of derivations from the hypotheses.

The assumption that the domain is governed by rules allows an elegant, powerful, and reasonably efficient algorithm, and firm theoretical results concerning the types of languages that may be discovered. The system has, in addition to discovering an axiomatization for simple arithmetic, synthesized logic programs for simple list-processing tasks, satisfiability of Boolean formulas, and tree isomorphism. These results though hinge on the assumption that the domain is describable in terms of (restricted) Horn clauses.

7. Conclusion

This work develops a formal, unified, and general (but basic) framework for investigating learning from examples. A primary goal was to keep the approach as general as possible and independent of any particular domain, representation scheme, or set of learning techniques. Hence, for example, there is no restriction placed on the ordering of the ground instances, nor is there any restriction with regard to introducing new ("known") predicate names during the learning process. Also, no agent is assumed to exist to help direct or focus the acquisition process. The major methodological stance is the adoption of the same basic ontology—individuals and relations—common to representation systems, as a starting point for investigating acquisition. In this way factors arising from our orientation toward learning emerge, presumably unobstructed by any *a priori* commitments to a representation scheme.

Formal systems are developed for introducing and maintaining the consistency of conjectures. An exact specification of what conjectures may potentially be formed is provided. Also it is shown how the consistency of a set of conjectures can be restored in the face of conflicting instances. The system illustrates that a reasonably rich and expressive set of conjectures can be derived using only a minimal set of assumptions. The expressiveness of the system is also indicated by the fact that it is as at least as general as a number of existing AI acquisition systems, including Brown (1973), Hayes-Roth (1978), Vere (1978) and Winston (1975). Results concerning decidability lend credence to the possibility that learning systems based directly on this approach and, in particular, incorporating the procedure for restoring efficiency, may be efficiently implementable. In addition, given the generality of the approach, it is possible that the framework could also provide an appropriate starting point for an investigation of other types of learning systems. That is, the approach could be conceivably be extended by incorporating further assumptions concerning the domain, underlying representation scheme, or an agent to assist in the learning.

The framework presented is intended not only as a basis for the development of systems that learn from examples, but also as a neutral point from which such systems may be viewed and compared. Presumably the issues addressed here are common to, and are relevant to, any system for learning from examples. However only a set of formal issues have been addressed, and the concern has been with what conjectures may potentially be formed, rather than with which of those conjectures should in fact be held. Pragmatic issues concerned with the justification of conjectures, strength of evidence, and degrees of confirmation (to name a few) are outside the scope of this work. Also outside the scope of this paper is the question of exceptions to general statements. The notion of constructive induction, or introducing new relations to simplify the overall structure of the hypothesis set, is not addressed. A further extension to this work, addressed in Delgrande (1985), concerns learning from examples, but where arbitrary information, in addition to the ground instances, is also permitted. Thus this extension addresses learning from examples where the learning now takes place in the presence of some arbitrary knowledge base.

The approach as it stands may have immediate practical applications. It could be used, for example, for verifying or completing definitional constraints in large databases or knowledge bases. Thus, as a specific example, database systems often use integrity constraints to partly maintain consistency and reliability. However, given a large number of relations, it is an arduous task to specify all integrity constraints and to ensure that the set is consistent. The approach then seems suited to the task of automatically proposing and verifying such constraints.

This work is part of my doctoral dissertation at the University of Toronto, supervised by Professor John Mylopoulos. Financial assistance from the Province of Ontario and the Department of Computer Science, University of Toronto, is gratefully acknowledged.

References

ANDERSON, A. R. & BELNAP JR, N. D. (1975). *Entailment: The Logic of Relevance and Necessity, Vol. I,* Princeton University Press.

ANGLUIN D. & SMITH, C. H. (1982). A survey of inductive inference: theory and methods, *Technical Report 250,* Department of Computer Science, Yale University.

BROWN, J. S. (1973). Steps toward automatic theory formation, *Proceedings of the Third International Conference on Artificial Intelligence,* Stanford, California, pp. 121–29.

BUNTINE, W. (1986). Induction of horn clauses: methods and the plausible generalisation algorithm, *Proceedings of the Knowledge Acquisition for Knowledge-Based Systems Workshop.* American Association for Artificial Intelligence, Banff, Canada.

CURRY, H. B. (1963). *Foundations of Mathematical Logic.* McGraw Hill Co.

DELGRANDE, J. P. (1985). A foundational approach to conjecture and knowledge, *Ph.D. thesis, Technical Report CSRI-173,* Department of Computer Science, University of Toronto.

DELGRANDE, J. P. (1986). A propositional logic for natural kinds. *Sixth Conference of the Canadian Society for the Computational Studies of Intelligence.*

DIETTERICH, T. G., LONDON, B. CLARKSON, K. & DROMEY, G. (1982). *Learning and inductive inference.* In COHEN, P. R. & FEIGENBAUM, E. A. Eds, *The Handbook of Artificial Intelligence.* William Kaufmann Inc.

T. G. DIETTERICH & R. S. MICHALSKI. (1983). A comparative review of selected methods for learning from examples. In MICHALSKI, R. S., CARBONELL, J. G. & MITCHELL, T. M., Eds, *Machine Learning: an Artificial Intelligence Approach.* Tioga.

FRAENKEL, A. A. BAR-HILLEL Y. & LEVY, A. (1973). *Foundations of Set Theory,* 2nd revised ed., North-Holland Publishing Co.

GOLD, E. M. (1967). Language identification in the limit. *Information and Control,* **10,** 447–474.

HAYES-ROTH, F. (1978). The role of partial and best matches in knowledge systems. In WATERMAN, D. A. & HAYES-ROTH, F. Eds, *Pattern-directed Inference Systems.* Academic Press.

HUGHES, G. E. & CRESSWELL, M. J. (1968). *An Introduction to Modal Logic.* Methuen and Co. Ltd.

KALMAN, J. A. (1958). Lattices with involution. *Transactions of the American Mathematical Society,* Vol. 87, pp. 485–491.

KLEENE, S. C. (1952). *Introduction to Metamathematics,* North Holland Publishing Co.

MICHALSKI, R. S. (1983). A theory and methodology of inductive learning. In MICHALSKI, R.

S., CARBONELL, J. G. & MITCHELL, T. M. Eds, *Machine Learning: an Artificial Intelligence Approach.* Tioga.
MITCHELL, T. M. (1977). Version spaces: a candidate elimination approach to rule learning. *Proceedings of the Fifth International Conference on Artificial Intelligence,* Cambridge, Massachusetts, pp. 305–310.
OSHERSON, D. N., STOB, M. & WEINSTEIN, S. (1983). Formal theories of language acquisition: practical and theoretical perspectives. *Proceedings of the Eighth International Conference on Artificial Intelligence,* Karlsruhe, West Germany.
SHAPIRO, E. Y. (1981). Inductive inference of theories from facts. *Research Report 192,* Department of Computer Science, Yale University.
SMITH, R. G. MITCHELL, T. M. CHESTEK, R. A. & BUCHANAN, G. B. (1977). A model for learning systems. *Proceedings of the Fifth International Conference on Artificial Intelligence,* Cambridge, Massachusetts, pp. 338–343.
SOLWAY, E. M. & RISEMAN, E. M. (1977). Levels of pattern description in learning. *Proceedings of the Fifth International Conference on Artificial Intelligence,* Cambridge, Massachusetts pp. 801–11.
VERE, S. A. (1978). Inductive learning of relational productions. In WATERMAN & HAYES-ROTH Eds, *Pattern-directed Inference Systems.* Academic Press.
WINSTON, P. H. (1975). Learning structural descriptions from examples. In WINSTON, P. Ed. *The Psychology of Computer Vision.* McGraw–Hill.

Knowledge base refinement by monitoring abstract control knowledge

DAVID C. WILKINS, WILLIAM J. CLANCEY AND BRUCE G. BUCHANAN
Department of Computer Science, Stanford University Stanford, CA 94305, U.S.A.

An explicit representation of the problem solving method of an expert system shell as abstract control knowledge provides a powerful foundation for learning. This paper describes the abstract control knowledge of the HERACLES expert system shell for heuristic classification problems, and describes how the ODYSSEUS apprenticeship learning program uses this representation to semi-automate "end-game" knowledge acquisition. The problem solving method of HERACLES is represented explicitly as domain-independent *tasks* and *metarules*. Metarules locate and apply domain knowledge to achieve problem solving subgoals, such as testing, refining, or differentiating between hypothesis; and asking general or clarifying questions.

We show how monitoring abstract control knowledge for metarule premise failures provides a means of detecting gaps in the knowledge base. A knowledge base gap will almost always cause a metarule premise failure. We also show how abstract control knowledge plays a crucial role in using underlying domain theories for learning, especially weak domain theories. The construction of abstract control knowledge requires that the different types of knowledge that enter into problem solving be represented in different knowledge relations. This provides a foundation for the integration of underlying domain theories into a learning system, because justification of different types of new knowledge usually requires different ways of using an underlying domain theory. We advocate the construction of a definitional constraint for each knowledge relation that specifies how the relation is defined and justified in terms of underlying domain theories.

1. Introduction

An apprenticeship period is the most effective means that human problem solvers use to refine domain-specific problem solving knowledge in expert domains. This provides motivation to give apprenticeship learning abilities to knowledge-based expert systems, since they derive their power from the quality and quantity of their domain-specific knowledge. By definition, apprentice learning programs improve an expert system in the course of *normal problem solving* and derive their power from the use of *underlying domain theories* (Mitchell *et al.*, 1985).

There are two principal apprenticeship learning scenarios used by human problem solvers in knowledge-intensive domains such as medicine and engineering. In the first scenario, an apprentice problem solver learns in the course of observing the problem solving behavior of another problem solver. A learning opportunity occurs when the apprentice fails to explain an observed problem solving action. At this point, the apprentice can often use the problem solving context and underlying domain theories to identify missing or wrong problem solving knowledge, or at worse be able to ask a pointed question that will isolate the knowledge discrepancy. Our past research focused on this type of scenario: the ODYSSEUS learning

KNOWLEDGE-BASED SYSTEMS Vol. 1
ISBN 0-12-273251-0

program improves HERACLES-based expert system in the course of watching a human expert solve problems (Wilkins *et al.*, 1986).

In the second apprenticeship learning scenario, an apprentice problem solver learns in the course of solving problems and monitoring his or her own problem solving failures. This paper described how the ODYSSEUS learning apprentice can perform this type of learning; the ODYSSEUS learning apprentice improves a HERACLES-based apprentice expert system by having ODYSSEUS monitor the expert system's normal problem solving.

This paper is organized as follows. Section 2 briefly describes the problem solving architecture of the HERACLES expert system shell. The key aspects of HERACLES that are crucial for learning are a separation of the domain knowledge from control knowledge and an explicit representation of the control knowledge using tasks and metarules. Section 3 describes the learning method used by ODYSSEUS, provides two learning examples, and discusses the generality and limitations of the learning approach. Section 4 covers related research, and Section 5 summarizes the contributions of this paper.

2. Heracles' problem solving architecture

HERACLES is an expert system shell for solving problems using the heuristic classification method; it provides the user with a vocabulary of knowledge relations for encoding domain knowledge, and a domain-independent body of control knowledge that solves problems using this domain knowledge. In HERACLES, control knowledge is represented as *task procedures* and *metarules,* which are invoked by a *task interpreter* (Clancey, 1986*b*).

A *task* is a procedure for accomplishing some well-defined problem-solving subgoal. Examples of tasks are to test a hypothesis, group and differentiate hypotheses, refine a hypothesis, forward reason, ask general questions, and process hard data. Each action within a task procedure for achieving the task procedure subgoal is called a *metarule,* Metarules, which might more precisely be called 'inference procedure rules', do not contain domain knowledge; they index the domain knowledge using a relational language.

The domain knowledge in HERACLES consists of MYCIN-like rules and facts and is encoded using the MRS relational language (Russell, 1985). This knowledge is accessed when metarules premises are unified with domain knowledge relations. There are approximately 120 knowledge relations, such as **subsumes($parm1, $parm2)**†, **trigger($rule)**, and **evidence . for($parm, $hypothesis, $rule, $cf)**‡. Tasks and metarules can be viewed as orchestrating the domain knowledge: they piece the domain knowledge together in order to achieve a problem solving goal. Examples of metarules are shown in Section 3. Currently HERACLES contains approximately thirty task procedures and eighty metarules.

The three main levels of organization in HERACLES are shown in Fig. 1. The bottom level of organization includes all domain-specific knowledge of the expert

† Throughout this paper, all variables start with a '$'.

‡ This last relation means that **$parm** contributes evidence for **$hypothesis** in **$rule** and the certainty factor or strength of this rule is **$cf**. If a rule has several parameters in the premise, and **evidence . for** tuple is constructed for each of them.

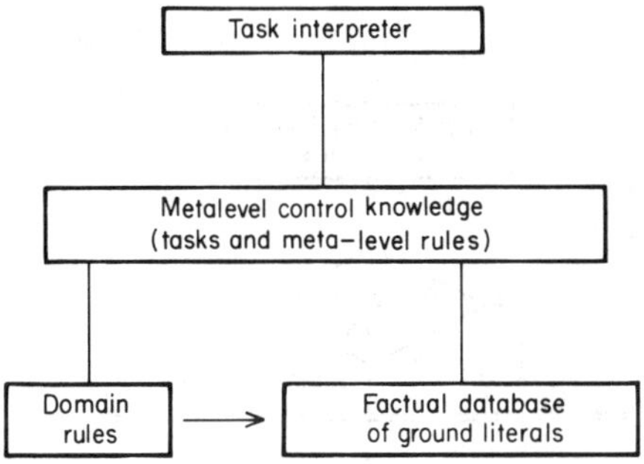

FIG. 1. Heracles' architecture.

domain, such as medical or engineering knowledge. The middle layer contains meta-level control knowledge, which encodes a problem-solving method such as heuristic classification or constraint propagation. Earlier shells such as EMYCIN did not have the middle layer of abstract control knowledge; rather, this knowledge was imbedded in the interpreter and the domain rules.

In the examples in this paper, the domain knowledge base to be refined is the NEOMYCIN knowledge base for diagnosing meningitis and neurological problems (Clancey, 1984). The NEOMYCIN knowledge base is a reorganization and extension of the MYCIN knowledge base, in which distinctions are made between different types of problem solving knowledge, and the control knowledge is more completely separated from the domain knowledge. The described HERACLES system was actually created by removing the domain knowledge from NEOMYCIN. Patient cases created for the NEOMYCIN domain are used as input. The ODYSSEUS induction theory uses the MYCIN library of solved patient cases (Buchanan & Shortliffe, 1984).

HERACLES metarules have the responsibility for locating and applying all domain knowledge. The form of the metarule provides a way to determine whether the premise of the rule is true by accessing dynamic state information and referencing and retrieving information from the domain knowledge base. ODYSSEUS monitors HERACLES metarule premises for failures. If the cause of the failure is missing domain knowledge, ODYSSEUS attempts to create this knowledge using underlying theories of the domain. If ODYSSEUS succeeds in finding the desired domain knowledge, the domain knowledge base in the expert system shell is automatically refined. The metarule for achieving a problem solving subgoal can now be successfully applied.

3. Odysseus' learning method

An overview of the learning method to be described is shown in Fig. 2. The first major task facing the learning system is global credit assignment, which is the determination of whether there is a potential gap in the knowledge base. The gap can be either a lack of factual or rule knowledge. The use of a relational language for all knowledge, including rules, provides a uniform approach to discovering both types of deficiencies. A gap in the knowledge base is suspected whenever the premise of a metarule fails. Given a failed metarule premise, the learning program

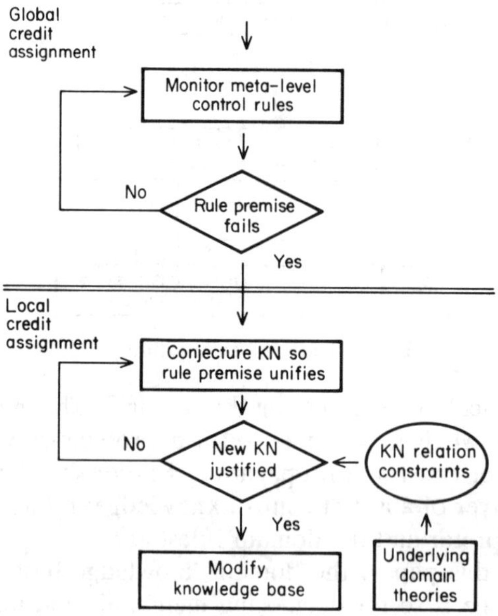

FIG. 2. Odysseus' learning method.

checks to see which conjuncts of the premise failed. If the failed conjunct indexes dynamic state information or is used to control the meta-level reasoning, then there is no learning opportunity, as there is no corresponding underlying domain theory. However, if the failed conjunct is the type that accesses the domain knowledge base, then this could be a learning opportunity.

After detecting the existence of a gap in the knowledge base, the next task is to pinpoint the gap; this is the local credit assignment problem. In our approach, there are two major parts to local credit assignment: generation of potential repairs and the testing of these repairs for validity.

The input to the ODYSSEUS candidate repair generator is the metarule that failed, the known bindings for variables in the clauses of the metarule premise that have been determined outside of the scope of the metarule, and a knowledge of the range of values that each variable in a metarule clause is allowed to assume. For example, the value of the variable **$finding** can be any finding in the domain vocabulary. The candidate repair generator focuses on the knowledge relations in the metarule and generates all allowable variable bindings for these relations. These instantiated relations are then passed on to the ODYSSEUS candidate tester.

The input to the ODYSSEUS candidate tester is a knowledge relation instance, such as **subsumes(visual-problems, double-vision)**. In order to test this candidate, two things are necessary. First, ODYSSEUS must have in hand a definition of all the constraints (empirical or otherwise) that determine whether an arbitrary instance of this knowledge relation is valid. Second, the learning program must have underlying theories of the domain that are capable of determining whether the constraints are satisfied, and hence whether the knowledge relation instance is valid. ODYSSEUS contains two underlying domain theories for testing of new knowledge: a strategy theory of heuristic classification problem solving and a confirmation theory based on induction over past cases.

In the remainder of this section, two learning examples will be described in detail to demonstrate the approach we are advocating. The first example, given in Section 3.1, illustrates the learning of factual knowledge for the knowledge relation **clarifying . questions**, using the ODYSSEUS strategy theory as the underlying domain theory. The second example, given in Section 3.2, illustrates the learning of rule knowledge for the knowledge relation **evidence . for**, using a confirmation theory based on induction over past cases as the underlying domain theory. These examples are based on the NEOMYCIN knowledge base, the MYCIN case library, and an actual medical case. Both sections assume that a metarule failure has occurred and that candidate repairs have been generated; they concentrate on the third stage of learning, wherein candidate repairs are tested.

3.1. LEARNING FACTUAL KNOWLEDGE

The focus of this example is the **clarifying . questions** knowledge relation in the **clarify . questions** metarule presented below. As an example of its use, suppose the doctor discovers that the patient has a **headache**. The **headache** finding is associated with many diagnostic hypotheses, so many that it is generally wise to narrow down this set of hypotheses by determining the severity and duration of the headache before pursuing a specific hypothesis. This is the process of *clarifying the finding,* and the questions about various subtypes of this finding (e.g., headache-duration, headache-severity) are called *clarifying questions.* In the HERACLES system, this is implemented by invoking the **clarify . finding** task whenever a new finding is derived by the system or provided by the user. In turn, the **clarify . finding** task invokes the **clarify . questions** metarule.

MetaRule 1: Clarify . questions

IF:	**goal(clarify . finding $finding1)** ∧ **clarifying . questions($finding1 $finding2)** ∧ **not(value-known $finding2)**
THEN:	**goal(findout $finding2)**
ENGLISH:	**If the current goal is to clarify finding1** **and finding1 can be clarified by finding2** **and finding2 is currently unknown** **then try to find out the value of finding2.**

Only one of the premise conjuncts of Rule 1 accesses domain knowledge, namely **clarifying . questions($finding1 $finding2)**. The first conjunct is for control purposes and the third conjunct checks the value of dynamic state knowledge.

The situation when learning may occur is when Rule 1 is passed a value for the variable **$finding1**, say 'headache', but the premise of Rule 1 fails because no bindings can be found for **$finding2**. In this situation, **$finding2** is a free variable at the time of failure. ODYSSEUS begins the learning process by invoking the candidate repair generator, which generates every possible candidate binding for

$finding2. Using information regarding the domain of **$finding2**, the learning critic is able to generate about 300 candidate relations.

In order to be able to validate candidate new domain knowledge for a particular knowledge relation, two steps must be taken beforehand. First, a *justification* for the knowledge relation must be constructed that specifies all the constraints that an instance of the knowledge relation must satisfy in order to be valid. In our example, this requires constructing a precise definition that captures the constraints on an instance of the **clarifying . questions** relation. Second, a way must be found to test these constraints using underlying theories of the domain. This two-step method constrasts with the current manual method of refining the NEOMYCIN knowledge base, which consists of asking physicians what clarifying questions to use.

Let us begin by giving an informal justification of **clarifying . questions**. One reasonable justification for asking clarifying questions is cognitive economy with respect to efficient diagnosis. Much of diagnosis involves the testing of specific hypotheses; however, sometimes a new piece of information is discovered that suggests a very large number of hypotheses. To reduce the number of relevant hypotheses, it is helpful to ask several clarifying questions that will add confirming or disconfirming evidence to many of the hypotheses associated with the new piece of information. After asking these questions, only a few of the numerous potential hypotheses will now be consistent with what is known.

We can now give a precise description of the constraints operating on **clarifying . questions**. This first-principles interpretation of a clarifying question is as follows: if a question is associated with many hypotheses, say more than six, and there exists a question that provides positive or negative evidence to many of these hypotheses, say between one-third and two-thirds, then always ask this question as a clarifying question. This can be formalized as follows.

Definition 1

For any finding f, let H_f be the set of all hypotheses h such that *relates-to* (f, h) is true. Let f_1 and f_2 be distinct findings, such that **subsumes** (f_1, f_2) is in the knowledge base. Let n be an empirically determined threshold indicating the minimum number of hypotheses that a finding must relate to in order to require the use of clarifying questions. Then

$$\textbf{clarifying . questions}(f_1, f_2) \leftrightarrow [(\|H_{f_1}\| \geq n) \wedge (\tfrac{1}{3}n \leq \|H_{f_1} \cap H_{f_2}\| \leq \tfrac{2}{3}n)].$$

The *relates-to*() relation is not part of the domain knowledge base; it is computed on the fly when a new piece of knowledge is validated, using a method which we will now describe. ODYSSEUS has two underlying domain theories that together can be used to check whether a new piece of knowledge satisfies all aspects of Definition 1. One underlying theory is a strategy theory for heuristic classification problem solving. A component of this theory is a line of reasoning explanation generator. Given a finding, all paths from that finding to reasonable possible diagnostic hypotheses via metarule applications can be determined. The generator can enumerate all the reasons that a question could possibly be asked, given the strategy and domain knowledge in HERACLES. The line of reasoning generator allows determination of all the hypotheses that are associated with any one question either directly or indirectly; it is used to compute *relates-to* (*f, h*).

We now describe the results of encoding Definition 1 and implementing our approach for the NEOMYCIN knowledge base. Currently, there are two clarifying questions for headache in the NEOMYCIN knowledge base: headache duration and headache severity. Our implemented metarule critic for the **clarify . questions** metarule considered the effect of all headache-related questions on the set of hypotheses associated with headache, and determined that one more clarifying question met the above described constraints: headache progression (i.e., is the headache getting better or worse). ODYSSEUS automatically modified a slot value under headache in the knowledge base to include this clarifying question; in the future, this question will always be asked when the patient complains of a headache.

3.2. LEARNING RULE KNOWLEDGE

All rule knowledge is represented within HERACLES using knowledge relations. This means that rules can be learned much as factual knowledge is learned. The example in this section involves learning an instance of the **evidence . for** relation in the **split . active . hypotheses** metarule. This rule is one of three invoked by the task **group . and . differentiate . hypotheses**. This metarule is useful during diagnosis when there are currently a large number of strong diagnostic hypotheses. The **split . active . hypotheses** metarule searches for a finding to ask about that will simultaneously provide strong positive evidence for some active hypotheses and strong negative evidence against other active hypotheses.

MetaRule 2: Split . active . hypotheses

IF:	**goal(group . and . differentiate . hyps $active . hypotheses)** ∧ **member($hypothesis1 $active . hypotheses)** ∧ **member($hypothesis2 $active . hypotheses)** ∧ **not(equal($hypothesis1 $hypothesis2))** ∧ **evidence . for($finding $hypothesis1 $rule1 $cf1)** ∧ **evidence . for($finding $hypothesis2 $rule2 $cf2)** ∧ **greater($cf1 . 2)** ∧ **less($cf2 − . 2)**
THEN:	**goal(findout $finding)**
ENGLISH:	**If the current goal is to group and differentiate a list of active hypotheses and a single finding provides positive evidence for one of the hypotheses and negative evidence for another of the hypotheses then try to find out the value of this finding.**

The metarule is passed a value for the variable **$active . hypotheses**. The interpreter attempts to find a unifier for all the clauses such that **$hypothesis1** is bound to one member in **$active . hypotheses, $hypothesis2** is bound to a different member of **$active . hypotheses**, and there is a single finding in the premise of a metarule that concludes that **$hypothesis1** is probably present and is also in the premise of a rule that concludes that **$hypothesis2** is probably absent. That is, a finding is asked that simultaneously provides evidence against some of the hypotheses and evidence for other hypotheses. Even though the NEOMYCIN knowledge base has been under development for several years, the **split . hypothesis . list** metarule is rarely invoked on any of the patient cases in the NEOMYCIN case library. Therefore implementing a learning critic for this metarule is useful.

In the example in which our learning critic was called into play, **$active . hypotheses** consisted of seven hypotheses: AV malformation, mycobacterium TB meningitis, viral meningitis, acute bacterial meningitis, brain aneurysm, partially treated bacterial meningitis and fungal meningitis. The metarule fails because a binding for **$finding** cannot be found in the two relations **positive . evidence . for** and **negative . evidence . for**. Other clauses establish bindings for **$hypothesis1** and **$hypothesis2**. Using information regarding the domain of **$finding**, the learning critic conjectures many potential missing rules. The number of conjectures can be quite large.

Given these conjectures, a confirmation theory determines whether any of them is true. This requires the use of a formal definition for each relation. In this case we need a formal definition of **$evidence . for**.

Definition 2

Let *r* be a justifiable domain rule. Let *f* be a finding that appears in the premise of *r*, and let *h* be a hypothesis that appears in the conclusion of *r*. Let *s* be the certainty factor strength of *r*, normalized to lie between ±1. Then

$$\textbf{evidence . for}(f, h, r, s) \leftrightarrow \textbf{general}(r) \wedge \textbf{specific}(r) \wedge \neg\textbf{complex}(r) \wedge \neg\textbf{colinear}(r)$$

To actually determine whether a domain rule is justifiable requires the use of an underlying domain theory. ODYSSEUS uses induction over a case library to determine whether the conjectured rule is valid. That is, ODYSSEUS does a statistical analysis of the cases and determines whether the rule has good generality, specificity, and economy, and satisfies other measures of rule fitness†.

The confirmation theory using the ODYSSEUS induction system found five rules that divide the list of active hypotheses, including:

Object-Level Rule 1

IF: **duration . of . symptoms ⩽ 1 day** ∧
evidence . for(meningitis) ⩾0·6

THEN: **suggests fungal . meningitis (cf = −0·8)** ∧
suggests mycobacterium . tb . meningitis (cf = −0·8) ∧
suggests acute . bacterial . meningitis (cf = 0·7)

† The library of test cases that we used to generate rules is the MYCIN case library (Buchanan & Shortliffe, 1984). Because diseases are defined in the NEOMYCIN knowledge base that are not defined in the MYCIN system (in this case, AV malformation, partially treated bacterial meningitis, and brain aneurysm), the values of the certainty factors (CFs) for some rules will be slightly inaccurate.

TABLE 1
Comparing apprenticeship scenarios

Learning Task	Scenario 1: watching other problem solver	Scenario 2: watching own problem solving
Global credit assignment	Attempt to construct an explanation of observed action fails.	Meta-level control rule premise fails.
Local credit assignment: generate repairs	Generate domain KN element that completes an explanation.	Generate domain KN element that allows rule to succeed.
Local credit assignment: test repairs	Check constraints on KN relation using underlying domain theories.	Check constraints on KN relation using underlying domain theories.

3.3. COMPARING APPRENTICE SCENARIOS

Table 1 contrasts the two different ODYSSEUS apprenticeship learning scenarios of watching another problem solver and watching one's own problem solving. Table 1 compares the way the two scenarios accomplish the three major learning tasks faced by an apprenticeship learning system: the realization that knowledge is missing, the generation of candidate repairs, and the testing of those repairs. Note that the latter two tasks, i.e. the local credit assignment process that involves the use of underlying domain theories and the construction of definitional constraints, are identical in the two scenarios. On the other hand, the global credit assignment process is easier when watching oneself, because there is none of the uncertainty connected with inferring another agent's line of reasoning. Generating repairs is also easier when watching oneself, as there is no uncertainty as to exactly which metarule and hence which knowledge relation is responsible for the failure.

Compared to watching another problem solver, one can learn from watching one's own problem solving earlier in the knowledge acquisition 'end-game'. When watching another problem solver, a relatively large knowledge base is required; otherwise it is impossible to follow the line of reasoning of an expert most of the time, which is a requirement of this scenario.

A disadvantage of watching oneself is a large number of false alarms. Metarules fail most of the time, and it is not clear what the failure rate would be for a really good knowledge base. Perhaps it would only be a little lower than with a fairly incomplete knowledge base. More experimentation is required to answer these questions.

4. Discussion

Monitoring abstract control knowledge appears to be a very promising lever for aiding apprenticeship learning. In showing two examples of the leverage obtained by this approach, we have only scratched the surface of the topic. This section discusses some of the remaining open issues.

As described in Sections 3.1 and 3.2, we have begun to implement constraint definitions to link knowledge relations to underlying theories. A key question that needs investigation is the reusability of these constraint definitions; are there sets of knowledge relations that can use the same or similar constraint definitions? As there are scores of different knowledge relations in the NEOMYCIN system, reuse of definitions could significantly reduce the amount of effort needed to create metarule critics for all metarules in the expert system shell. Further, it is not yet known whether all types of knowledge relations will be amenable to formal constraint definitions.

The best method of gauging the improvement produced by the addition of new knowledge is another open question. The heuristic knowledge that the examples of Section 3 added to the knowledge base is clearly helpful for the example cases, because it allows several hypotheses to be confirmed or disconfirmed with a single question. However, a complete validation should show improvement in performance on a validation set of cases. The measure of performance should be diagnostic accuracy and efficiency.

Another issue involves the control of the learning process. When should this type of learning be invoked? Not every metarule failure signals missing knowledge; how can learning opportunities be distinguished from routine failures?

Another open problem relates to the quantity of new knowledge introduced into the system. For example, in Section 3.2 five new rules were found that would divide the current hypothesis list. More generally, an open problem in the induction of rule bases is how to adequately *bias* the selection of rules (Fu & Buchanan, 1985; Michalski *et al.*, 1983). There may be very many good candidate rules, but having too many rules is injurious to an expert system—efficiency is decreased, debugging is complicated, and explanations of actions become harder to follow. Of course, learning knowledge in the context of normal problem solving increases the likelihood that the rules produced by the induction system are going to be useful for problem solving. Only adding rules that are needed by the metarules of the inference procedure is a good step towards introducing a sufficient bias on rule selection.

5. Related work

Two major apprenticeship learning systems are LEAP and DIPMETER ADVISOR (Mitchell *et al.*, 1985, Smith *et al.*, 1985). In both of these systems there is a single *type* of knowledge. In LEAP, all knowledge is implementation rules. In DIPMETER ADVISOR all knowledge is heuristic rules. In contrast, there are dozens of types of knowledge in HERACLES—each knowledge relation corresponds to a different type of knowledge. The key to automatic learning seems to be the definition of constraints to tie each knowledge relation individually to one or more underlying domain theories.

There has been a great deal of research on failure driven learning that monitors control and planning knowledge (Mitchell *et al.*, 1983; Korf, 1985; Minton, 1985). The goal of these research effort is to create better control knowledge so as to speed up problem solving, rather than to learn domain-specific factual knowledge. This compliments our approach, as we do not address the learning of abstract control knowledge for a problem-solving method; in other words, we do not learn tasks and metarules.

ODYSSEUS has a separate definitional constraint for each knowledge relation. This allows ODYSSEUS to determine whether the candidate new knowledge relation instance is valid. This is reminiscent of the approach taken in AM (Lenat, 1976), where each slot of a concept has a set of associated heuristic rules that can be used to validate the contents of the slot.

6. Summary

It is well known that expert systems derive much of their power from the quality and quantity of their domain specific knowledge. The method described in this paper provides a method of partially automating the acquisition of some of this knowledge.

The construction of expert system shells for generic tasks has become a common practice. There is a growing awareness that the power of a knowledge acquisition system for an expert system shell is bounded by the complexity and explicitness of the inference procedure (Eshelman & McDermott, 1986; Kahn *et al.*, 1985). There is also a growing awareness that *automated* knowledge acquisition must be grounded in underlying domain theories (Mitchell *et al.*, 1985; Smith *et al.*, 1985). Using the HERACLES expert system shell and the ODYSSEUS apprenticeship learning program, we have demonstrated how underlying theories of a problem solving domain can be effectively used by a learning method centered around an explicit representation (i.e., tasks and metarules) of the problem solving method.

The learning method described in this paper has three stages. The first stage is global credit assignment, the process of determining that there is a gap in the knowledge base. This is accomplished by monitoring metarule premise failures in the expert system shell, since all knowledge base gaps cause these. The second stage of learning is generating candidate repairs. Candidate repairs are generated by locating the knowledge relation in the failed metarule premise, and generating all values of the relation for the free variables in the relation. The last stage of learning is evaluation of candidate repairs. The ODYSSEUS method involves constructing a constraint definition for each different type of knowledge, to describe how an underlying domain theory can be used to validate the repair. In the described experiments, we used the NEOMYCIN knowledge base for the HERACLES expert system shell. The underlying domain theories are a strategy theory and a confirmation theory based on induction over past cases.

A major open question is to determine how many of the knowledge relations in the expert system shell can be grounded in underlying theories of the domain. In particular, we are investigating the extent to which the different knowledge relations can be grounded in the two underlying theories that are part of ODYSSEUS. However, for certain types of domain knowledge used in the metarules, such as definitional and causal knowledge, we currently have no underlying theory; construction of such theories to allow *automated* knowledge acquisition will be difficult and perhaps impossible.

The type of learning demonstrated in this paper is more powerful than most forms of failure-driven learning, because the definition of failure is weaker. Failure to solve the overall problem is not necessary; rather, failure to satisfy a metarule premise for achieving a problem solving subgoal is sufficient for learning to take place.

We express our gratitude for helpful comments provided by Haym Hirsh and Marianne Winslett for several draft versions of this paper.

This work was supported in part by NSF grant MCS-83-12148, ONR/ARI contract N00014-79C-0302, Advanced Research Projects Agency (Contract DARPA N00039-83-C-0136), the National Institute of Health (Grant NIH RR-00785-11), National Aeronautics and Space Administration (Grant NAG-5-261), and Boeing (Grant W266875). We are grateful for the computer time provided by the Intelligent Systems Lab of Xerox PARC and SUMEX-AIM.

References

BUCHANAN, B. G. & SHORTLIFFE, E. H. (1984). *Rule-Based Expert Systems: The MYCIN Experiments of the Stanford Heuristic Programming Project.* Reading, MA: Addison-Wesley.

CLANCEY, W. J. (1984). NEOMYCIN: reconfiguring a rule-based system with application to teaching. In CLANCEY, W. J. & SHORTLIFFE, E. H. Eds. *Readings in Medical Artificial Intelligence.* Reading, MA.: Addison-Wesley. pp. 361–381.

CLANCEY, W. J. (1986*a*). From GUIDON to NEOMYCIN to HERACLES in twenty short lessons. *AI Magazine,* **7**, 40–60.

CLANCEY, W. J. (1986*b*). Representing control knowledge as abstract tasks and metarules. In COOMBS, M. & BOLC, L. Editors. *Computer Expert Systems* Heidelberg: Springer Verlag. Also, *Knowledge Systems Lab Report KSL*-85-16, Stanford University, April 1985.

ESHELMAN, L. & MCDERMOTT, J. (1986). MOLE: a knowledge acquisition tool that uses its head. In *Proceedings of the Fifth National Conference on Artificial Intelligence.*

FU, L. & BUCHANAN, B. G. (1985). Inductive knowledge acquisition for rule based expert systems. *Technical Report KSL* 85-42, Stanford University, Computer Science Department.

KAHN, G., NOWLAN, S. & MCDERMOTT, J. (1985). MORE: an intelligent knowledge acquisition tool. In *Proceedings of the 1985 IJCAI.* pp. 573–580.

KORF, R. (1985). *Learning to solve problems by searching for macro-operators.* Marshfield, MA: Pitman.

LENAT, D. B. (1976). AM: An artificial intelligence approach to discovery in mathematics as heuristic search. Ph thesis, Stanford University.

MICHALSKI, R. S., CARBONELL, J. G., & MITCHELL, T. M. Eds (1983). *Machine Learning: An Artificial Intelligence Approach.* Pato Alto: Tioga Press.

MINTON, S. (1985). Selectively generalizing plans for problem solving. In *Proceedings of the* 1985 *IJCAI,* pages 596–599.

MITCHELL, T., UTGOFF, P. E. & BANERJI, R. S. (1983). Learning by experimentation: acquiring and refining problem-solving heuristics. In MICHALSKI, T. M., CARBONELL, J. G. & MITCHELL, T. M., Eds. *Machine Learning: An Artificial Intelligence Approach.* Palo Alto: Tioga Press. pp. 163–190.

MITCHELL, T. M., MAHADEVAN, S. & STEINBERG, L. I. (1985). LEAP: a learning apprentice for VLSI design. In *Proceedings of the* 1985 *IJCAI.* pp. 573–580.

RUSSELL, S. (1985). The Complete Guide to MRS. *Technical Report KSL*-85-108, Stanford University.

SMITH, R. G., WINSTON, H. A., MITCHELL, T. M. & BUCHANAN, B. G. (1985).

Representation and use of explicit justifications for knowledge base refinement. In *Proceedings of the* 1985 *IJCAI*. pp. 673–680.
WILKINS, D. C., CLANCEY, W. J. & BUCHANAN, B. G. (1986). An overview of the ODYSSEUS learning apprentice. In MITCHELL, T. M., MICHALSKI, R. S. AND CARBONELL, J. G., Eds. *Machine Learning*: *A Guide to Current Research,* New York: Kluwer Academic Press. pp. 332–340.

A conceptual clustering program for rule generation

EDWARD WISNIEWSKI,† HOWARD WINSTON, REID SMITH,‡ AND MICHAEL KLEYN
Schlumberger–Doll Research, Old Quarry Road, Ridgefield, CT 06877-4108, U.S.A.

We present an *Interesting Situation Generator* (ISG) that assists in the synthesis of interpretation rules from basic domain knowledge. The ISG is a hierarchical clustering program that discovers equivalence classes of situations (e.g. types of geological formations) that give rise to qualitatively distinct manifestations (e.g. different patterns of geophysical measurements corresponding to types of geological formations). The equivalence classes can be used by a rule generator to construct an initial set of interpretation rules of the form manifestation ⇒ situation.

1. Introduction

Traditionally, the design of knowledge-based systems has involved a process of interaction between a domain expert and a computer scientist, or knowledge engineer. During this interaction, the knowledge engineer formalizes the domain knowledge of the expert into inference rules and data structures (Buchanan & Shortliffe, 1984). Once the system has been implemented, failures in its performance may necessitate additional modification. As a result, the development of knowledge-based systems often involves a lengthy interaction between computer scientist and domain expert. This has been called the *knowledge acquisition bottleneck* (Mitchell, 1983). To relieve this bottleneck, recent research in AI has focused on partially automating the knowledge acquisition process. For example, one approach has been to develop interactive Learning Apprentice Systems (LASs). Mitchell, Mahadevan, and Steinberg (1985) define such systems as "interactive knowledge-based consultants that directly assimilate new knowledge by observing and analysing the problem-solving steps contributed by their users through their normal use of the system". In these systems, knowledge acquisition is viewed as *refinement* of an *existing* knowledge-based system. Refinement consists of adding rules to the existing knowledge base or modifying incorrect ones.

In this paper, we address the problem of *initial* knowledge base construction in the design of interpretation systems. We describe a program called the Interesting Situation Generator (ISG) that assists in the synthesis of interpretation rules from basic domain knowledge. The ISG is a *hierarchical clustering program* that discovers equivalence classes of *situations* (e.g. types of geological formations) that give rise to qualitatively distinct *manifestations* (e.g. patterns of qualitatively different geophysical measurements). Michalski and Stepp (1983) define conceptual clustering as a process of constructing a concept network characterizing a collection of objects, with nodes marked by concepts describing object classes, and links marked by relationships between the classes. The ISG automates this process. It differentiates a *general* situation class into a hierarchical tree of more *specific* situation classes. The most

† Current address: Honeywell Systems and Research Center, 3660 Technology Drive NE, Minneapolis, MN 55418.

‡ Current address: Schlumberger Palo Alto Research, 3340 Hillview Avenue, Palo Alto, CA 94304.

KNOWLEDGE-BASED SYSTEMS Vol. 1
ISBN 0-12-273251-0

specific classes are equivalence classes whose members produce similar manifestations. As a conceptual clustering program, the ISG constructs a network of nodes (the situation classes) linked by ISA-relations.

We have tested the program in the context of geological interpretation systems. The task for our systems is to infer the geological structures penetrated by an oil-well borehole, given a variety of petrophysical data recorded by logging tools (e.g. the tilt, or dip, of rock formations penetrated by the borehole, indexed by depth, as measured by a dipmeter tool—see Smith & Young, 1984). Our approach is to construct a set of initial *linkage rules* that embody fundamental, "textbook" knowledge of a domain (Winston, Smith, Kleyn, Mitchell & Buchanan, 1986; Swartout, 1983). This knowledge forms the underlying justification for a specialist's conclusions. A log interpreter, for example, may see a series of patterns and conclude that the data indicate the presence of an overturned antiform—a particular type of fold. The justification for such a conclusion can be explained by some set of basic principles, or deeper domain knowledge, combined with simplifying assumptions and approximations of the domain.

Linkage rules describe the way in which individual situations (e.g. geologic structures such as folds, or pathological states in medicine such as infections) produce particular manifestations of sensed data (e.g. dip of rock formations in geology, or symptoms like fever in medicine). Our formalization of the domain theory into linkage rules is one of the explicit inputs to the ISG.†

The ISG discovers a situation class by first selecting a situation and using the linkage rules to determine its corresponding manifestation. During this process, the ISG records dependencies between parameters of the manifestation and situation.‡ A dependency, in the context of the ISG, describes a sensitivity relationship between parameters. For example, increasing the height of a rectangle increases its area, all other things being equal. Next, the ISG determines the largest variations in the manifestation's parameters such that the manifestation retains its qualitative features. These are the boundaries of the manifestation class. Using the parameter dependencies, the ISG determines the corresponding variations in the situation's parameter values. These variations define ranges of the situation parameter values such that situations possessing values within these ranges will produce qualitatively similar manifestations. This process amounts to *backpropagating* the boundaries of the manifestation class to find the corresponding boundaries of the situation class. In this way, the ISG determines an equivalence class of situations that produce manifestations having the same qualitative features.

The ISG solves the *aggregation* and *characterization* problems of conceptual clustering programs. Fisher and Langley (1985) define the aggregation problem as the process of determining useful subsets of an object set. In our case, the object set is the general situation class initially presented to the ISG to differentiate. The useful subsets found by the ISG are the equivalence classes that produce qualitatively

† The ISG is most useful in these cases where it is easier to formalize a domain theory as a set of linkage rules than it is to construct a complete and consistent set of interpretation rules. In such cases a rule generator can construct the interpretation rules from the output of the ISG.

‡ Dependencies in the ISG were implemented with error propagation information (described in Smith, Winston, Mitchell & Buchanan, 1985) that specifies how errors propagate across data-dependency links connecting beliefs matched by rule conditions to beliefs created by rule actions.

similar manifestations (the leaves of the tree described above). Fisher and Langley define the characterization problem as the process of determining a useful characterization, or concept, for a group of objects. Such a characterization defines a group of objects that are similar to each other in some respect. The ISG determines useful characterizations of equivalence classes that are defined in terms of *ranges of parameter values*.

The ISG *efficiently* solves the aggregation and characterization problems. The program contains an *internal instance selector* (Buchanan, Mitchell, Smith & Johnson, 1978). That is, it does not rely on an externally supplied choice of training instances. The ISG selects and generalizes a *single* situation into an equivalence class (as described above). By generalizing situations into an equivalence class, the ISG reduces the unexplored (or uncharacterized) set of situations. The ISG then selects a situation *outside* this class and repeats the process. In this way, the ISG *optimizes* the aggregation process by choosing a representative sample of the possible situation instances. Further, the ISG defines a situation class in terms of parameter ranges by generalizing from a single, training instance. In this sense, it efficiently characterizes a situation class.

Using an example from geology, we will show how the ISG utilizes linkage rules and parameter dependencies to construct a hierarchical tree of situation classes. These equivalence classes of situations can be used by a *rule generator* to synthesize interpretation—or diagnostic—rules of the form manifestation⇒situation (e.g. sequence of dip patterns⇒overturned antiform fold, or fever⇒meningitis). This set of rules forms an initial knowledge base for an interpretation system.

2. Related work

Two other examples of learning systems that employ internal instance selectors are LEX (Mitchell, Utgoff & Banerjii, 1983) and AM (Lenat, 1982). Given a set of integral transformations, LEX could learn the heuristics that suggest the conditions under which particular transformations are useful. Its instance selector generated practice problems for the performance element to solve. The solutions to these problems were then analysed in order to propose and refine problem solving heuristics. LEX's problem generator autonomously looked for problems that could lead to refinement of existing, but partially learned heuristics, and problems that could lead to discovery of new heuristics.

AM could discover new mathematical concepts and relations from an initial set of facts, together with a collection of heuristics for synthesizing and evaluating new facts. Its instance selector autonomously generated new tasks that either suggested new concepts to explore or new extensions to existing concepts. Tasks were ordered by a measure of how "interesting" they appeared to be. Given that far more tasks could be proposed than explored, this ordering ensured that AM explored the most promising tasks first, effectively executing a best-first search through a space of mathematical concepts.

One of AM's heuristics for identifying interesting new concepts is very similar to the method our ISG uses to partition a set of situations by backpropagating the boundaries of manifestation equivalence classes. This heuristic states:

If f is a function which transforms elements of A into elements of B, and B is ordered, then consider just those members of A which are transformed into *extremal* elements of B. This set is an interesting subset of A.

If a domain theory in the form of linkage rules that map situations into manifestations is substituted for f, a set of situations is substituted for A, a set of corresponding manifestations is substituted for B, and a manifestation partition is substituted for the order relation on B, then this heuristic expresses the ISG's basic strategy for partitioning a set of situations. The interesting subset of A—discovered by the above heuristic—corresponds to the boundaries of the situation equivalence classes.

3. A structural model of geological interpretation

We will discuss the ISG in terms of a geological domain—dipmeter interpretation. The method, however, is domain-independent, assuming a domain theory that maps situations into manifestations and certain constraints (see next session) on domain parameters. The dipmeter records the slope, or dip, of rock layers as it ascends an oil-well borehole. These recordings can be characterized in terms of primitive patterns. Recordings of increasing dip† with borehole depth are known as red patterns, while those of decreasing dip with depth are referred to as blue patterns. The length of such a pattern is the depth at which the pattern ends minus the depth at which it begins. The rock layers of geological objects like folds and faults manifest themselves as *composites* of such patterns. One manifestation might consist of a red pattern, followed in depth by a blue pattern (a red/blue composite). Another manifestation might consist of a blue pattern followed in depth by a red pattern (a blue/red composite).

For example, Fig. 1 shows a cross-section of a geologic situation, called a fold, with its associated manifestation consisting of a red/blue/red (R/B/R) composite pattern. The heavy curved line is a representative bedding layer within the fold, and the horizontal and vertical lines establish a coordinate system such that the horizontal line is parallel to the earth's surface. The entire fold is composed of layers parallel to the representative layer. If a vertical borehole penetrates this fold, it initially encounters rock layers parallel to the top flank part of the representative bedding layer. These layers all have a local slope parallel to the slope of the top tangent (tangent-1). As the borehole passes through the fold's hinge, the slope of the bedding layers it encounters first increases from the slope of tangent-1 to the slope of the vertical line (i.e. 90°), then decreases from 90° to 0°, and finally increases from 0° to the slope of tangent-2. The borehole exits the fold through rock layers parallel to the bottom flank part of the representative bedding layer. These layers all have a local slope parallel to the slope of the bottom tangent (tangent-2). The top borehole interval within which the slope is *increasing* produces a *red* pattern in the corresponding dipmeter log, the middle interval within the slope is *decreasing* produces a *blue* pattern, and the bottom interval within which the slope is *increasing* produces another *red* pattern.

† Patterns of monotonically increasing, decreasing, or constant dip are required to have uniform azimuth. That is, the dip direction must remain constant.

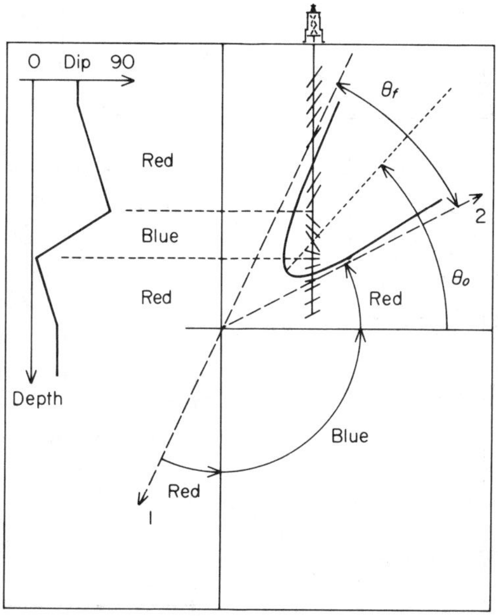

FIG. 1. Change in dip as a function of depth.

4. Equivalence classes and parameters

A set of manifestations can be partitioned into a collection of equivalence classes by specifying an equivalence relation over the manifestations. For example, let two manifestations be equivalent iff they consist of the same sequence of primitive patterns. Then one equivalence class would be the set of all manifestations that consist of red/blue composites. In this equivalence class, the length of the primitive patterns is not a determining factor for membership, because the equivalence relation does not require the lengths of corresponding primitive patterns within two red/blue composites to be equal in order for the two composities to be equivalent. For example, manifestations consisting of red/blue composites with short red patterns are equivalent, or qualitatively similar, to manifestations consisting of red/blue composites with long red patterns.

Associated with any situation or manifestation is a set of parameters or properties. A situation like a fold, for example, can be described by its fold angle θ_f (the angle between the flanks of the fold) and its orientation θ_o with respect to the earth's surface (see Fig. 1). Any fold may possess any value in the range of these parameters. The ISG uses such parameters to discover new subclasses of a given situation class. For example, the ISG discovers acute, right, and obtuse fold subclasses based on the value of their fold angle parameter. With these subclasses, values of the fold angle parameter are relevant to determining class membership.

Manifestation parameters are defined with respect to some partition of the set of manifestations. They are manifestation properties that can be used to distinguish, or index, individual members of the partition's equivalence classes. For the manifesta-

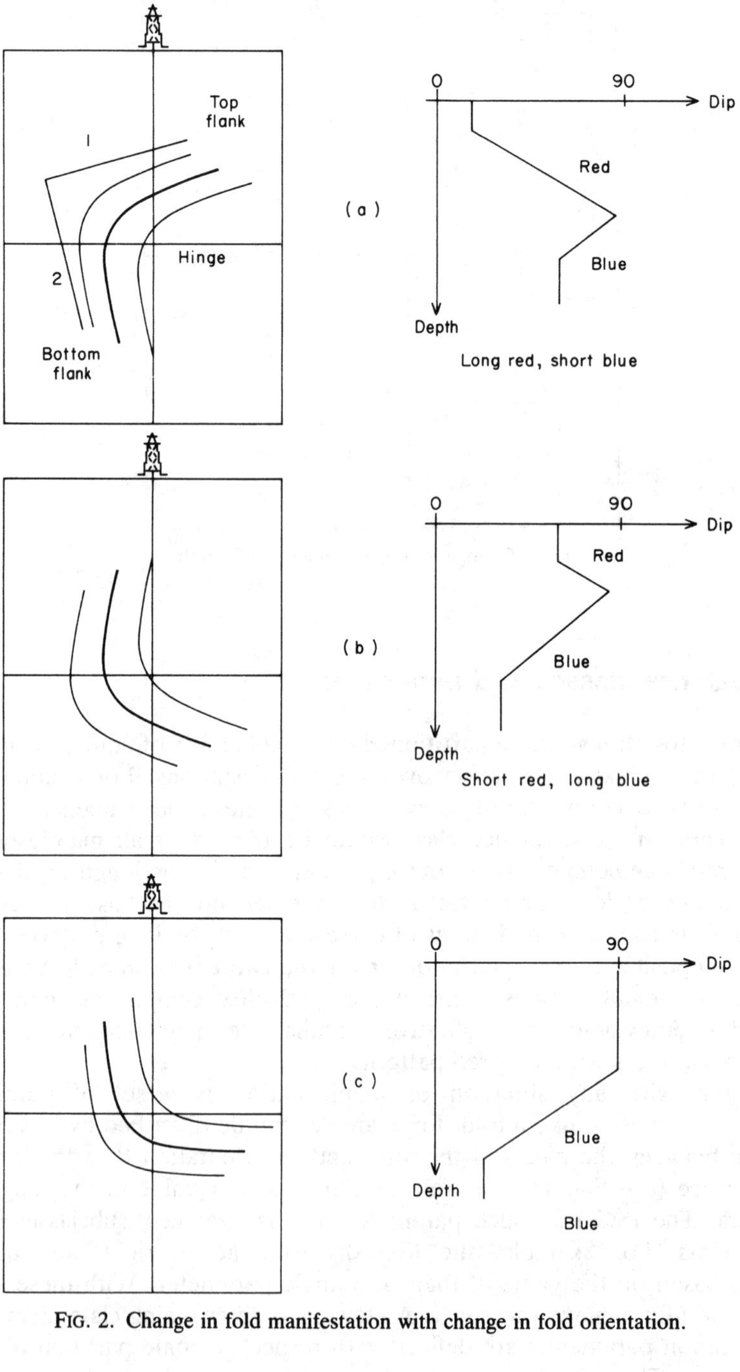

FIG. 2. Change in fold manifestation with change in fold orientation.

tion partition described above, we chose primitive pattern lengths to be the manifestation parameters. For example, in this case, a red/blue composite is a manifestation whose parameters are the lengths of its component red and blue patterns. Note that we could have alternatively chosen the top and bottom dips of the primitive patterns to be the manifestation parameters because they can also be used to distinguish manifestations within the same equivalence class. In the worked example, presented in section 5, top and bottom dips were used instead of pattern lengths because dip values have a lower bound (0°) and an upper bound (90°) whereas pattern lengths only have lower bounds (0). As explained below, this provides more constraints on determining the boundaries of situation equivalence classes.

The ISG algorithm requires that the *datatype* of each situation and manifestation parameter be an ordered set of values for which a metric is defined. That is, the domains of situation parameters and the ranges of manifestation parameters must be partially ordered sets (posets), and the distance between any two values of a parameter must be defined. In our example, the domain of fold angle θ_f is the interval (0°, 180°), the domain of fold orientations θ_o is the interval (0°, 90°), and the domain of primitive red and blue pattern lengths is the interval $(0, \infty)$. The order relation on these intervals is the standard ordering of the real numbers (i.e. $a \leq b$ iff $b - a$ is non-negative), and the metric is the usual metric for the real numbers (i.e. $d(a, b) = |b - a|$).

For the ISG to be applicable, the domain theory must define dependencies or causal relations between situation and manifestation parameters. Suppose that some value of a situation parameter, called a *critical value,* (e.g. $x = x^c$) causes a manifestation parameter to assume a value that represents the lower or upper bound of its range (e.g. $y = y_{min}$ or $y = y_{max}$). By definition, the manifestation having this extreme value for one of its parameters lies on the boundary of a manifestation equivalence class. Thus, situations with x less than the critical value x^c must give rise to a different set of manifestations than situations with x greater than x^c.

For example, suppose that a fold manifests itself as a red/blue composite [Fig. 2(a)]. As Fig. 2(b) shows, increasing the orientation of the fold causes a decrease in the length of the red pattern associated with the fold's manifestation. By continuing to increase the orientation to a critical value (θ_o^c) the length of the red pattern will eventually decrease to zero [Fig. 2(c)]. As a consequence, the fold now produces a *different* manifestation that consists of a single blue pattern.

In the course of applying rules to the knowledge base, the ISG generates dependency information about the way in which a change in a situation parameter affects a manifestation parameter. Returning to Fig. 2(a), suppose that increasing the orientation of the fold by one degree causes the length of the red pattern in the fold's manifestation to decrease by one meter (i.e. $\partial R_L/\partial \theta_o = -1$). Using this dependency, the initial values of the red pattern length (R_L^i) and fold orientation (θ_o^i), and the lower bound of the length of a red pattern (R_L^c), the ISG could determine the critical value of the orientation that produces the different manifestation in Fig. 2(c). That is $(R_L^c - R_L^i)/(\theta_o^c - \theta_o^i) = -1$ (assuming linearity). Therefore, $\theta_o^c = R_L^i - R_L^c + \theta_o^i$. In this way, the ISG discovers the boundaries of situation equivalence classes.

5. The ISG algorithm

5.1. INPUT/OUTPUT BEHAVIOR

This section explains how the ISG works by showing how it generates the tree of fold classes in Fig. 3. As shown in the hierarchy, Folds are classified as Acute, Right, or Obtuse depending on the value of their fold angle θ_f. Each of these subclasses can in turn be divided according to the fold's orientation θ_o.† Each of the terminal situation classes has a single manifestation m. For example, the manifestation of class Acute-2 (folds with fold angles between 0° and 90° and orientations between a and b) is a composite R/B/R pattern. The input/output behavior of the ISG is summarized in Table 1.

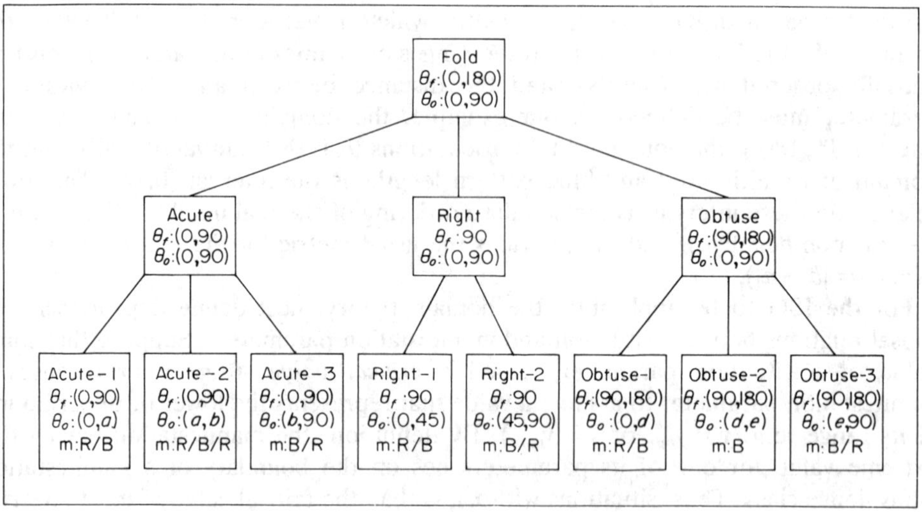

FIG. 3. Differentiated folds

TABLE 1
The ISG problem

Given:
Situation class S
Ordering of S parameters
Set of M parameters
Manifestation partition π_M
Linkage rules L
Find:
Situation partition π_S

Given a domain theory of linkage rules L that map situations in S into manifestations in M, a set of manifestation parameters, and an ordering of S's situation parameters, the ISG finds the partition π_S of S induced by a given partition

† As explained later, the variables a, b, c, d and e are functions of θ_f.

π_M of M:

$$S \xrightarrow{L} M$$

$$\pi_S \xleftarrow{\text{ISG}} \pi_M$$

In our example, the situation class S is the set of folds parameterized by fold angle θ_f and orientation θ_o (i.e. the Fold class shown in Fig. 3). The fold angle lies between 0° and 180°, and the fold orientation lies between 0° and 90°—the domains of these parameters. We will order these parameters such that the situation partition π_S produced by the ISG first differentiates S with respect to θ_o and then with respect to θ_f.† The manifestation class M is the set of primitive red and blue patterns together with composite patterns made up of sequences of red and blue patterns. The set of linkage rules L is a set of geologic principles that map folds in S into their corresponding manifestations in M. The set of manifestations is given a partition π_M, such that any two manifestations are equivalent iff they are composed of the same sequence of red and blue patterns. In this example, the top and bottom dips of the red and blue patterns comprising a manifestation will be used as the manifestation parameters. Given the above information, the ISG will find the partition π_S of S shown in Fig. 3.

The ISG will explore the situation parameter space shown in Fig. 4. Regions of the parameter space are labelled with the manifestations that they produce, and the situations corresponding to some representative points are illustrated in Fig. 5; where only the orientations of the top and bottom tangents with respect to the horizontal and vertical axes are shown. For example, Fig. 5(g) is a representation of the situation depicted in Fig. 1.

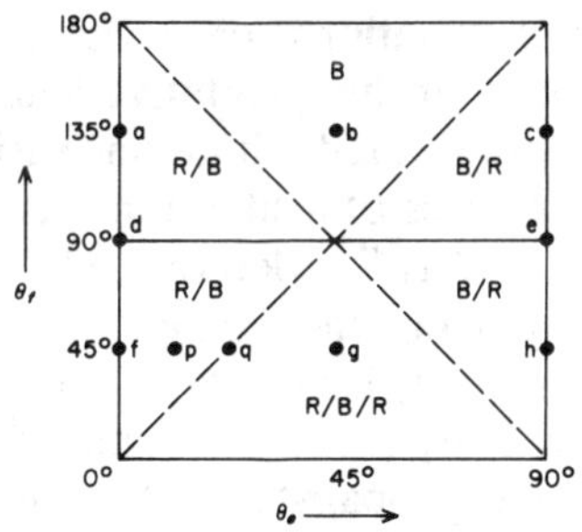

FIG. 4. Fold parameter space.

5.2. DEPENDENCIES BETWEEN PARAMETERS

As described in section 4, the ISG uses dependency information to find situation equivalence classes that correspond to given manifestation equivalence classes. Given an initial situation–manifestation pair (S_0, M_0), the ISG determines how a change in a situation parameter affects manifestations parameters. Using these

† If the order of these parameters was reversed, the ISG would grow a tree of fold classes such that the top-level differentiated folds with respect to θ_o and the bottom-level differentiated folds with respect to θ_f. The particular ordering selecting for our example was chosen so that the situation equivalence classes discovered by the ISG correspond to those typically used by geologists (see e.g. *Schlumberger Dipmeter Interpretation*, 1981).

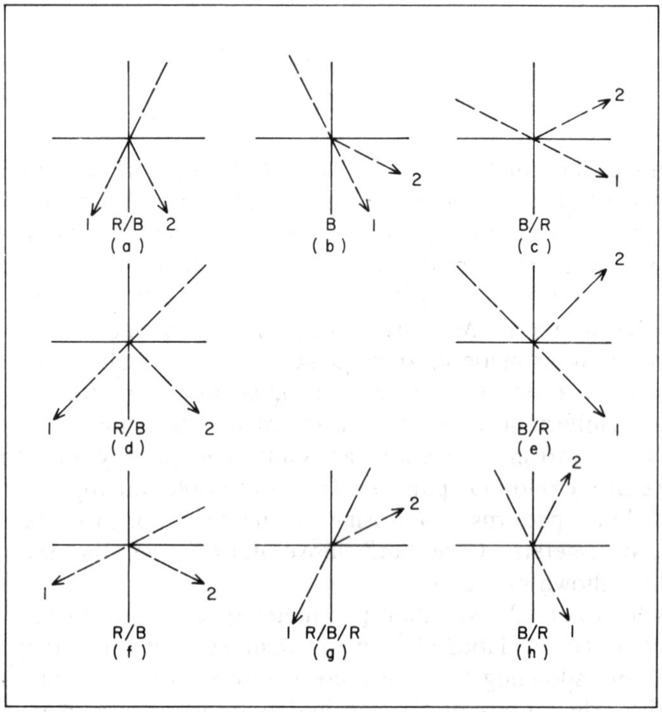

FIG. 5. Fold situations.

relationships, the ISG calculates the situation parameter variations, or deviations, that correspond to the largest variations, or deviations, that can be made in the manifestation parameters, such that the qualitative features of M_0 are preserved. By changing S_0 by these backpropagated situation variations, the ISG finds the situations that define the boundaries of a situation class.†

Dependency information is *locally* determined with respect to the values referenced and created by applying the linkage rules. That is, if a linkage rule computes a value of $y(x_1, x_2, \ldots x_n) = y_0$ based on a value of $x_i = x_{i,o}$, then the associated dependency is essentially $(\partial y/\partial x_i)$ evaluated at $x_i = x_{i,o}$, and tells us how deviations of x_i from $x_{i,o}$ would propagate to deviations in y from y_o; with other variables (i.e. $x_{j \neq i}$) held constant. For example, consider a rule for computing the area of a rectangle, $A = H \times W$, where the area is a function of height and width. Here, the dependency information relating H and A states that a change δH in H would cause a change δA in A such that $\text{sign}[\delta A] = \text{sign}[(\partial A/\partial H)\delta H] = \text{sign}[W]\,\text{sign}[\delta H] = \text{sign}[\delta H]$, and $|\delta A| = |(\partial A/\partial H)\delta H| = W\,|\delta H|$. Note that the change in A with respect to the change in H depends upon the value of W that obtained when the rule was applied.

5.3. GETTING STARTED

In the following example, for any manifestation M_i, the set of manifestations that are members of the same equivalence class as M_i will be denoted $[M_i]$ and will be

† As explained in section 5.4, the ISG also forward propagates maximum situation parameter variations in an effort to extend the boundaries of the equivalence class that are discovered by backpropagation.

called the equivalence class generated by M_i. Similarly, for any situation S_i, the set of situations that are members of the same equivalence class as S_i will be denoted $[S_i]$ and will be called the equivalence class generated by S_i. We define the boundary of $[S_i]$ as the pair of elements of $[S_i]$ possessing the extreme values (relative to the other members in $[S_i]$) of the parameter currently being differentiated. We denote this boundary as $\partial[S_i]$. One element will have the minimum value and the other element the maximum value of the parameter currently being differentiated. The boundary of $[M_i]$ (i.e. those elements of $[M_i]$ having the extreme values of the manifestation parameters) will be denoted as $\partial[M_i]$.

The ISG starts by randomly selecting an initial situation S_0 from the Fold class. Suppose this situation corresponds to point $p = (\theta_o^0, \theta_f^0) = (10°, 45°)$ in the situation parameter space. By applying the linkage rules L to S_0, the ISG determines that this situation produces a particular R/B manifestation M_0 with its associated manifestation parameter values. (In this example, we will consider four manifestation parameters $y_1 = R_{top}$, $y_2 = R_{bot}$, $y_3 = B_{top}$, and $y_4 = B_{bot}$.) The ISG also determines the dependencies between situation and manifestation parameters.

The ISG begins to grow (depth-first) the tree in Fig. 3 by creating the structure shown in Fig. 6. Node fold-1 represents what will become the Acute class and node fold-1-1 represents what will become the Acute-1 subclass. For now, node fold-1-1 represents the initial situation S_0.

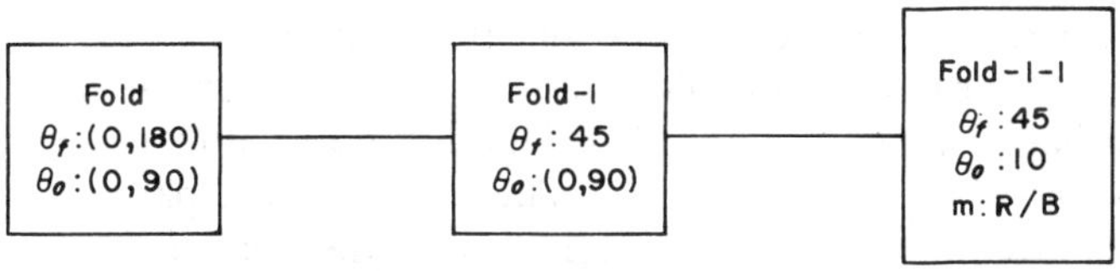

FIG. 6. ISG intermediate results—1.

5.4. FINDING THE FIRST SITUATION EQUIVALENCE CLASS

Given S_0 and M_0, the ISG will backpropagate the largest possible variations in M_0 and forward propagate the largest possible variations in S_0 to determine $\partial[S_0]$. *Maximum manifestation variations* are the largest changes that can be made to M_0's paarameter values such that the resulting manifestation is still qualitatively similar to M_0 (i.e. these changes correspond to the distance between M_0's parameter values and the parameter values of the manifestations that lie on $\partial[M_0]$). *Maximum situation variations* are the largest changes that can be made to S_0's parameter values such that the resulting situation is not an element of a *previously determined equivalence class.*

Figure 7 illustrates these definitions. Here x is a situation parameter whose domain is $[x_-, x_+]$, and y is a manifestation parameter whose range is $[y_-, y_+]$. Let situation S_0 be characterized by $x = x_0$ and let its corresponding manifestation M_0 be characterized by $y = y_0$. Maximum manifestation variations correspond to changes in y from y_0 to values of y on $\partial[M_0]$ (i.e. $y = y_\pm$). Maximum situation variations correspond to changes in x from x_0 to x_{min} and x_{max}. Here $[x_{min}, x_{max}]$ is the largest interval surrounding x_0 that does not overlap other equivalence classes. (If we are just starting to partition the x domain, and have not determined any equivalence classes, then $x_{min} = x_-$ and $x_{max} = x_+$).

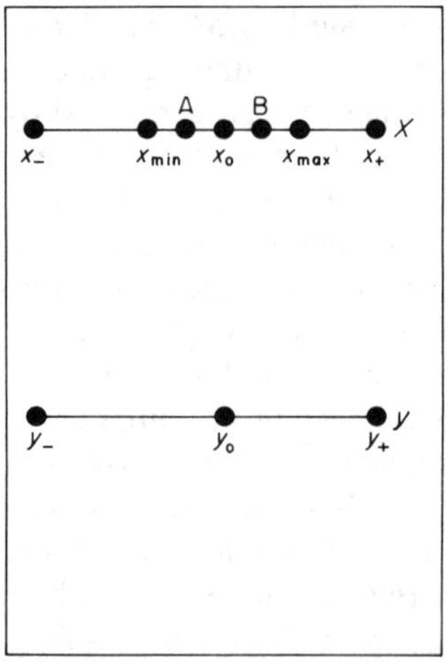

FIG. 7. ISG algorithm definitions.

Returning to our example, dependency information (generated when S_0 was mapped by linkage rules into M_0) records the way in which changes in the situation parameters are related to changes in the manifestation parameters. This information will be used to find the interval of orientations surrounding the orientation of S_0 (i.e. $\theta_o^0 = 10°$) such that any situation having an orientation within this interval and fold angle $\theta_f^0 = 45°$ will produce a manifestation qualitatively similar to M_0 (i.e. a R/B manifestation that is a member of $[M_0]$). Figure 8 shows how this is accomplished. The situation parameters are θ_o and θ_f. The manifestation parameters, R_{top}, R_{bot}, B_{top}, and B_{bot}, are the top and bottom dips of the red and blue patterns that comprise the R/B manifestation. We will hold θ_f constant at $\theta_f = 45°$ while θ_o is partitioned into an interval corresponding to $[S_0]$.

In Fig. 8, the situation parameter values of S_0, $(\theta_f^0, \theta_o^0) = (45°, 10°)$, and M_0's

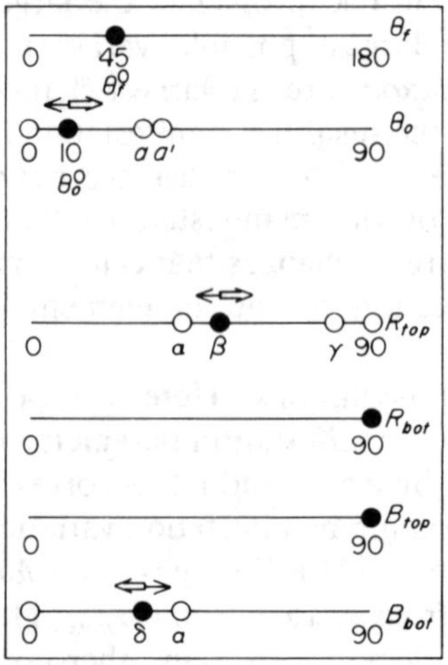

FIG. 8. Extending fold-1-1.

corresponding manifestation parameter values $(R_{top}, R_{bot}, B_{top}, B_{bot}) = (\beta, 90, 90, \delta)$ are indicated by the solid circles (●). The arrows above these points show how decreasing (←) or increasing (⇒) θ_o affects the manifestation parameter values. That is, they illustrate how variations in situation parameter values propagate to variations in manifestation parameter values. The manifestation parameters change in the direction indicated by the single arrows (double arrows) when θ_o is decreased (increased). For example, *increasing* θ_o from θ_o^0 causes R_{top} to *increase* from β and B_{bot} to *decrease* from δ. Note that R_{bot} and B_{top} are not affected by altering θ_o. As explained below, the open circles (○) show the results of forward propagating extreme situation parameter variations and backpropagating extreme manifestation parameter variations.

The equivaluence class $[S_0]$ will be determined in two steps.

(1) Elements of $\partial[M_0]$ will be *backpropagated,* using dependency information, to find elements of $\partial[S_0]$. By definition, manifestations on $\partial[M_0]$ are manifestations one of whose parameters has either the largest or smallest value of all manifestations in $[M_0]$. Situations on $\partial[S_0]$ are defined to be those situations in $[S_0]$ that correspond to manifestations on $\partial[M_0]$.

In Fig. 7, let the lower bound of the x interval that corresponds to $[S_0]$ be denoted by A and the upper bound of that interval be denoted by B. That is, these points characterize situations on $\partial[S_0]$. Then B is the minimum of the set of backpropagated y values that correspond to manifestations on $\partial[M_0]$, (i.e. the minimum of the set of backpropagated $y_\pm$ values). The minimum is taken over x values that lie in the interval $[x_0, x_{max}]$. Recall that this the largest interval of x values greater than x_o that do not already characterize previously determined situation classes. We take B as the minimum of the set of backpropagated values because if two extreme manifestation values $y_1 = y_1^*$ and $y_2 = y_2^*$ map into B_1 and B_2 such that $x_0 < B_1 < B_2$, then values of x between B_1 and B_2 would map into invalid values of y_1 (i.e. a value outside of the range of the manifestation parameter). By definition as x is increased from x_0 to B_1, y_1 goes from y_0 to an endpoint of its range y_1^*. Further increasing x from B_1 to B_2 would force y_1 to exceed y_1^*.

Similarly, A is the maximum of the set of backpropagated y values that correspond to manifestations on $\partial[M_0]$ (i.e. the maximum of the set of backpropagated $y_\pm$ values). The maximum is taken over x values that lie in the interval $[x_{min}, x_o]$. This is the largest interval of x values smaller than x_0 that do not already characterize previously determined situation classes. We take A as the maximum of the set of backpropagated values because if two extreme manifestation values $y_1 = y_1^*$ and $y_2 = y_2^*$ map into A_1 and A_2 such that $A_1 < A_2 < x_0$, then values of x between A_1 and A_2 would map into invalid values of y_2. By definition as x is decreased from x_0 to A_2, y_2 goes from y_0 to an endpoint of its range y_2^*. Further decreasing x from A_2 to A_1 would force y_2 to exceed y_2^*.

(2) The ISG tries to extend $\partial[S_0]$ (obtained by backpropagating extreme M_0 parameter variations in step (1)) by *forward propagating* extreme S_0 parameter variations. This is accomplished by checking to see if the largest or smallest values a situation parameter can assume (i.e. the endpoints of the largest interval surrounding a situation parameter value that does not include values characterizing other equivalence classes) map into valid values of all the manifestation parameters. If so, the extreme situation parameter value is taken as belonging to a situation on $\partial[S_0]$.

Again, referring to Fig. 7, forward propagation will extend A to x_{min} if this point maps into the interval of valid y values $[y_-, y_+]$ of all manifestation parameters. Similarly, forward propagation will extend B to x_{max} if this point also maps into the interval of valid y values.

Step 1—Backpropagation
Returning to our example (Fig. 8), the maximum manifestation parameter variations, corresponding to manifestations on $\partial[M_0]$, occur when R_{top}, R_{bot}, B_{top}, and B_{bot} equal 0° or 90°. These values are backpropagated into the interval of unexplored θ_o values [0°, 90°] as follows.

R_{bot}, $B_{top} = 0°, 90°$. Using dependency information, the ISG determines that these parameters remain *constant* with variations in θ_o. Therefore, forward and backward propagation of maximal variations does not take place between θ_o and these manifestation parameters. Thus, these manifestation parameters do not play a role in determining $[S_0]$.

$R_{top} = 0°$, $B_{bot} = 90°$. However, the elements in $[M_0]$ can be distinguished by their R_{top} and B_{bot} values. By definition, the elements of $[M_0]$ that lie on the $\partial[M_0]$ have the largest or smallest values of these parameters. If $R_{top} = 0°$ characterizes one of these boundary points, then it will backpropagate into a valid value of θ_o characterizing a situation on $\partial[S_0]$. That is, if $R_{top} = \beta$ suffers the largest possible parameter variation such that it should equal 0°, then the corresponding change in θ_0 should put its value between 0° and 90°. However, $R_{top} = 0°$ would cause θ_o to be less than 0° and lie outside the domain of valid fold orientations. Thus, a manifestation with $R_{top} = 0°$ does not lie on $\partial[M_0]$ and cannot be used to determine a θ_o value of a situation on $\partial[S_0]$. Similarly, if $B_{bot} = 90°$ for an element of $\partial[M_0]$, then θ_o would also be less than 0° for the corresponding element of $\partial[S_0]$. Because this lies outside the domain of θ_o, there is no manifestation with $B_{bot} = 90°$ on $\partial[M_0]$, and this point is discarded in looking for $\partial[S_0]$.

$R_{top} = 90°$, $B_{bot} = 0°$. However, $R_{top} = 90°$ backpropagates into $\theta_o = a'$, and $B_{bot} = 0°$ backpropagates into $\theta_o = a$; where $\theta_o^0 < a < a'$. The *smallest* of these two backpropagated values *greater* than θ_o^0 is selected as the upper bound of θ_o values for situations in $[S_0]$. This is because values of θ_o between a and a' would map into *valid* values of R_{top} between some value γ and 90°, but would also map into *invalid* values of B_{bot} less than 0°.

The ISG has at this point discovered the upper bound of the interval surrounding θ_o^0 that contains values of θ_o belonging to elements of $[S_0]$. This upper bound is $\theta_0 = a$, and characterizes a situation on $\partial[S_0]$. Although backpropagation did not extend the interval below θ_o^0, the next step will.

Step 2—forward propagation
The maximum situation parameter variations, corresponding to situations on $\partial[S_0]$, occur when θ_0 equals 0° or 90°. These values are forward propagated into the range of manifestation parameter values as follows.

$\theta_0 = 0°$. The smallest value of θ_0 less than 10° that is not already included in another situation equivalence class is 0°. Using dependency information, the ISG determines that this value maps into $R_{top} = \alpha$, and $B_{bot} = \alpha$. These are valid values for manifestation parameters. Therefore, the interval $(10°, a)$ is extended to $(0°, a)$. We now know that $[S_0]$ situations with θ_0 in $(0°, a)$ and $\theta_f = 45°$ map into the same equivalence class of R/B patterns. Note that, in general, a will depend on θ_f.

$\theta_0 = 90°$. The largest value of θ_o greater than 10° that is not already included in another situation equivalence class is 90°. We do not have to forward propagate this value because we have already extended the interval to the right of θ_o^0 by a smaller amount. That is, we already determined that a situation with $\theta_o = a$; $\theta_o^0 < a < 90°$ lies on $\partial[S_0]$. If we forward propagated $\theta_o = 90°$, it would have mapped into invalid values of the manifestation parameters.

5.5. FINDING OTHER PROGENCY OF FOLD-1

At this point, we can extend the tree of partial results shown in Fig. 6 into that shown in Fig. 9 by extending the orientation of the fold-1-1 class from $\theta_o = 10°$ to $\theta_o \in (0°, a)$.

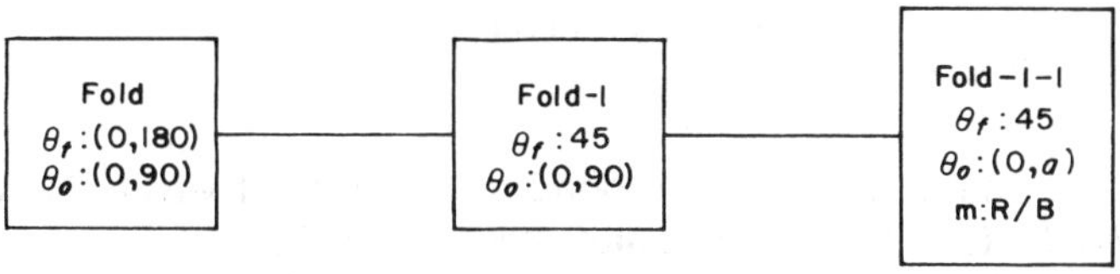

FIG. 9. ISG intermediate results—2.

To continue partitioning the domain of θ_o, a new child of the fold-1 class (i.e. fold-1-2) is created having a θ_o value that lies in the unexplored region outside of $(0°, a)$. For example, this new situation S_1 might correspond to $(\theta_o^1, \theta_f^1) = (40°, 45°)$; where $40° > a$. If the method by which situation S_0 was extended to $[S_0]$ is applied to situation S_1, the ISG will find that members of $[S_1]$ have values of θ_o in the interval (a, b); where $a < b < 90°$. Here, b (like a) depends upon θ_f. Because the fold-1-1 and fold-1-2 subclasses of fold-1 do not yet cover the entire domain of fold-1 orientations, a third child of the fold-1 class (i.e. fold-1-3) is created that represents a situation S_2 having a θ_o value in the interval $(b, 90°)$. Node fold-1-3 is expanded into $[S_2]$. It turns out that values of θ_o for members of this equivalence class lie in the interval $(b, 90°)$. At this point, the ISG has found a complete set of fold-1 subclasses such that each subclass corresponds to a single manifestation (see Fig. 10). The intervals $(0°, a)$, (a, b), and $(b, 90°)$ cover the entire domain of θ_o.

5.6. EXTENDING THE FOLD-1 CLASS

The ISG now tries to extend the fold-1 class by looking for an interval surrounding $\theta_f^0 = 45°$ such that as θ_o is varied from 0° to 90° the same *sequence* of manifestations (R/B, R/B/R, B/R) is produced. That is, the ISG tries to enlarge fold-1 to include situations with fold angles other than 45° that have the same set of subclasses fold-1-1, fold-1-2, and fold-1-3. The technique for enlarging the fold-1 class is similar to the method we used to enlarge the fold-1-1, fold-1-2, and fold-1-3 subclasses. In this case, θ_f is the situation parameter and the *lengths of the θ_o intervals* occupied by the R/B, R/B/R, and B/R equivalence classes are the manifestation parameters.

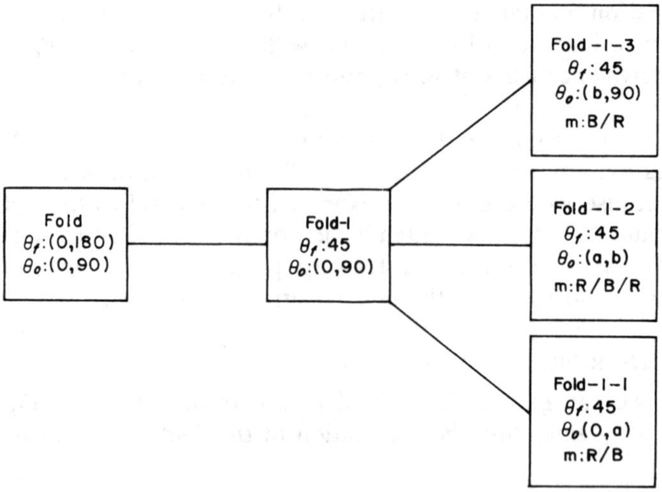

FIG. 10. ISG intermediate results—3.

The fold-1 class boundaries will correspond to θ_f values at which the length of a θ_o interval of a *R/B, R/B/R,* or *B/R* manifestation vanishes.†

After extending the fold-1 class, the ISG's intermediate results are shown in Fig. 11. The interval of θ_f values (0°, 90°) characterize a situation equivalence class that

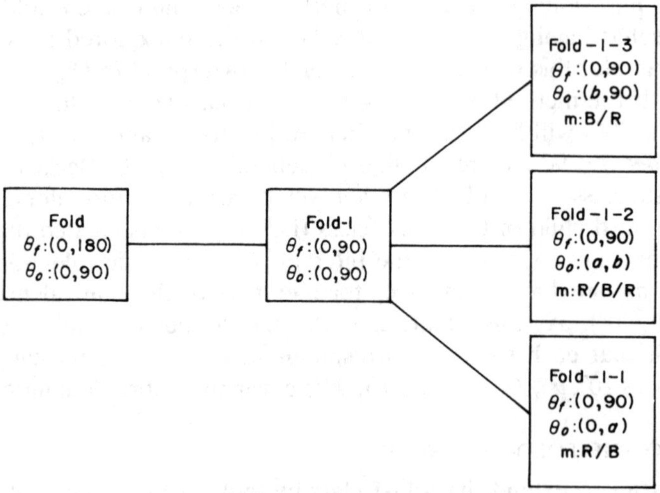

FIG. 11. ISG intermediate results—4.

produces the same sequence of manifestations (R/B, R/B/R, B/R) as θ_o is varied through its domain from 0° to 90°. In other words, we have distinguished Acute folds as the set of situations whose manifestations go through the sequence R/B→R/B/R→B/R as their orientation is increased from 0° to 90°.

† Wisniewski, Winston, Smith & Kleyn (1986) describe this part of the algorithm in more detail.

5.7. COMPLETING THE EXAMPLE

At this point, the ISG has grown the branch of Acute folds in Fig. 3; where fold-1 corresponds to Acute, fold-1-1 corresponds to Acute-1, fold-1-2 corresponds to Acute-2, and fold-1-3 corresponds to Acute-3. These folds have fold angles θ_f between 0° and 90°. It remains for the ISG to complete the differentiation of situations in the top-level fold class that have fold angles in the interval (90°, 180°).

This is accomplished by randomly selecting a fold angle outside the fold-1 class and creating a new child (i.e. fold-2) with this value of θ_f. Suppose this fold angle is 100°. An instance (i.e. fold-2-1) of fold-2 is also created that has a specific orientation θ_o between 0° and 90°. This branch of the fold tree is extended by the ISG into the branch of Obtuse folds in the same way as the original branch shown in Fig. 3 was extended into the branch of Acute folds. It turns out that this new branch includes situations having fold angles between 90° and 180°.

A third child of the fold class (i.e. fold-3) is created having the remaining unexplored fold angle $\theta_f = 90°$. An instance (i.e. fold-3-1) of fold-3 is also created with a specific orientation between 0° and 90°. The ISG extends this branch of fold situations into the branch of Right folds shown in Fig. 3.

At this point the ISG has finished partitioning the initial set of fold situations. The output of the algorithm is the tree structure shown in Fig. 3; except that the progency of fold have labels of the form fold-k, and the progeny of each of these children have labels of the form fold-k-1.

6. Conclusion

6.1. SUMMARY

Given a situation class in a domain model, the ISG produces conceptual clusterings of its members that constitute equivalence classes corresponding to distinct manifestations. That is, a given situation class is divided into more specific subclasses. In this sense, the ISG assists in completing a domain model. If a given situation class already maps into a single manifestation, the ISG makes no further divisions in the situation class hierarchy. Thus, the ISG picks up where the domain mode leaves off and continues to differentiate situation classes until each of them corresponds to a single manifestation.

When a complete partitioning of the situations already exists, the ISG can *verify* its correctness. In particular, the output of the ISG might be used to look for cases where two or more situation classes correspond to the same manifestation. Here, classification of a situation instance is ambiguous based on its manifestation. In this way, the ISG focuses attention on those ambiguous situation classes that require further differentiation.

6.2. CURRENT STATUS

The ISG can succesfully differentiate folds with orientations between 0° and 90° and fold angles between 0° and 180° into three equivalence classes. It discovers that: (i) *acute* folds (those with fold angles less than 90°) produce red/blue, red/blue/red, and blue/red patterns; (ii) *right angle* folds (those with folds angles equal to 90°) produce red/blue and blue/red patterns; and (iii) *obtuse* folds (those with fold angles greater than 90°) produce red/blue, blue, and blue/red patterns. In another test of the algorithm, the ISG successfully differentiated folds with orientations between 90° and 180° and fold angles between 0° and 180°.

In its current form, the ISG program is implicitly tied to the fold domain. It has not been applied to other problems, because its inputs (see Table 1) are not explicitly specified. For example, when applied to the fold problem, the program assumes the specific manifestation partition that distinguishes, for example, R/B from B/R patterns. The program will not run with a manifestation partition that only distinguishes, for example, complosite patterns based on the number of their constituent primitive patterns so that R/B and B/R both belong to the same two-primitive-pattern equivalence class.

6.3. LIMITATIONS

The ISG starts with dependency information that describes how local changes in situation parameters affect manifestation parameters. During forward and backward propagation, it uses this information to extrapolate the value of the manifestation parameters that result from larger changes in the value of situation parameters, and vice versa. This information is the basis for determining an equivalence class of situations that produce equivalent manifestations. The extrapolation is correct only in those cases where the situation and manifestation parameters are linearly related. Thus, the accuracy of the situation partition produced by the ISG depends upon the extent to which manifestation parameter values are linear functions of situation parameter values in the context of backpropagating each manifestation equivalence class.

The ISG is exponential in the number of undifferentiated situation parameters. That is, it is $O(P^N)$; where P is the average number of intervals into which a situation parameter domain is partitioned, and N is the number of situation parameters. This result implies that the ISG can usefully be employed as a way of elaborating an almost-complete domain theory. It is not a practical way of partitioning situation classes that have more than a few undifferentiated parameters.

6.4. FUTURE PLANS

We plan to isolate the ISG algorithm from the context of the folding problem. Once this is done, more of the ISG's inputs can be explicitly specified and independently varied. For example, we plan to experiment with the ISG by checking the results of varying: (i) the order of situation parameters; (ii) the definitions of manifestation partitions; and (iii) the content of the domain theory that maps situations into manifestations.

We have only applied the ISG to situations involving two parameters (e.g. folds with the parameters orientation and fold angle). After generalizing the program to handle any number of situation parameters, we plan to reapply it to the fold problem by adding another situation parameter—the orientation of the borehole. In this paper, we assumed the borehole had a vertical orientation.

As a weak-method, the ISG suffers from the limitations mentioned in section 6.3. We need to understand better the extent to which these limitations constrain its utility. In particular, more experience and analysis is required in order to find out whether the ISG always terminates its search for interesting situations, and if it does not—under what conditions would one expect to observe divergent behavior. We need to have a better understanding of how the local nature of dependency information compromises the accuracy with which the ISG partitions situation classes. If this will usually be the case, then we need to know whether it will be possible to and under what circumstances can we recover from such initial partitioning errors.

References

BUCHANAN, B. G., MITCHELL, T. M., SMITH, R. G. & JOHNSON, C. R. (1978). Models of learning systems. *Encyclopedia of Computer Science and Technology,* **11,** 24–51.

BUCHANAN, B. G. & SHORTLIFFE, E. H. (Eds) (1984). *Rule-based Expert Systems: the MYCIN Experiments of the Stanford Heuristic Programming Project.* Reading, MA: Addison–Wesley.

FISHER, D. & LANGLEY, P. (1985). Methods of conceptual clustering and their relation to numerical taxonomy, *Technical Report 85-26,* Department of Information and Computer Science, Irvine, CA.

LENAT, D. B. (1982). AM: discovery in mathematics as heuristic search. In DAVIS, R. & LENAT, D. B. Eds, *Knowledge-based Systems in Artificial Intelligence,* **1-225**. New York, N.Y: McGraw–Hill, Inc.

MICHALSKI, R. S. & STEPP, R. E. (1983). Learning from observation: conceptual clustering. In MICHALSKI, R. S., CARBONELL, J. G. & MITCHELL, T. M. Eds, *Machine Learning,* Vol. 1. Palo Alto, CA: Tioga Publishing Company.

MITCHELL, T. M. (1983). Learning and problem solving. *Proceedings of the Eighth International Joint Conference on Artificial Intelligence,* pp. 1139–1151.

MITCHELL, T. M., MAHADEVAN, S. & STEINBERG, L. I. (1985). LEAP: a learning apprentice system for VLSI design. *LCSR Technical Report 64,* Rutgers University Department of Computer Science, New Brunswick, N.J.

MITCHELL, T. M., UTGOFF, P. E. & BANERJII, R. B. (1983). Learning by experimentation: acquiring and refining problem-solving heuristics. In MICHALSKI, R. S., CARBONELL, J. G. & MITCHELL, T. M. Eds, *Machine Learning,* Vol. 1. Palo Alto, CA: Tioga Publishing Company.

SMITH, R. G. & YOUNG, R. L. (1984). The design of the dipmeter advisor system. *Proceedings of the ACM Annual Conference,* pp. 15–23.

SMITH, R. G., WINSTON, H. A., MITCHELL, T. M. & BUCHANAN, B. G. (1985). Representation and use of explicit justifications for knowledge base refinement. *Proceedings of the Ninth International Joint Conference on Artificial Intelligence,* pp. 673–680.

SWARTOUT, W. R. (1983). XPLAIN: a system for creating and explaining expert consulting programs. *Artificial Intelligence,* **21,** 285–325.

WINSTON, H. A., SMITH, R. G., KLEYN, M., MITCHELL, T. M. & BUCHANAN, B. G. (1986). Learning apprentice systems research at Schlumberger. In MITCHELL, T. M., CARBONELL, J. G. & MICHALSKI, R. S. Eds, *Machine Learning: a Guide to Current Research.* Booston, MA: Kluwer Academic Publishers.

WISNIEWSKI, E. J., WINSTON, H. A., SMITH, R. G. & KLEYN, M. (1986). Case generation for rule synthesis. *Proceedings of the First Annual Knowledge Acquisition for Knowledge-Based Systems Workshop.*

Explanation-based learning for knowledge-based systems

MICHAEL J. PAZZANI
UCLA Artificial Intelligence Laboratory, 3531 *Boelter Hall, Los Angeles, CA 90024 and The Aerospace Corporation, P.O. Box* 92957, *Los Angeles, CA* 90009, *U.S.A.*

We discuss explanation-based learning for knowledge-based systems. First, we identify some potential problems with the typical means of acquiring a knowledge base: interviewing domain experts. Next, we review some examples of knowledge-based systems which include explanation-based learning moduluses and discuss two systems in detail: ACES (Pazzani, 1986*a*) which learns heuristics for fault diagnosis from device descriptions, and OCCAM (Pazzani, 1986*b*) which learns to predict the outcome of economic sanction episodes from simple economic theories. We conclude that explanation-based learning is a promising approach to constructing knowledge-based systems when the required information is available but not in the form of heuristic rules. In this case, the role of explanation-based learning is to explicate heuristics which are only implicit in deep models.

Introduction

The typical image of the development of a knowledge-based system is that a domain expert is interviewed to explain how he solves a problem in a given area. In the interview, the expert reveals heuristics which can be encoded as rules in a knowledge-based system. For some applications, there are a number of potential difficulties with the development of knowledge bases in this manner. Many of these difficulties could be avoided if the rules for a knowledge-based system could be learned. Potential problems which must be addressed in the development of knowledge-based systems include:

Availability of experts

Domains which are selected for development of knowledge-based systems are often those domains with few experts. By definition, an expert is someone who performs extremely well in a problem area. It may be more important for the expert to continue to solve problems in his area than to devote time to the development of a knowledge-based system.

Existence of experts

In some areas, there may not exist experts with enough experience to possess the heuristics required by knowledge-based systems. Consider how knowledge-based systems in medicine are developed: typically a physician will report on the symptoms of various diseases and possibly the certainty of having a disease given a set of symptoms. This empirical association between symptoms and diseases summarizes knowledge gained through years of experience. Now, consider if the same

KNOWLEDGE-BASED SYSTEMS Vol. 1
ISBN 0-12-273251-0

methodology could be applied to a system to diagnose faults in a new satellite. The experts in this case are not really experienced in diagnosis of this particular satellite. The expert most probably will be someone who is experienced with diagnose of a different satellite, or a designer of the satellite. It will be a difficult task for either type of expert to give empirical associations between atypical behavior and faulty components. This difficulty will be present in any new system. The problem is especially serious if only a few systems are manufactured. There may never be enough failures to develop a human expert to "knowledge engineer".

EXPERTS' ABILITY TO REPORT ON THE PROCESSES USED IN MAKING A DECISION

A critical assumption in developing a knowledge-based system by interviewing an expert is that an expert is able to reveal the heuristics he uses to make decisions. In some areas, such as vision and natural language understanding, people do not even attempt to explain their decisions. For example, consider the sentence "Many words have more than one sence". How did you know the word "sence" is "sense" misspelled and means "meaning" rather than "capacity to perceive stimuli" or any other meaning? For those who answer the previous question: were you consciously aware of this when making the decision or are you really answering the question "How could one (rather than how did I) make the decision"?

Even when an expert is willing to give an explanation for making a decision, can he really tell us how he made the decision? When a stock broker recommends investing in United Hotcakes Preferred does he have conscious access to the processes he used to come to the conclusion that this stock is a good investment? The psychological evidence is mixed. For example Berl, Lewis & Morrison (1976) students verbal reports of their decision criteria for deciding on a college to attend matched well with their apparent actual criteria. However a number of discrepancies have been found between the information reportedly used to influence a decision and the information actually used:

Failure to report an influential factor

For example, in one experiment (Nisbett & Wilson, 1977), subjects were given electric shocks of increasing intensity. Some subjects were given a placebo pill which they were told would produce hand tremors, butterflies in the stomach (i.e. the symptoms of electric shock). This experiment was performed to demonstrate that those subjects who attributed the symptoms of the shock to the pill would tolerate more shock than those who did not. In fact, the subjects who received the placebo pill withstood an average of four times the amperage of those who did not. However, 75% of the subjects who received the placebo pill and tolerated an extreme amount of shock did not report the pill as the explanation.

Erroneous reporting of non-influential factors

In one experiment (Nisbett & Ross, 1978), some subjects viewed a film while a power saw made a distracting noise and the control subjects viewed the film with no distraction. Subjects then related the film according to how interesting they thought it to be, how much they thought others would be affected by it and how sympathetic they found the main character to be. Although there was no detectable difference of the rating between the two groups, 55% of the subjects who were distracted by the power saw felt that the distraction had lowered their ratings.

Ericson & Simon (1984) have proposed a model which accounts for some of the conditions which lead to excluding relevant information from protocols as well as including irrelevant information. At best, however, information arrived at by interviewing domain experts is only the tip of the iceberg. At worst, this information can be misleading when the expert is actually answering the questions "How might one solve this sort of problem?" rather than "How did I solve this problem?".

CHANGES IN THE UNDERLYING THEORY REQUIRE CHANGES IN THE KNOWLEDGE-BASE

Consider the problem of several companies which built financial knowledge-based systems. After spending a considerable amount of time and effort interviewing tax experts, the tax laws have changed. More interviews with tax experts are now needed to encode the implications of the new tax code as heuristics which indicate financial strategies. In general, when changes in the underlying theory occur, the knowledge-acquisition process may need to be repeated to encode changes as new heuristics.

One way to avoid the above potential problems is to eliminate the knowledge engineer's task of interviewing an expert by having the machine learn its own heuristics. We are exploring this issue in a number of contexts: ACES (Pazzani, 1986*a*) which identifies faults in the attitude control system of the DSCS-III satellite† and OCCAM (Pazzani, 1986*b*) which predicts the outcome of economic sanction episodes. Both of these programs use explanation-based learning techniques. For example, in ACES, rather than *inducing* diagnosis heuristics (i.e. empirical associations between symptoms and device failures) from a number of training examples (Michie, 1983), diagnosis heuristics are *deduced* as needed from device models which describe the system functionality and connectivity.

The idea of explanation-based learning of heuristic rules fits in well with Steels' notion of Second Generation Expert Systems (Steels & Van de Veide, 1986). For a variety of reasons, including enhanced explanation capabilities, and robustness, Steels argues that expert systems should contain both "shallow" associations (i.e. heuristics) which allow for rapid solutions to commonly encountered problems and "deep" causal models which enable the system to solve uncommon or even unanticipated problems with general search techniques. An important advantage of second generation expert systems lies in the area of knowledge acquisition. After searching for a solution to a problem in the deep model, a heuristic can be created. Explanation-based learning is one such technique for creating heuristics.

Explanation-based learning

Let us briefly examine the problem of how explanation-based learning might be used to create heuristics for financial expert systems. In the proposed new tax law, the interest expense on a home mortgage is tax deductible, but the interest on an

† The attitude control system is responsible for detecting and correcting deviations from the desired orientation of the satellite.

automobile loan is not. A suitable representation of the tax law would serve as a deep model. Let us assume that the problem that the financial expert system is intended to solve is to maximize net income. One aspect of maximizing net income is minimizing income taxes. A general search program, with operators such as take out an automobile loan, and take out a second mortgage, would find that taking out a second mortgage on one's home to pay for an automobile would result in lower income taxes. After this expensive general search is done once, the results of the search can be saved as a heuristic:

IF the goal is to minimize taxes,
AND the client wants to buy an automobile
THEN take out a second mortgage to pay for the car

In addition, the conditions which are necessary to perform the operator (take out a second loan), can be propagated to the heuristic as additional preconditions (Minton, 1984):

IF the goal is to minimize taxes,
AND the client wants to buy an automobile
AND the client owns a home
AND the equity in the home is greater than the price of the car
THEN take out a second mortgage to pay for the car

Explanation-based learning systems (DeJong, 1983, Mitchell, Kedar-Cabelli & Keller, 1986) share a common approach to generalization. First, an example problem is solved producing an explanation (occasionally called a justification, or a proof) which indicates what information (e.g. features of the example and inference rules) was needed to arrive at a solution. Next, the example is generalized by retaining only those features of the example which were necessary to produce the explanation. Various systems differ according to the problem solved, and who does the problem-solving. For example, in LEAP (Mitchell, Mahadevan & Steinberg, 1986) a user designs a VLSI circuit to achieve some specified functionality. LEAP produces a justification which indicates how the circuit implements the specified function. In OCCAM, the reason that an economic sanctions incident failed or succeeded to achieve the desired effect determines which features of the incident should be generalized. In ACES, an explanation is produced which indicates why a fault proposed by a fault diagnosis heuristics was not confirmed by device models.

Fault diagnosis

Two different approaches have been used for fault diagnosis. In one approach (Davis, Shrobe *et al.* 1982, Genesereth, Bennett & Hollander 1981, Scan, 1985), the observed functionality of devices are compared with their predicted functionality which is specified by a quantitative or qualitative model of the device (de Kleer & Brown, 1984; Kuipers, 1984; Forbus, 1984). For a large system, such as a satellite, with a number of rapidly changing data values, comparing observed to predicted functionality can be inefficient. The alternative approach (Shortliffe, 1976; Nelson, 1982; Wagner, 1983) encodes empirical associations between unusual behavior and faulty components as heuristic rules. This approach requires extensive debugging of the knowledge base to identify the precise conditions which indicate the presence of a particular fault. In previous work, (Pazzani & Brindle, 1985, 1986) we have described the Attitude Control Expert System (ACES) in which these two approaches are integrated. Heuristics examine the atypical features and hypothesize potential faults. Device models confirm or deny hypothesized faults. Thus, heuristics focus diagnosis by determining which device in a large system might be at fault. Device models determine if that device is indeed responsible for the atypical features.

Here, we address the problem of revising the fault-diagnosis heuristics when they hypothesize a fault which is later denied. This occurs when all of the possible exceptions to a heuristic are not explicitly stated. When a fault is proposed, and later denied by device models, the heuristic which suggested the fault is revised so that the hypothesis will not be proposed in future similar cases.

ACES learns how to avoid a hypothesis failure after just one example. It does this by finding the most general reason for the hypothesis failure. Device models provide an explanation for the hypothesis failure. The device models indicate which features would have been needed to be present (or absent) to confirm the hypothesis. Explanation-based learning improves the performance of ACES by creating fault diagnosis heuristics from information implicit in the device models.

EXPLANATION-BASED LEARNING OF FAULT-DIAGNOSIS HEURISTICS

The basic idea behind the learning of fault-diagnosis heuristics is that simulating a fault with device models will result in a number of predictions. If these predictions are not present in a particular system, the fault which was simulated can be ruled out. When a hypothesized fault is not confirmed, in addition to ruling out the fault in the current case, the heuristic which suggested the fault can be revised to not propose the fault in future similar cases. The top-level diagnosis algorithm is presented in Fig. 1.†

```
To diagnose a fault in circuit,
First, suggest a fault with diagnosis heuristics,
Then confirm (or deny) the fault with device models.

diagnose (Circuit, Fault):-
        problem (Fault, Rule_id),
        confirm (Circuit, Fault, Rule_id).
```

FIG. 1. Top-level diagnosis algorithm.

A simple example of explanation-based learning of diagnosis heuristics should help the presentation of the learning algorithm. Consider the circuit in Fig. 2. The circuit consists of two light bulbs (a parking light and a taillight) in parallel protected by a fuse. ACES utilizes three types of data:

(1) *Diagnosis heuristics*: shallow associations between atypical data values and faults. Two simple diagnosis heuristics which define a blown fuse and a burnt out light bulb are illustrated in Fig. 3. These diagnosis heuristics are

† This version of ACES is implemented entirely in PROLOG. By providing English descriptions in addition to PROLOG rules, I hope that anyone will be able to read this section, and that there is enough detail for those who wish to experiment with ACES.

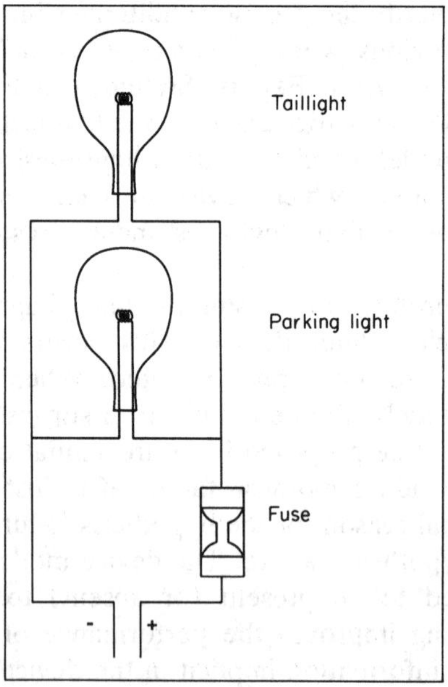

FIG. 2. A simple circuit consisting of two light bulbs in parallel protected by a fuse.

interpreted by a PROLOG meta-interpreter. In the second rule, the term:

device__functionality (L,__Any__input, not__working),

gives the modified qualitative functionality for a burnt-out light bulb: Given any input, the light bulb does not work. The normal functionality for a light bulb:

device__functionality (L,__Any__input,__Any__input),

indicates that if the input to a light is working, then the light bulb is working, and if the input is not working, the light bulb will not work.

(2) *Device models*: a description of the connectivity of a circuit and the functionality of the components. Figure 4 illustrates the definition for the circuit in Fig. 2.

```
IF a device is not working
THEN the fuse of the device's circuit is blown.
symptom (device__functionality (F,__Any__input, not__working), rule__001) ⇐
        value (D, not__working) &
        isa (D, device) &
        connected (F, D) &
        isa (F, fuse).

IF a light is not working
THEN the light is burnt out.
symptom (device__functionality (L,__Any__input, not__working), rule__002) ⇐
        value (L, not__working) &
        isa (L, light).
```

FIG. 3. Initial diagnosis heuristics.

```
circuit (c1, [battery_1, fuse_1, taillight, parking_light, ground_1)].
connected battery_1, fuse_1).
connected (fuse_1, taillight).
connected (fuse_1, parking_light).
connected (taillight, ground_1).
connected (parking_light, ground_1).
isa (battery_1, battery).
isa (ground_1, ground).
isa (fuse_1, fuse).
isa (taillight, light).
isa (parking_light, light).
isa (light, device).
isa (fuse, device).
device_functionality (fuse, Input, Input).
device_functionality (light, Input, Input).
```

FIG. 4. Description of functionality and connectivity for the circuit in Fig. 2.

(3) *Measurements*: Observed data from the device. For this example, let us assume that the battery is working, the taillight is working and the parking light is not working (see Fig. 5).

```
value (taillight, working).
value (parking_light, not_working).
value (battery_1, working).
```

FIG. 5. Initial data.

When diagnosis starts, the first rule in Fig. 3 (rule_001) will suggest that the fault is a blown fuse (fuse_1) since a device (parking_light) protected by the fuse is not working. The next step is to confirm or deny the fault using device models. A qualitative simulation of the circuit with the fuse blown yields a prediction that the taillight will also not be working. This prediction is denied since the taillight is in fact working. Therefore, the hypothesis that the fuse is blown can be ruled out. In addition, the rule which proposed the hypothesis can be revised so that the hypothesis failure does not occur in future similar cases. (After rule_001 is revised, rule_002 suggests that the bulb of the parking light is burned out. This fault is confirmed by the device models.) To illustrate how this learning occurs and what exactly is meant by "similar cases", it is necessary to describe the confirmation process in more detail (see Fig. 6).

```
To confirm a fault:
First generate a prediction,
Next see if the prediction is false
If the prediction is false,
        revise the rule which generated the prediction
        and rule out the fault
        (otherwise, try to generate a new prediction)
Succeed when there are no more predictions are generated.

confirm (Circuit, Fault, Rule_id):-
        prediction (Circuit, Fault, Prediction, Justification),
        denied (Prediction)
        fix_rule_failure (Rule_id, Fault, Prediction, Justification)
        !,
        fail.
confirm (_,_,_).
```

FIG. 6. Confirmation: a failed prediction initiates the learning process.

When the qualitative simulation generates a prediction, it also returns a *justification* of how the prediction was arrived at. This justification consists of the conditions which were needed to establish the prediction. In the current example, the justification consists of the fact that the taillight is a light, and the taillight is connected to the fuse (see Fig. 7). The justification indicates that no specific knowledge about the taillight or fuse_1 was needed to make the prediction. The same prediction would be made for any light which was connected to any blown fuse.

fault:	device_functionality (fuse_1,_Any_input, not_working)
prediction:	value (taillight, not_working)
from database:	value (taillight, working)
justification:	connected (fuse_1, taillight) & isa (taillight, light)

FIG. 7. Violated prediction of a blown fuse and its justification.

When a prediction fails, the learning process is initiated (Schank, 1982). The fault-diagnosis heuristic which proposed the fault is modified to include a test for the violated prediction in all future cases which would generate the same prediction for the same reason (i.e. have the same justification). To continue our example, rule_001 will be modified so that in future cases, to propose a blown fuse, if there is a light connected to the fuse, the light must not be working. The revised version of rule_001 is shown in Fig. 8.

symptom (device_functionality (F,_Any_input, not_working), rule_001) ⇐
value (D, not_working) &
isa (D, device) &
connected (F, D) &
isa (F, fuse) &
if connected (F, V037) & isa (V037, light)
then value (V037, not_working)
else true.

FIG. 8. Revised heuristic which proposes a blown fuse-changes in italics.

A DEFINITION OF FAULT DIAGNOSIS WITH HEURISTICS AND DEVICE MODELS

The goal of diagnosis is to find a hypothesis H which accounts for the abnormal functionality of a device. The process of diagnosis can be viewed as applying the following inference rule to conclude H.

$$\frac{F \text{ and } F \rightarrow H}{H}, \qquad \text{if consistent } (H, M, O)$$

where F is a set of features, $F \rightarrow H$ is a fault diagnosis heuristic, H is a fault hypothesis, M is a set of implications which represent a device model, O is the set of observed data (F is a subset of O), and *consistent*(H, M, O) is true if H is consistent with the device model and the observed data. In the discussion of the blown fuse in the previous section, F is the feature indicating that the taillight is not working:

value (D, not_working) &
isa (D, device) &
connected (F, D) &
isa (F, fuse).

H is the hypothesis that the fuse is blown:

device_functionality (fuse_1,_Any_input, not_working)

$F \rightarrow H$ is the diagnosis heuristic which indicates a blown fuse if a light is not working (see Fig. 3), M is the device model of the circuit (see Fig. 4), and O is the set of observed features (see Fig. 5). Note that the above inference rule is similar to *modens ponens* except that the conclusion H must be consistent with the model and the observed data. This is necessary because the implication $F \rightarrow H$ may not be correct.

In our approach to learning and fault diagnosis, *consistent*(H, M, O) corresponds to confirming a hypothesis with device models. It is computed by first finding M_H, a model of a new device which is identical to M except that a faulty component has a different functionality. In the discussion of the blown fuse M_H, corresponds to a device similar to two lights in parallel protected by a fuse, but the fuse is replaced by a broken wire:

device_functionality (fuse_1,_Any_input, not_working).

This new model, M_H, has a number of predictions, P_I. M_H can be viewed as a number of implications:

$$
\begin{aligned}
&H \text{ and } M \rightarrow M_H \\
&J_1 \text{ and } M_H \rightarrow P_1 \\
&J_i \text{ and } M_H \rightarrow P_i \\
&J_n \text{ and } M_H \rightarrow P_n.
\end{aligned}
$$

In the blown fuse example, P_i might be the feature indicating that the taillight is not working:

value (taillight, not_working)

and J_i, the justification for deducing P_i would correspond to:

connected (fuse_1, taillight) & isa (taillight, light).

A hypothesis H is not consistent (i.e. *consistent*(H, M, O) is false) if there is an implication J_i and $M_H \rightarrow P_i$ and P_i is false. Here, the inconsistency arises because from J_i and $M_H \rightarrow P_i$ and M_H and J_i, it is possible to deduce P_i. However, P_i is known to be false.

To summarize, we view diagnosis as consisting of two processes: generating a fault hypothesis H from diagnosis heuristics of the form $F \rightarrow H$, and then confirming or denying H by evaluating *consistent*(H, M, O).

A DEFINITION OF LEARNING FAULT-DIAGNOSIS HEURISTICS

When a diagnosis heuristic proposes a fault hypothesis H which is denied because *consistent*(H, M, O) is false, the diagnosis heuristic can be revised to not propose the fault in future similar conditions. By similar, we mean those conditions which would result in the same inconsistency. Recall that an inconsistency is detected when P_i and not (P_i) can be derived. The following is an example of an inconsistent theory:

$$F$$
$$F \rightarrow H$$
$$M$$
$$H \text{ and } M \rightarrow M_H$$
$$J_i$$
$$J_i \text{ and } M_H \rightarrow P_i$$
$$\text{not } (P_i).$$

We blame $F \rightarrow H$ for the inconsistency, because we are assuming that the observed data (not (P_i) and F), the justification J_i and the implication from the device model J_i and $M_H \rightarrow P_i$ are correct. To avoid this inconsistency the diagnosis heuristic is revised to:†

$$F \text{ and } J_i \text{ and } P_i \rightarrow H$$
$$F \text{ and not } (J_i) \rightarrow H.$$

The new theory, given below, no longer allows H and, therefore, P_i to be derived:

$$F$$
$$F \text{ and } J_i \text{ and } P_i \rightarrow H$$
$$F \text{ and not } (J_i) \rightarrow H$$
$$M$$
$$H \text{ and } M \rightarrow M_H$$
$$J_i$$
$$M_H \rightarrow P_i$$
$$\text{not } (P_i).$$

In the broken fuse example, J_i is

connected (F, V037) & isa (V037, light)

and P_i is:

value (V037, not_working).

The point of failure-driven learning of diagnosis heuristics is that it is simpler to rule out a hypothesis by testing for P_i than proving *consistent*(H, M, O). In the above example, after rule_001 is modified, only one fault hypothesis rather than two are required to correctly identify the failure as a blown light bulb for the parking light. The number of logical inferences required drops from 451 to 305.

LEARNING HEURISTICS TO DIAGNOSIS ATTITUDE CONTROL ANOMALIES

Explanation-based learning has proved useful in satellite diagnosis. Figure 9 presents the definition of two fault diagnosis rules. The rules in this version of ACES have a LISP-like syntax since the satellite diagnosis system is written in a version of PROLOG implemented in LISP. The first element of a list is the predicate name. Variables are preceded by "?". The part of the rule preceded by ":-" is a fault hypothesis, and the part of the rule after ":-" is the conditions which

† We use the equivalent if-then-else construct to avoid proving J_i again.

```
1: (problem (problem wheel-tach ?from
                        (broken-wheel-tach ?wheel ?from))):-
;there is a tachometer stuck at 0
(feature (value-violation ?sig ?from ?until 0))
(measurement ?sig ?wheel speed ?tach)
(isa ?wheel reaction-wheel)
;if the speed of a wheel is 0

2: (problem (problem wheel-drive ?from
                   (broken-wheel-drive ?wheel ?from ?sig))):-
;there is a wheel drive motor not responding to the drive signal
(feature (value-violation ?sig ?from ?until 0))
(measurement ?sig ?wheel speed ?tach)
(isa ?wheel reaction-wheel)
;if the speed of a wheel is 0
```

FIG. 9. Initial fault-diagnosis heuristics.

are necessary to be proved to propose the hypothesis. These rules implement two very crude diagnosis heuristics: "if the speed of a reaction wheel is 0, then the tachometer is broken" and "if the speed of a reaction wheel is 0, then the wheel drive is broken".

Some information about the attitude control system should help to follow the example. The attitude control system consists of a number of sensors which indicate the attitude (i.e. the orientation) of the satellite by the position of the sun and the earth. In addition, there are a set of four reaction wheels whose speed can be varied to adjust the momentum (and therefore the attitude) of the satellite. The speed of the reaction wheels as measured by a number of tachometers is also used as an estimate of attitude to provide finer control of the attitude.
In this example, five atypical features are detected:

(1) value-violation WSPR+ 9:40 15:00 0): The value of the pitch-roll+ wheel speed is 0 from 9:40 until the end of the sample period at 15:00.
(2) (rate-violation WSPR+ 8:00 9:40 −200 0): The value of the pitch-roll+ wheel speed changed an atypical amount (from −200 to 0) between 8:00 and 9°40.
(3) (rate-violation WSPR− 8:00 9:40 −300 −100): The value of the pitch-roll+ wheel speed changed an atypical amount (from −300 to 100) between 8:00 and 9:40.
(4) (rate-violation WSPY+ 8:00 9:40 −200 −400): The value of the pitch-yaw+ wheel speed changed an atypical amount (from −200 to −400) between 8:00 and 9:40.
(5) (rate-violation WSPY− 8:00 9:40 −100 −300): The value of the pitch-yaw− wheel speed changed an atypical amount (from −100 to −300) between 8:00 and 9:40.

The first rule in Fig. 9 suggests that the cause of the value violation is a broken tachometer of the pitch-yaw+ wheel. Next, the device models of the tachometer and the attitude control system confirm or deny this hypothesis. If the tachometer were stuck at 0, the attitude control system would react by adjusting the speeds of the other wheels to compensate for the perceived loss of momentum. However, since the wheel speed is not actually 0, the attitude control system would not have a correct estimate of the momentum. Changing the speed of the other wheels would result in disturbing the attitude of the satellite. Therefore, the hypothesis that the tachometer is faulty is denied and the heuristic which suggested the hypothesis is modified to prevent suggesting the hypothesis in future similar cases. The conditions which were necessary to explain why the hypothesis failed are exactly those which should be added to the heuristic to avoid a hypothesis failure in similar cases. In this example, the heuristic is modified to test that the attitude is disturbed before postulating a faulty tachometer. Figure 10 contains the modified heuristic.

```
(problem (problem wheel-tach ?from
                 (broken-wheel-tach ?wheel ?from))):-
(feature (value-violation ?sig-31 ?from-32 ?end-33 ?value-34))
(isa ?sig-31 attitude-signal)
;make sure the attitude has been disturbed
(feature (value-violation ?sig ?from ?until 0))
(after ?from-32 ?from)
;make sure the attitude disturbance is after the value violation
(measurement ?sig ?wheel speed ?tach)
(isa ?wheel reaction-wheel)
```

FIG. 10. Revised faulty tachometer heuristic-changes in italics.

The primary difference between explanation-based learning and inductive learning is how the class of "similar" failures is discovered. The goal is to define those failures which have a wheel speed of 0 but are not a broken tachometer. In explanation-based learning, this class is found to be those failures which do not have an attitude disturbance after the wheel speed reaches 0. This conclusion is deduced from a model of the connectivity and functionality of the components of the satellite. In contrast, in inductive learning, this class would be arrived at by correlating features of a number of positive and negative training examples. Since the satellite has over 300 signals in the attitude control system alone which are sampled every second, the time and space demands of correlating these features may make inductive learning impractical.

(After the tachometer rule is revised, ACES correctly identifies the fault to be in the wheel drive motor. The second rule in Fig. 9 suggests that the pitch-roll+ wheel drive motor is not responding to its drive signal. The device model confirms this fault. This fault accounts for the pitch-roll+ wheel speed going from −200 to 0 as the results of friction. The speed of the other three wheels are automatically adjusted by the attitude control system to conserve the momentum. This occurs without any disturbance of the attitude.)

Results

There are two standards for evaluating the effects of learning in ACES. First, there is the performance of ACES using the rules in Fig. 9. We call this version naive-ACES. Additionally, there is the performance of ACES using rules hand-coded from information provided by an expert. We call this version of the system expert-ACES.

The data in Table 1 demonstrate that the learning technique presented in this paper improves the simple fault diagnosis heuristics to the extent that the performance of ACES using the learned heuristics is comparable to the system using

TABLE 1
Number of fault hypotheses

Case	Fault	Naive-ACES	Naive-ACES after learning	Expert-ACES
1	Tachometer	21	1	1
2	Wheel drive-1	4	1	2
3	Wheel unload	1	1	1
4	Wheel drive-2	2	1	1

the rules provided by an expert. In one case, the performance of the learned rules is even better than the expert provided rules.

Economic sanctions

Our objective in constructing OCCAM is to build a knowledge-based system which predicts the outcome of applying economic sanctions. However, instead of heuristics rules, OCCAM relies on a memory of generalizations built when analyzing the outcome of previous incidents. OCCAM starts with general knowledge about coercion, and domain knowledge about political and economic relationships. OCCAM builds specializations of coercion which represent such events as blackmail, kidnapping† and economic sanctions. These specializations serve as schemata‡ which allow OCCAM to easily make a prediction when a situation is recognized as similar to previous situations (Schank, 1982). For example, consider answering the following question:

What might happen if the United States refuses to sell computer equipment to South Korea unless South Korea improves its record on human rights?

An expert on economic sanctions or someone familiar with this sort of incident night reply:

South Korea will probably buy computers from some other country such as Japan in the same manner that the USSR imported grain from Argentina when the US refused to sell grain to the USSR.

On the other hand, someone with no knowledge of previous incidents might arrive at a similar conclusion through a long inference chain: South Korea will have an increased demand for computers; South Korea would be willing to pay a higher price for computers; There are several countries which export computers; South Korea may buy the computers from one of them. In OCCAM, this sort of reasoning is done when a new event is added to memory, saved as a new schema and relied on to answer questions. For example, OCCAM should generate an answer to the South Korea question by accessing a generalization which represents those economic

† In the kidnapping domain, the domain knowledge includes interpersonal relationships (Pazzani, 1985).

‡ OCCAM's schemata can be thought of as a special type of chunk (Laird, Rosenbloom & Newell, 1984, Larkin, McDermott, Simon & Simon 1980) which record causal and motivational information for events.

sanction incidents which have failed because the threat was to refuse to sell a product which was easily obtainable elsewhere. A primary question then is how should a schema be generalized from examples of economic sanctions. How can one determine which features of an economic sanctions incident are useful predictors of the outcome?

One way to determine which features are predictive is to use a similarity-based learning mechanism (Lebowitz, 1980) to determine which features are always present for each outcome. The problem with this approach is that when there are a large number of features, it requires a large number of examples to distinguish those similarities which are coincidental from those which are relevant. For example, consider the following two kidnapping examples:

Kidnapping-1
John was abducted. His father, a wealthy, fair-skinned man, received a note which stated John would be killed unless he paid a $100 000 *ransom.*

Kidnapping-2
Mary was abducted. Her mother, a wealthy, fair-skinned woman, received a note which stated Mary would be killed unless she paid a $50 000 *ransom.*

There are a number of similarities between these two kidnappings. In both instances, the parent who received the random note is wealthy and has fair skin. It is relevant that the ransom note be sent to a parent rather than just any person, since parents have a goal of preserving the health of their children. It is relevant that the parent is wealthy, since wealthy persons have the ability to pay the ransom. However, it is just a coincidence that the parents in these two kidnappings have fair skin.

The distinction between relevant and coincidentally similar features is important in making inferences and predictions. For example, if the only examples of kidnapping encountered are Kidnapping-1 and Kidnapping-2 and an intelligent person hears about another kidnapping he might want to infer that the parent of the hostage is wealthy. On the other hand, he would not want to infer that the parent has fair skin. Similarly, if presented with a new kidnapping case where the parent has fair skin but is not wealthy, an intelligent person would not want to predict that the ransom would be paid. To avoid making erroneous inferences, a schema should not include all features which are common to all experienced events.

One way to avoid the problem of coincidentally similar features is to have an *a priori* set of features which are relevant (Kolodner, 1984). However, in general, the relevance of features is dependent on the type of event. Consider the following examples:

Cancer-1
John, a wealthy, fair-skinned man, was advised by his doctor to wear a 15 *SPF sunscreen at the beach.*

Cancer-2
Mary, a wealthy, fair-skinned woman, was advised by her doctor to wear a 15 *SPF sunscreen at the beach.*

In these examples, fair-skinned would be considered relevant since fair-skinned persons are more prone to skin cancer. The fact that Mary and John are both wealthy is incidental. These examples illustrate that the relevance of feature is not absolute. A key task for a learner, either human or computer, is identifying which features of a particular event should be expected to appear in future events.

When a proper domain theory exists, explanation-based learning offers a solution to the problem of identifying relevant features. It can determine the relevant features from analysing only one example. The relevant features are precisely those features which enable the causal and intentional relationships to be inferred. The explanation-based learning approach used in OCCAM is very similar to that used by ACES. An explanation is constructed which explains why a certain event occurs. For OCCAM, the explanation indicates why an economic sanctions incident succeeded or failed. In ACES, the explanation indicates why a hypothesized fault has been ruled out. The explanation is generalized by retaining only those parts of the example which were needed to produce the explanation. The generalized example and the explanation are saved so that a single step (or chunk) can be substituted for a longer inference chain.

For some problems, it is important that a schema be created to avoid extremely complex searches. One situation in which this occurs is when the motivation for a particular course of action is to avoid a potential goal failure. It is difficult, if not impossible to determine why the action was decided upon without knowing about the potential failure. For example, OCCAM notices a similarity between a number of kidnappings whose victims are all infants with blond hair. However, it is not capable of finding an explanation for this similarity until it finds a kidnapping case in which there is a goal failure when the victim was not an infant. After the ransom was paid, the victim was able to give evidence which led to the arrest and conviction of the kidnapper (which is, of course, a goal failure for the kidnapper.) With this new information, one explanation for the kidnapping of infants is found to be avoiding this goal failure since infants cannot give police evidence nor testify. Since the explanation does not reference the infants' hair color, this similarity is treated as a coincidence.

In the next section we give an example of OCCAM creating a schema with explanation-based learning. In addition to the explanation-based learning module, OCCAM consists of a memory indexing scheme which enables appropriate examples and generalization to be found in memory, and a similarity-based learning mechanism which can learn some of the domain theories used by the explanation-based learning system. The interested reader is referred to Pazzani, (1986*b* and 1985). Here, we concentrate on only the explanation-based learning module.

ECONOMIC SANCTIONS: AN EXAMPLE

Consider the following example of economic sanctions:

Economic-Sanctions-1
In 1983, *Australia refused to sell uranium to France unless France ceased nuclear testing in the South Pacific. France paid a higher price* ($300 000 000) *to buy* 1500 *tons of uranium from South Africa.*

What lesson should be learned from Economic-Sanctions-1? That economic sanctions never achieve their desired goal? This conclusion would be overly general. That economic sanctions never work against countries that export wine? This conclusion would be wrong. That economic sanctions will not work when Australia refuses to sell a commodity which is sold by South Africa. This is close to being true, but is probably too specific to apply to many future cases. The problem here is to identify which features of Australia, South Africa, France, and uranium as well as the features of demand (to stop nuclear testing in the South Pacific) and the threat (to not sell uranium) were necessary to explain why Australia did not achieve its goal. To get an idea of the magnitude of the problem consider the representation of Economic-Sanctions-1 which is used in OCCAM (see Fig. 11).

In Fig. 11, the Conceptual Dependency representation of Economic-Sanctions-1 is displayed. The top level concept is coerce. In addition to economic sanctions, *coerce* underlies the representation of blackmail and kidnapping. *coerce* has a number of roles: an *actor* who performs the coercion; a *target* which is the victim of the coercion; an *object* which is the focus of the coercion (in kidnapping, the *object* is typically a relative of the *target*, in economic sanctions, the *object* is usually a commodity); a demand which is an action the *actor* wants the *target* to perform; a *threat* which the actor will *perform* if the *target* does not meet the demand; a *response* which the *target* performs in response to the *threat*; and a *result* which is the outcome of the coercion. In Fig. 11, the notation (**role* target*) indicates that this component is identical to the *target* of the coerce and the notation SOUTH-AFRICA indicates that the representation for South Africa (see Fig. 12) should be substituted for the token SOUTH-AFRICA.

The information in OCCAM's representation about individual countries is derived from the World Almanac. For example, for South Africa, which is the *actor* of the

```
(coerce object (commodity type (uranium))
        actor AUSTRALIA
        target FRANCE
        demand (act type (explode)
                    actor (*role* target)
                    object (weapons type (nuclear))
                    location (southern-hemisphere)
                    mode (neg)
        threat (act type (sell)
                   actor (*role* actor)
                   object (*role* object)
                   to (*role* target)
                   mode (neg))
        response (act type (sell)
                     actor SOUTH-AFRICA
                     object (*role* object)
                     price (money dollars (300000000)
                     value (>market))
                     amount (weight number (1500)
                                    unit (tons)
                     to (*role* target))
        result (statetype (possess)
                   actor (*role* target)
                   value (yes)
                   object (*role* object)))
```

FIG. 11. Conceptual dependency representation of Economic-Sanctions-1.

```
(country name (south-africa)
         language (english)
         location (southern-hemisphere)
         business-relationship (*set* US
                                      JAPAN
                                      (*role* actor)
                                      (*role* target)
                                      UK)
         government (parliamentary)
         continent (africa)
         exports (*set* (*role* object)
                        (commodity type (gold))
                        (commodity type (chromium))
                        (commodity type (diamond))
         imports (commodity type (oil))
```

FIG. 12. OCCAM's representation for South Africa.

response, the language is English, the location is the southern hemisphere, it has a business relationship with the US, Japan, Australia (the *actor*), and France (the *target*), the continent is Africa, it exports uranium (the *object*), gold, chromium and diamonds and it imports oil. In general, the approach in OCCAM has been to provide as many features of each action and entity as possible. It is an important part of the generalization process to determine which of the features should be included in a generalization.

When generalizing Economic-Sanctions-1, some features of entities such as South Africa are relevant, other's are not. It is the simple economic theories used by OCCAM which indicate the relevancy of features. Only those features which are needed to establish an explanation of why France was able to buy uranium from South Africa are considered relevant. A closer examination of the generalization procedure in OCCAM should help clarify how relevancy is determined.

The first step in generalizing is deciding if an event should be generalized at all. There are a number of criteria to determine if a single event is worth generalizing. De Jong (1983) gives a number of criteria to determine if a single event is worth generalizing (e.g. does the event achieve a goal in a novel manner). In this case, let us assume that there are no previous examples of economic sanctions in memory so when Economic-Sanctions-1 is encountered, a goal has failed in a novel manner. This goal failure initiates the explanation and generalization process. The goal failure for Australia was accomplished when France purchased uranium from South Africa. The explanation process tries to determine why France was able to purchase the uranium. The rules in Fig. 13 are used by OCCAM to create an explanation how France was able to purchase the uranium causing a goal failure for Australia.

The first rule in Fig. 13 indicates that the result of Australia refusing to sell uranium to France is that France will have an increased demand for uranium. The second rule allows OCCAM to infer that the increased demand for uranium will enable a country that exports uranium and has a business relationship with France (i.e. South Africa) to sell France the uranium at a price greater than the market price. Finally, the third rule indicates that selling the uranium to France will result in France having the uranium. The components of the representation of Economic-Sanctions-1 which correspond to the components of the rules will be included in the

```
(def-rule refuse-to-sell- > demand-increase
        (act type (sell)
              actor (country exports ?y)
              TO ?x
              object ?y
              mode (neg))
   result
   (state type (demand-increase)
         actor ?x
         value (yes))
         object ?y))
(def-rule demand-increase- > price-increase
         (state type (demand-increase)
               value (yes)
               actor ?x
               object ?y)
   enables
   (act type (sell)
       actor (country exports ?y
                      business-relationship ?x)
       to ?x
       object ?y
       price (money value (>market))
       mode (yes)))
(def-rule sell- > possess
         (act type (sell)
              to ?x
              object ?y
              mode (yes))
   result
   (state type (possess)
         object ?y
         value (yes)
         actor ?x))
```

FIG. 13. OCCAM's economic rules.

```
(coerce result (state type (possess)
                      actor (*role* target)
                      object (*role* object)
                      value (yes))
        response (act type (sell)
                      actor (country exports (*role* object)
                                     business-relationship
                                     (*role* target))
                      object (*role* target)
                      to (*role* target)
                      price (money value (>market)))
        threat (act type (sell)
                    actor (*role* actor)
                    object (*role* object)
                    mode (neg))
        target (country imports (*role* object))
        object (commodity)
        actor (country exports (*role* object)))
```

FIG. 14. Generalization of Economic-Sanctions-1 produced by OCCAM.

generalization of Economic-Sanctions-1. Everything else is irrelevant and will be discarded. The generalization produced by OCCAM is illustrated in Fig. 15.

The generalization in Fig. 15 indicates that if a country that exports a commodity tries to coerce a country which imports the commodity by refusing to sell them the commodity, then a response might be to buy the commodity at a higher price from another country. Although this seems like a simple conclusion, there are many examples where economic sanctions have failed for this reason (Hufbauer & Schott, 1985) (e.g. in 1961, the USSR refused to sell grain to Albania, who purchased it instead from China, in 1980, the US refused to sell grain to the USSR who purchased it instead from Argentina, and in 1981, the US refused to sell pipeline equipment to the USSR who purchased it instead from France).

A similar conclusion about the effectiveness of sanctions was arrived at by Ian Smith, the former Prime Minister of Rhodesia which was the target of a decade of economic sanctions following its independence from Great Britain (cited in Renwick, 1981):

> *We find that we are compelled to export at discount and import at a premium.*

It is interesting to note which features of South Africa were generalized. In Fig. 15, the only features of the *actor* of the *response* are that it exports the *object* and it has a business relationship with the *target.* Notice that in the individual episode in Fig. 13, the fact that South Africa has a business relationship with the *actor* was also included. However, this feature was not necessary to produce the explanation, so it is not included in the generalization. The features of South Africa which were included are those that matched against the features of the second rule in Fig. 14.

The generalizations produced by OCCAM serve as a means of recognizing when economic sanctions will fail or succeed at achieving the desired goal. These generalizations summarize an inference chain by retaining only those surface features of an example which were required to establish an explanation for the outcome using simple economic theories.

Conclusion

We have demonstrated that explanation-based learning is a viable technique for obtaining expert-like performance in an knowledge-based system. Explanation-based learning can be used in knowledge-based systems where the heuristics to be learned are implicit in a different representation of the same knowledge. For example, in fault diagnosis, device models are a natural way of expressing the functionality of a component. However, they are not the most natural or efficient representation for diagnosis (Sembugamoorthy & Chandraskaran, 1985). The ACES system explicates associations between failures and atypical behavior which are implicit in device models. In OCCAM, explanation-based learning creates schemata which recognize when economic sanctions will fail or succeed at achieving the desired goal. These schemata explicate the implications of simple economic theories such as supply and demand on economic sanctions incidents.

The research on ACES was supported by the Aerospace Sponsored Research Program. The research on OCCAM was supported in part by the UCLA-RAND Artificial Intelligence

Fellowship. Carl Kesselman implemented the PROLOG meta-interpreter and commented on an earlier draft of this paper.

References

BERL, J., LEWIS, G. & MORRISON, R. (1976). Applying models of choice to the problem of college selection. In CARROLL, J. & PAYNE, J. Ed., *Cognition and Social Behavior*. Hillsdale, New Jersey: Erlbaum.

DAVIS, R., SHROBE, H., *et al.* (1982). Diagnosis based on description of structure and function. *Proceedings of the National Conference on Artificial Intelligence, Pittsburgh, PA*. American Association for Artificial Intelligence.

DEJONG, G. (1983). Acquiring schemata through understanding and generalizing plans. *Proceedings of the Eighth International Joint Conference on Artificial Intelligence*, Karlsruhe, West Germany.

ERICSSON, K. & SIMON, H. (1984). *Protocol Analysis: Verbal Reports as Data*. MIT Press.

FORBUS, K. (1984). Qualitative process theory. *Artificial Intelligence*, **24**.

GENESERETH, M., BENNETT, J. S. & HOLLANDER, C. R. (1981) DART: expert systems for automated computer fault diagnosis. *Proceedings of the Annual Conference*. Baltimore, Maryland: Association for computing Machinery.

HUFBAUER, G. C. & SCHOTT, J. J. (1985). *Economic Sanctions Reconsidered: History and Current Policy*. Washington, D.C.: Institute For International Economics.

DE KLEER, J. & BROWN, J. (1984). A Qualitative physics based on confluences. *Artificial Intelligence*, **24**, 7–83.

KOLODNER, J. (1984). *Retrieval and Organizational Strategies in Conceptual Memory: a Computer Model*. Hillsdale, New Jersey: Lawrence Erlbaum Associates.

KUIPERS, B. (1984). Commonsense reasoning about causality: deriving behavior from structure. *Artificial Intelligence* **24**, 169–203.

LAIRD, J., ROSENBLOOM, P. & NEWELL, A. (1984). Towards chunking as a general learning mechanism. *Proceedings of the National Conference on Artificial Intelligence*. Austin, Texas: American Association for Artificial Intelligence.

LARKIN, J., MCDERMOTT, J., SIMON, D. & SIMON, H. (1980). Expert and novice performance in solving physics problems. *Science*, **208**, 1335–1342.

LEBOWITZ, M. (1980). *Generalization and memory in an integrated understanding system. Computer Science Research Report* 186. Yale University.

MICHIE, D. (1983). Inductive rule generation in the context of the fifth generation. *Proceedings of the International Machine Learning Workshop*. Monticello, Illinois.

MINTON, S. (1984). Constraint-based generalization: learning game-playing plans from single examples. *Proceedings of the National Conference on Artificial Intelligence*. Austin, Texas: AAAI.

MITCHELL, T., KEDAR-CABELLI, S. & KELLER, R. (1986). Explanation-based learning: a unifying view. *Machine Learning*, **1**.

MITCHELL, T., MAHADEVAN, S. & STEINBERG, L. (1986). LEAP: a learning apprentice for VLSI design. *International Meeting on Advances in Learning*, Les Arc, France.

NELSON, W. R. (1982) REACTOR: an expert system for diagnosis and treatment of nuclear reactor accidents. *Proceedings of the National Conference on Artificial Intelligence*. Pittsburgh, Philadelphia: AAAI.

NISBETT, R. & ROSS, L. (1978) *Human Inference: Strategies and Shortcomings of Social Judgements*. Engelwood Cliffs, New Jersey: Prentice-Hall, Inc.

NISBETT, R. & WILSON, T. (1977) Telling more than we can know: verbal reports on mental processes. *Psychological Review*, (1977), **84**, 231–259.

Pazzani, M. (1985). Explanation and generalization-based memory. *Proceedings of the Seventh Annual Conference of the Cognitive Science Society*. Irvine, California: Cognitive Science Society.

Pazzani, M. (1986) Refining the knowledge base of a diagnostic expert system: an application of failure-driven learning. *Proceedings of the National Conference on Artificial Intelligence.* American Association for Artificial Intelligence.

PAZZANI, M., DYER, M. & FLOWERS, M. (1986). The role of prior causal theories in generalization. *Proceedings of the National Conference on Artificial Intelligence.* American Association for Artificial Intelligence.

PAZZANI, M. & BRINDLE, A. (1985). An expert system for satellite control. *Proceedings of ITC/USA/85, the International Telemetering Conference.* Las Vegas, Nevada: International Foundation for Telemetering.

PAZZANI, M. & BRINDLE, A. (1986) Automated diagnosis of attitude control anomalies. *Proceedings of the Annual AAS Guidance and Control Conference.* Keystone, Colorado: American Astronautical Society.

RENWICK, R. (1981). *Economic Sanctions.* Center for International Affairs, Harvard University.

SCARL, E. A., JAMIESON, J. & DELAUNE, C. (1985). A fault detection and isolation method applied to liquid oxygen loading for the space shuttle. *Proceedings of the Ninth International Joint Conference on Artificial Intelligence,* Los Angeles, California.

SCHANK, R. (1982). *Dynamic Memory: a Theory of Reminding and Learning in Computers and People.* Cambridge University Press.

SEMBUGAMOORTHY, V. & CHANDRASKARAN, B. (1985). Functional representation of devices and compilation of diagnostic problem solving systems. *Technical Report,* Ohio State University.

SHORTLIFFE, E. H. (1976). *Computer-based Medical Consultation: MYCIN.* New York: American Elsevier.

STEEL, L. & VAN DE VELDE, W. (1986). Learning in second generation expert systems. In KOWALIK, J. S. Ed., *Knowledg-Based Problem Solving.* Englewood Cliffs, New Jersey: Prentice-Hall, Inc.

WAGNER, R. E. (1983). Expert system for spacecraft command control. In *Computers in Aerospace IV Conference.* Hartford, Connecticut: American Institute of Aeronautics and Astronautics.

Inductive Rule Generation

Simplifying decision trees

J. R. QUINLAN†

Artificial Intelligence Laboratory, Massachusetts Institute of Technology, 545 Technology Square, Cambridge, MA 02139, U.S.A.

Many systems have been developed for constructing decision trees from collections of examples. Although the decision trees generated by these methods are accurate and efficient, they often suffer the disadvantage of excessive complexity and are therefore incomprehensible to experts. It is questionable whether opaque structures of this kind can be described as knowledge, no matter how well they function. This paper discusses techniques for simplifying decision trees while retaining their accuracy. Four methods are described, illustrated, and compared on a test-bed of decision trees from a variety of domains.

1. Introduction

Since people began building knowledge-based systems, it has become painfully obvious that the ability to function at an expert level in some task domain does not necessarily confer a corresponding ability to articulate this know-how. The knowledge for most early and many current expert systems has been amassed by an interview process in which a *knowledge engineer* interacts with a domain expert to extract and refine a set of *rules*. This process can be taxing for all concerned because the expert, as Waterman (1986) puts it,

> has a tendency to state [his] conclusions and the reasoning behind them in general terms that are too broad for effective machine analysis . . . the pieces of basic knowledge are assumed and combined so quickly that it is difficult for him to describe the process.

Consequently, the productivity of the interview method is usually low. This led Feigenbaum (1981) to identify knowledge acquisition as the 'bottleneck' problem in building knowledge-based systems.

One way around this bottleneck, long advocated by Donald Michie (1983) and others, uses inductive methods to extract general rules from concrete examples. The expert is not asked to articulate his skill but instead to provide a framework of important concepts in the task domain, augmented perhaps by a collection of tutorial examples; the hard work is carried out by a suitable induction engine. Most researchers in Machine Learning will be familiar with Meta-DENDRAL and its synthesis of chemical knowledge (Buchanan & Mitchell, 1978) and with AQ11's results on soybean diagnosis (Michalski & Chilausky, 1980). The feasibility of this inductive approach to knowledge acquisition has also been confirmed in several industrial projects, such as British Petroleum's recent successful construction of a 2500-rule expert system for the design of hydrocarbon separation vessels in just one man-year (*Expert Systems User*, August 1986, pp. 16–19).

Many current commercial induction packages (including *Ex-Tran*, *RuleMaster* and

† Permanent address: Basser Department of Computer Science, University of Sydney, Sydney 2006, Australia.

KNOWLEDGE-BASED SYSTEMS Vol. 1
ISBN 0-12-273251-0

1*st-Class*) express the derived rules in the form of *decision trees*. From the standpoint of execution efficiency this is a simple and economical representation, but the trees can become complex and thus opaque (Michie, 1986). If a decision tree that measures up very well on the performance criterion is nevertheless totally incomprehensible to a human expert, can it be described as *knowledge*? Under the common-sense definition of this term as material that might be assimilated and used by human beings, it is not, in just the same way that a large program coded in assembly language is not knowledge.

This paper examines four methods for improving the intelligibility of decision trees and thereby making them more knowledge-like. Three of the methods involve *pruning* the decision tree by replacing one or more subtrees with leaves, while the remaining method reformulates the decision tree as a set of production rules. Section 2 introduces the methods and illustrates their operation with respect to a small but real example. Section 3 presents an empirical comparison of the methods using sets of decision trees from six task domains.

2. Methods for simplifying decision trees

Induction algorithms that develop decision trees view the task domain as one of *classification*. The underlying framework consists of a collection of *attributes* or properties which are used to describe individual *cases*, each case belonging to exactly one of a set of *classes*. Attributes may be either continuous or discrete. A case's value of a continuous attribute is always a real number while its value of a discrete attribute is one of a small set of possible values for that attribute. In real-life tasks it is also important to recognise that a case may have *unknown* values for one or more of the attributes.

A *decision tree* may be either a leaf identified by a class name, or a structure of the form

$$\begin{array}{ll} C_1: & D_1 \\ C_2: & D_2 \\ \vdots & \vdots \\ C_n: & D_n \end{array}$$

where the C_is are mutually exclusive and exhaustive logical conditions and the D_is are themselves decision trees. The set of conditions involves only one of the attributes, each condition being

$$A < T \quad \text{or} \quad A > T$$

for a continuous attribute A, where T is some threshold, or

$$A = V \quad \text{or} \quad A \ \text{in} \ \{V_i\}$$

for a discrete attribute A, where V is one of its possible values and $\{V_i\}$ is a subset of them. To improve legibility, the non-leaf subordinate decision trees above will be indented when the trees are printed.

Such a decision tree is used to classify a case as follows. If the tree is a leaf, we simply determine the case's class to be the one nominated by the leaf. If the tree is a structure, we find the single condition C_i that holds for this case and continue with the associated decision tree. The only complexity arises when the value of the attribute appearing in the C_is is unknown. In this eventuality we explore all the decision trees associated with the structure and combine their findings with weights proportional to the estimated probability of the associated condition being satisfied. Quinlan (1986) discusses the procedure in more detail.

Figure 1 shows such a decision tree for the diagnosis of hypothyroid conditions with classes {*primary hypothyroid, secondary hypothyroid, compensated hypothyroid, negative*}. Some attributes such as *TSH* and *FTI* are continuous and have real values, while attributes like *thyroid surgery*, with possible values {t, f}, are discrete. To classify a case with this tree, we would first enquire whether the value of *TSH* was greater than 6·05. If the value was below this threshold we would continue with the decision tree commencing with *T4U measured* = t, while a value above this threshold would lead us to the decision tree headed *FTI* < 64·5. In either case we would continue in similar fashion until a leaf was encountered.

The set of cases with known classes from which a decision tree is induced is called the *training set*. Other collections of cases not seen while the tree was being developed are known as *test sets* and are commonly used to evaluate the performance of the tree.

This paper focusses on simplifying decision trees, not with the inductive methods used to construct them in the first place. Various ways of developing trees from training sets may be found in (Breiman *et al.*, 1984), (Kononenko, Bratko & Roškar, 1984) and (Quinlan, 1986).

```
TSH < 6·05:
|   T4U measured = t: negative (1918)
|   T4U measured = f:
|   |   age > 43·5: negative (58)
|   |   age < 43·5:
|   |   |   query hypothyroid = f: negative (41)
|   |   |   query hypothyroid = t: secondary hypothyroid (1)
TSH > 6·05:
|   FTI < 64·5:
|   |   thyroid surgery = f:
|   |   |   T3 < 2·3: primary hypothyroid (51)
|   |   |   T3 > 2·3:
|   |   |   |   sex = M: negative (1)
|   |   |   |   sex = F: primary hypothyroid (4)
|   |   thyroid surgery = t:
|   |   |   referral source = SVI: primary hypothyroid (1)
|   |   |   referral source = ⟨other⟩: negative (2)
|   FTI > 64·5:
|   |   on thyroxine = t: negative (32)
|   |   on thyroxine = f:
|   |   |   thyroid surgery = t: negative (3)
|   |   |   thyroid surgery = f:
|   |   |   |   TT4 < 150·5: compensated hypothyroid (120)
|   |   |   |   TT4 > 150·5: negative (6)
```

FIG. 1. Sample decision tree.

2.1. COST-COMPLEXITY PRUNING

Breiman *et al.* (1984) describe a two-stage process in which a sequence of trees $T_0, T_1, \ldots, T_k$ is generated. T_0 is the original decision tree and each T_{i+1} is obtained by replacing one or more subtrees of T_i with leaves until the final tree T_k is just a leaf. The second stage evaluates these trees and selects one of them as the final pruned tree.

Consider a decision tree T used to classify each of the N cases in the training set from which T was generated, and let E of them be misclassified. If $L(T)$ is the number of leaves in T, Breiman *et al.* define the *cost-complexity* of T as the sum

$$\frac{E}{N} + \alpha \times L(T)$$

for some parameter α. Now, suppose we were to replace some subtree S of T by the best possible leaf. In general, the new tree would misclassify M more of the cases in the training set but would contain $L(S) - 1$ fewer leaves. This new tree would have the same cost-complexity as T if

$$\alpha = \frac{M}{N \times (L(S) - 1)}$$

As before, let T_0 be the original tree. To produce T_{i+1} from T_i we examine each non-leaf subtree of T_i to find the minimum value of α above. The one or more subtrees with that value of α are then replaced by their respective best leaves.

To illustrate the process, consider the decision tree of Fig. 1. This was generated from 2514 cases, where the number in parentheses after each leaf shows how many of these cases are covered by that leaf.† Consider the subtree

```
T4U measured = t: negative (1918)
T4U measured = f:
|   age > 43·5: negative (58)
|   age < 43·5:
|   |   query hypothyroid = f: negative (41)
|   |   query hypothyroid = t: secondary hypothyroid (1)
```

The vast majority of cases at the leaves of this subtree are of class *negative* which is clearly the best leaf. If the subtree were replaced by the leaf *negative* the new tree would misclassify the lone non-*negative* case, so M is 1. The new tree would also have three fewer leaves, giving a value for α of 0·00013 at which the cost-complexity of the original and modified trees would be equal. This is the lowest such value for any subtree, so the tree T_1 would be formed by replacing this subtree as above.

The second stage of this process abandons the cost-complexity model and attempts to select one of the T_is on the basis of reliability alone. We cannot assess this simply from the proportion of cases in the original training set that are misclassified. Whatever induction algorithm was employed has almost certainly built the original tree to fit the training set and thus the error rate on these cases would be expected to understate the error rate on unseen cases. We therefore assume some

† The counts do not sum to 2514 because cases with unknown values of tested attributes cannot be associated with any one leaf and are therefore not included.

test set containing N' cases and use each T_i to classify all of them. Let E' be the minimum number of errors observed with any T_i, with the standard error of E' being given by

$$\mathrm{se}(E') = \sqrt{\frac{E' \times (N' - E')}{N'}}.$$

The tree selected is the smallest T_i whose observed number of errors on the test set does not exceed $E' + \mathrm{se}(E')$.

In this example, a test set containing 629 cases gave a sequence of eight trees, T_0 being the original tree and T_7 the leaf *negative.* The selected tree was T_4 which appears in Fig. 2. This tree is indeed a great deal simpler than the original and would qualify as 'knowledge' under the most stringent criterion. Notice that the class *secondary hypothyroid,* which is represented by just a single case in the training set, has sensibly been omitted.

Nevertheless, cost-complexity pruning raises several problematic issues. First, it is unclear why the particular cost-complexity model used above is superior to other possible models such as the product of error rate and number of leaves. Second, it seems anomalous that the cost-complexity model used to generate the sequence of subtrees is abandoned when the best tree is selected. Last, the procedure requires a test set distinct from the original training set; the authors show, however, that a cross-validation scheme can be employed to generate these estimates at the time the original tree is constructed, but at the expense of a substantial increase in computation.

2.2. REDUCED ERROR PRUNING

Rather than form a sequence of trees and then select one of them, a more direct procedure suggests itself as follows. We again assume a separate test set, each case in which is classified by the original tree. For every non-leaf subtree S of T we examine the change in misclassifications over the test set that would occur if S were replaced by the best possible leaf. If the new tree would give an equal or fewer number of errors and S contains no subtree with the same property, S is replaced by the leaf. The process continues until any further replacements would increase the number of errors over the test set.

Using the same example of Fig. 1 and the same test set as before, reduced error pruning generates the tree shown in Fig. 3.

- *TSH* < 6·05: negative (2018)
- *TSH* > 6·05:
    - *FTI* < 64·5: primary hypothyroid (62)
    - *FTI* > 64·5:
        - on thyroxine = t: negative (32)
        - on thyroxine = f:
            - thyroid surgery = t: negative (3)
            - thyroid surgery = f:
                - *TT4* < 150·5: compensated hypothyroid (120)
                - *TT4* > 150·5: negative (6)

FIG. 2. Decision tree after cost-complexity pruning.

```
TSH < 6·05: negative (2018)
TSH > 6·05:
|   FTI < 64·5:
|   |   thyroid surgery = f: primary hypothyroid (59)
|   |   thyroid surgery = t:
|   |       referral source = S/I: primary hypothyroid (1)
|   |       referral source = ⟨other⟩: negative (2)
|   FTI > 64·5:
|   |   on thyroxine = t: negative (32)
|   |   on thyroxine = f:
|   |   |   thyroid surgery = t: negative (3)
|   |   |   thyroid surgery = f:
|   |   |   |   TT4 < 150·5: compensated hypothyroid (120)
|   |   |   |   TT4 > 150·5: negative (6)
```

FIG. 3. Decision tree after reduced error pruning.

As with cost-complexity pruning, this process generates a sequence of trees. Its rationale is clearer, though, since the final tree is the most accurate subtree of the original tree with respect to the test set and is the smallest tree with that accuracy. The disadvantages of the method are, first, that it again requires a separate test set and second, that parts of the original tree corresponding to rarer special cases not represented in the test set may be excised.

2.3. PESSIMISTIC PRUNING

When the original tree T is used to classify the N cases in the training set from which it was generated, let some leaf account for K of these cases with J of them misclassified. As observed before, the ratio J/K does not provide a reliable estimate of the error rate of that leaf when unseen cases are classified, since the tree has been tailored to the training set. A more realistic error rate might be obtained using the continuity correction for the binomial distribution (Snedecor & Cochran, 1980, pp. 117ff) in which J is replaced by $J + 1/2$.†

Let S be a subtree of T containing $L(S)$ leaves and let $\sum J$ and $\sum K$ be the corresponding sums over the leaves of S. A more pessimistic view of S is that it will misclassify $\sum J + L(S)/2$ out of $\sum K$ unseen cases, where the standard error of this number of misclassifications can be determined as before. If S were replaced by the best leaf, let E be the number of cases from the training set that it misclassifies. The pessimistic pruning method replaces S by the best leaf whenever $E + 1/2$ is within one standard error of $\sum J + L(S)/2$. All non-leaf subtrees are examined just once to see whether they should be pruned but, of course, sub-subtrees of pruned subtrees need not be examined at all.

To illustrate the idea we return to the subtree of Fig. 1 that commences with the condition $T4U$ *measured* = *t*. As before, $\sum K$ is 2018, $L(S)$ is 4, $\sum J$ is 0, so the estimate of the number of errors due to S is 2·0 with standard error 1·41. If the subtree is replaced by the leaf *negative* it will give one error, so E is 1. Since $1 + 1/2 < 2{\cdot}0 + 1{\cdot}41$, pessimistic pruning would indeed replace this subtree. Repeating this evaluation on all subtrees of T gives a pruned tree identical to that of Fig. 2.

† This makes the unsurprising assumption that $J/K < 0{\cdot}5$.

This method has two advantages. It is much faster than either of the preceding methods since each subtree of T is examined at most once. Unlike these methods, it does not require a test set separate from the cases in the training set from which the tree was constructed.

2.4. SIMPLIFYING TO PRODUCTION RULES

This form of simplification does not give a smaller decision tree at all but instead develops an 'equivalent' set of *production rules,* a representation medium widely used in expert systems (Winston, 1984). The process has two stages: individual production rules are first generated and polished, and then the rules produced are evaluated as a collection.

Whenever a decision tree is used to classify a case, a path is established between the top of the tree and one of its leaves. In order for the case to reach that leaf, it must have satisfied all the conditions along the path. For example, any case that is classified as *negative* by the last leaf of the decision tree in Fig. 1 must satisfy all the conditions

$$\begin{aligned}&TSH > 6{\cdot}05,\\&FTI > 64{\cdot}5,\\&on\ thyroxine = f,\\&thyroid\ surgery = f,\\&TT4 > 150{\cdot}5.\end{aligned}$$

Every leaf of a decision tree thus corresponds to a production rule of the form

$$\textbf{if } X_1 \wedge X_2 \wedge \cdots \wedge X_n \textbf{ then } class\ c$$

where the X_is are conditions as before and c is the class of the leaf.

Merely rewriting a tree as the collection of these equivalent production rules would not represent any simplification at all. Instead, the first stage examines each production rule to see whether it should be generalised by dropping conditions from its left-hand side. Let X_i be one of the conditions and consider those cases in the training set that satisfy all the other conditions in the rule. With respect only to these cases, the relevance of X_i to determining whether a case belongs to class c (given that the other conditions are satisfied) can be summarised by the 2×2 contingency table

	Class c	Not class c
satisfies X_i	sc	$s\bar{c}$
does not satisfy X_i	$\bar{s}c$	$\overline{sc}$

where sc is the number of these cases that satisfy X_i and belong to class c, $s\bar{c}$ is the number that satisfy X_i but belong to some class other than c, and so on. Fisher's exact test (Finney *et al.*, 1963) can then be invoked to assess the probability that the division by X_i arises merely from chance or, in other words, the significance level at which we can reject the hypothesis that X_i is irrelevant to whether a case belongs to class c.† Each X_i is examined in turn to find the one that has the least relevance to classification and, unless the hypothesis that this X_i is not significant can be rejected at the 0·1% level or better, the condition is discarded and the process repeated.

Consider the rule above. When the training cases that satisfy all conditions other than the first are examined, the table for the condition $TSH > 6{\cdot}05$ comes out to be

	Class *negative*	Not class *negative*
$TSH > 6{\cdot}05$	6	0
$TSH < 6{\cdot}05$	154	0

which shows that this condition is entirely irrelevant. On the other hand, the table of cases satisfying all conditions other than the last is

	Class *negative*	Not class *negative*
$TT4 > 150{\cdot}5$	6	0
$TT4 < 150{\cdot}5$	0	120

which is significant at better than the 0·1% level. Repeated application of the above process reduces the original rule to one with a single condition

if $TT4 > 150{\cdot}5$ **then** *class negative*

The final step in this first stage is to estimate a *certainty factor* for the simplified rule, using a device similar to that of pessimistic pruning. If the left-hand side of a rule is satisfied by V cases in the training set, W of which belong to the class indicated by the right-hand side, the certainty factor of the production rule is taken as $(W - 1/2)/V$. In the example above, the training set contains 246 cases that match the left-hand side, all of them being class *negative*, so this rule's CF is 99·8%.

Note that we need not develop one rule for each leaf of the decision tree. Some leaves give rise to identical rules while other leaves generate vacuous rules from which all conditions have been dropped. The number of rules is generally smaller than the number of leaves.

The second stage of this process looks how well the rules will function as a set. This evaluation depends on the way in which the rules will be used. A simple strategy has been adopted here: To classify a case, find a rule that applies to it; if there is more than one, choose the rule with the higher certainty factor; if no rule applies, take the class by default to be the most frequent class in the training set.

For each rule in turn, we now determine how the remaining rules would perform

† I am indebted to Donald Michie of the Turing Institute for making me aware of this test and its advantages over the approximate χ^2 test.

on the training set if this rule were omitted. If there are rules whose omission would not lead to an increased number of errors classifying the cases in the training set, or would even reduce it, the least useful such rule is discarded and the process repeated.

Continuing the example, the decision tree of Fig. 1 is reduced by this method to just three rules:

if *TSH* $< 6{\cdot}05$ **then** *class negative* [99·9%]

if *thyroid surgery* $= f \wedge$
TSH $> 6{\cdot}05 \wedge$
FTI $< 64{\cdot}5$ **then** *class primary hypothyroid* [97·5%]

if *on thyroxine* $= f \wedge$
thyroid surgery $= f \wedge$
TSH $> 6{\cdot}05 \wedge$
*TT*4 $< 150{\cdot}5 \wedge$
FTI $> 64{\cdot}5$ **then** *class compensated hypothyroid* [99·6%]

As with pessimistic pruning, this method does not require a set of test cases apart from the original training set. In its current implementation it is the slowest of the four tree-simplifying methods. The method should be able to be improved by adopting a more sophisticated condition-elimination strategy than the simple hill-climbing approach used above, and by employing a better production rule interpreter.

2.5. OTHER METHODS

The four methods of simplifying decision trees certainly do not exhaust all possibilities. The cross-validation method of Breiman *et al.* (1984) has already been mentioned. Kononenko *et al.* (1984) present an information-based heuristic used in their ASSISTANT system, but this is now being changed to another form of cross-validation (Lavrač, Mozetič & Kononenko, 1986). I have previously experimented with a form of pruning based on the path lengths in the decision tree and observed error rates (Quinlan, Compton, Horn & Lazarus, 1986).

3. Empirical comparison

The performance of a simplification method can be assessed in terms of the clarity and accuracy of its final product. Ideally, the pruned decision tree or set of production rules should be much more comprehensible than the original decision tree but should not be significantly less accurate when classifying unseen cases.

To test how well the methods of the previous section measure up to these two criteria, they were compared using decision trees developed for six task domains. For each domain, the available data was shuffled, then divided into a training set containing approximately two-thirds of the data and two equal-sized test sets. This division was carried out so as to make the proportion of cases belonging to each class as even as possible across the three sets. The training set was used to induce ten decision trees for the domain. Each simplification method was applied to each tree and the resulting classifier evaluated on both test sets.

The six domains include both real-world tasks and synthetic tasks constructed to provide some particular challenge. They are:

- Diagnosis of hypothyroid conditions (*Hypothyroid*): This domain has been encountered in the running example of the previous section. The data comes from the archives of the Garvan Institute of Medical Research, Sydney, and covers all 3772 thyroid assays carried out by Garvan's clinical laboratory between January and November 1985. The data uses seven continuous and sixteen discrete attributes with quite high rates of missing information—values of four of the attributes are unknown in more than 10% of the cases. The 3772 cases, each belonging to one of four classes, were split into a training set of 2514 and two test sets of 629. This domain is a good starting point because it uses 'live' data from which, warts and all, extremely accurate classifiers can be constructed.
- Discordant assay results (*Discordant*): This domain is taken from the same Garvan data, this time looking to detect anomalous combinations of thyroid hormone values. There are two classes and the 3772 cases were divided as above. The percentage of discordant cases is very low (about 1·5%) and, in contrast with the first domain, the decision trees generated from this training set perform comparatively poorly on unseen cases.
- Recognising faulty digits (*LEDDigits*): The third domain comes from (Breiman *et al.*, 1984). Imagine a seven-element representation for a decimal digit such as is commonly found on LED or LCD displays. Each element of a faulty display is subject to a 10% random error, i.e. with probability 0·1 its correct status is inverted. The data consists of 3000 randomly-generated cases, each described in terms of the seven binary attributes, with ten equiprobable classes. The training set contains 2000 cases, the test sets 500 each. This artificial domain is interesting because it tests the ability of the simplification methods to deal with the complex decision trees commonly obtained from noisy training sets.
- Assessing consumer credit applications (*Credit*): The data for this domain were provided by a large bank. Each case concerns an application for credit card facilities described by 9 discrete and 6 continuous attributes, with two decision classes. The 690 cases making up the data are divided into a training set of 460 and two test sets of 115. Some discrete attributes have large collections of possible values (one of them has 14) resulting in broad, shallow decision trees. These data are also both scanty and noisy, giving decision trees that are extremely complex and not very accurate on unseen cases.
- King and rook versus king and knight (*Endgame*): This domain from a chess endgame seeks to decide whether the rook's side can capture the opposing knight and/or checkmate in 3 ply. Positions are described by 39 binary attributes, with all possible board positions giving rise to 551 distinct cases. This domain models an idealised noise-free environment with no missing information in which the accuracy of the decision tree depends only on the completeness of the training set. Here the training set contains 367 cases, the test sets 92 cases each.
- Probabilistic classification over disjunctions (*Prob–Disj*): The last domain is an artificial one designed to model tasks in which only probabilistic classification is possible and which contains explicit disjunctions. There are ten boolean attributes a_0 through a_9 and the criterion used to generate the data can be expressed as: if

$a_0 \wedge a_1 \wedge a_2$ or $a_3 \wedge a_4 \wedge a_5$ or $a_6 \wedge a_7 \wedge a_8$ then the class is Y with probability 0·9, N with probability 0·1; otherwise, the class is N with probability 0·9, Y with probability 0·1. (The remaining attribute a_9 is irrelevant.) Because the class of a case is determined probabilistically, no classification procedure can achieve more than 90% accuracy on this task. Six hundred cases with random values for each attribute were generated and classified as above. Of these, 400 are used as the training set, leaving test sets of 100 cases each.

The results of these experiments are summarised in the following tables. The effectiveness of the simplification methods in reducing the size of the original decision trees is shown in Table 1, each entry being the average over the ten decision trees in that domain. As a general observation, all the methods achieve significant simplification in all domains. Cost-complexity pruning tends to produce smaller decision trees than either reduced error or pessimistic pruning, especially in the *Credit* domain. While the complexity of decision trees and sets of production rules cannot be compared directly, it would appear that the last method achieves the greatest reduction overall, its advantages being particularly noteworthy in the *Prob–Disj* domain.

The other side of the coin is the effect of simplification on classification accuracy. Table 2 shows the results in each domain of using the ten original decision trees and their simplified counterparts to classify cases in the two test sets, expressed as the average percentage of misclassifications over each set. Perhaps surprisingly, the simplified trees on the whole are of superior or equivalent accuracy to the originals, so pruning has been beneficial on both counts. Note, though, that both the cost-complexity and reduced error methods have 'seen' the first training set in performing their respective simplifications. The slight superiority of reduced error pruning, coupled with the fact that cost-complexity pruning produces smaller trees, suggests that the latter may be slightly over-pruning. Despite *not* having seen the first test set, the performance of pessimistic pruning is marginally better then cost-complexity pruning averaged over all domains. Simplification to production rules, though, scores pretty clear wins in the last two domains. In the *Prob–Disj* domain in particular, this can be explained by observing that disjunctive concepts tend to scatter cases from some disjuncts throughout the decision tree. Pruning the tree is unable to re-collect these cases, but simplification of rules can.

TABLE 1
Average size before and after simplification

	Original decision trees	Cost-complexity pruning	Reduced error pruning	Pessimistic pruning	Production rule form
Hypothyroid	23·6 nodes	11·4 nodes	14·4 nodes	11·0 nodes	3·0 rules
Discordant	52·4 nodes	11·8 nodes	12·4 nodes	13·6 nodes	1·8 rules
LED Digits	92·2 nodes	45·6 nodes	59·0 nodes	56·0 nodes	15·8 rules
Credit	248·0 nodes	9·7 nodes	26·3 nodes	32·5 nodes	7·8 rules
Endgame	88·8 nodes	51·0 nodes	55·6 nodes	62·6 nodes	11·6 rules
Prob-Disj	190·0 nodes	30·4 nodes	43·0 nodes	42·6 nodes	4·2 rules

TABLE 2
Average error rates on test sets

	Original decision trees	Cost-complexity pruning	Reduced error pruning	Pessimistic pruning	Production rule form
Hypothyroid					
Test 1	0·3%	0·4%	0·3%	0·5%	0·3%
Test 2	0·8%	0·7%	0·8%	0·6%	1·0%
Discordant					
Test 1	1·6%	1·1%	1·0%	1·0%	1·1%
Test 2	2·1%	1·6%	1·7%	1·5%	1·5%
LED Digits					
Test 1	30·0%	29·9%	27·8%	28·8%	31·3%
Test 2	27·9%	28·7%	28·0%	27·4%	28·3%
Credit					
Test 1	20·2%	14·4%	12·9%	15·8%	15·2%
Test 2	21·0%	17·1%	17·4%	16·4%	17·8%
Endgame					
Test 1	11·8%	13·8%	10·0%	13·1%	11·1%
Test 2	10·5%	13·4%	11·6%	12·1%	7·3%
Prob–Disj					
Test 1	17·0%	14·2%	10·1%	14·0%	10·0%
Test 2	18·4%	17·2%	17·8%	15·8%	10·0%

One further possibility has been explored. There is no obvious way to merge distinct decision trees, so pruned trees from different originals cannot be combined to form a composite tree that reflects the various strengths of its components. No such limitation applies to the production rule representation, though, because the union of sets of rules is itself a set. This line of thought led to a final experiment in which, for each domain, the rule sets produced from all ten decision trees were amalgamated and the collection winnowed as before. The composite rule set was then used to classify all cases in the test sets. The results in Table 3 show that these composite sets of production rules are both compact and accurate classifying

TABLE 3
Error rates of composite rule sets

	Number	Error rates	
	of rules	Test 1	Test 2
Hypothyroid	3	0·3%	1·0%
Discordant	2	0·6%	1·4%
LED Digits	23	28·2%	25·8%
Credit	11	13·0%	15·7%
Endgame	12	9·8%	5·4%
Prob–Disj	4	10·0%	10·0%

mechanisms, matching or outperforming the best of all other methods on nine of the twelve test sets.

4. Conclusion

The intention of this paper has been to investigate methods for simplifying decision trees without compromising their accuracy. The motivation behind this drive towards simplicity is the desire to turn decision trees into knowledge for use in expert systems.

Four methods have been discussed, all of which managed to achieve significant simplification when put to the test on sets of decision trees from six task domains. This simplification was often coupled with an actual improvement in classification accuracy on unseen cases. Two of the four methods needed a separate set of test cases in order to carry out the simplification and, since these did not perform noticeably better than the remaining two methods, the requirement of additional test data is a weakness. The last method, in which decision trees are reformulated as sets of production rules, has proved especially powerful.

I am grateful to the Garvan Institute of Medical Research for providing access to the thyroid data, and to Les Lazarus and Paul Compton in particular for their help. This work has been supported in part by grants from the Australian Research Grants Scheme and the Westinghouse Corporation.

References

Breiman, L., Friedman, J. H., Olshen, R. A. & Stone, C. J. (1984). *Classification and Regression Trees.* Belmont: Wadsworth.

Buchanan, B. G. & Mitchell, T. M. (1978). Model-directed learning of production rules. In Waterman, D. A. & Hayes-Roth, F. Eds. *Pattern Directed Inference Systems.* New York: Academic Press.

Feigenbaum, E. A. (1981). Expert systems in the 1980s. In Bond, A. Ed. *State of the Art Report on Machine Intelligence.* Maidenhead: Pergamon-Infotech.

Finney, D. J., Latscha, R., Bennett, B. M. & Hsu, P. (1963). *Tables for Testing Significance in a 2×2 Contingency Table.* Cambridge: Cambridge University Press.

Kononenko, I., Bratko, I., & Roškar, E. (1984). Experiments in automatic learning of medical diagnostic rules. *Technical Report,* Jozef Stefan Institute, Ljubljana, Yugoslavia.

Lavrač, N., Mozetič, I. & Kononenko, I. (1986). An experimental comparison of two learning programs in three medical domains. *Proceedings of ISSEK Workshop* 86. Turing Institute, Glasgow.

Michalski, R. S. & Chilausky, R. L. (1980). Learning by being told and learning by examples: an experimental comparison of the two methods of knowledge acquisition in the context of developing an expert system for soybeam disease diagnosis. *International Journal of Policy Analysis and Information Systems,* **4,** 2.

Michie, D. (1983). Inductive rule generation in the context of the Fifth Generation. *Proceedings of the Second International Machine Learning Workshop.* University of Illinois at Urbana-Champaign.

Michie, D. (1986). Current developments in expert systems. *Proceedings of the Second Australian Conference on Applications of Expert Systems.* Sydney.

Quinlan, J. R. (1986). Induction of decision trees. *Machine Learning,* **1,** 1.

Quinlan, J. R., Compton, P. J., Horn, K. A. & Lazarus, L. (1986). Inductive knowledge

acquisition: a case study. *Proceedings of the Second Australian Conference on Applications of Expert Systems,* Sydney.
SNEDECOR, G. W. & COCHRAN, W. G. (1980). *Statistical Methods* (7th edition). Iowa State University Press.
WATERMAN, D. A. (1986). *A Guide to Expert Systems.* Reading, MA: Addison-Wesley.
WINSTON, P. H. (1984). *Artificial Intelligence* (2nd edition). Reading, MA: Addison-Wesley.

PRISM: An algorithm for inducing modular rules

JADZIA CENDROWSKA
c/o The Faculty of Mathematics, The Open University, Walton Hall, Milton Keynes, MK7 6AA, U.K.

(Received 29 May 1987)

The decision tree output of Quinlan's ID3 algorithm is one of its major weaknesses. Not only can it be incomprehensible and difficult to manipulate, but its use in expert systems frequently demands irrelevant information to be supplied. This report argues that the problem lies in the induction algorithm itself and can only be remedied by radically altering the underlying strategy. It describes a new algorithm, PRISM which, although based on ID3, uses a different induction strategy to induce rules which are modular, thus avoiding many of the problems associated with decision trees.

1. Introduction

Considerable effort has recently been devoted to the development of efficient knowledge acquisition techniques for expert systems, with rule induction algorithms coming under the scrutiny of a substantial number of researchers. Particular attention has been paid to Ross Quinlan's ID3 algorithm (Quinlan, 1979*a*, 1979*b*, 1983*a*) which, having performed well in the domain of chess end-games, was soon adopted for use in a number of commercial applications. However, despite this apparent success, some major limitations to the ID3 algorithm have been identified (Bundy, Silver & Plummer, 1984; Cendrowska, 1984; Hart, 1985; O'Rorke, 1982), which makes its use unsuitable for many domains. The algorithm's inability to deal with noisy input data is an area for much current research and new improved variants of ID3 are constantly being reported in the technical press (A-Razzak, Hassan & Pettipher, 1985; Hart, 1985; Lavrac *et al.* 1986; Michie, 1983; Quinlan, 1983*b*), but concern has been shown about the way in which the results of the induction process are expressed.

This report discusses the second of these two limitations. ID3 produces its output in the form of a decision tree, which can be incomprehensible (to humans), difficult to manipulate (by humans and computers) and complicates the provision of explanations (by computers for humans). In addressing this subject, it is argued that current research aimed at modifying the decision tree output of ID3 is misplaced, that the decision tree output is an inherent weakness in the algorithm itself and that this can only be remedied by radically altering the underlying induction strategy.

The first part of this report explains the problem in more detail, highlighting it by means of a simple example which is introduced in Section 2. Section 3 describes how ID3 tackles the induction task using an information theoretic approach, and the inherent weaknesses of this approach are discussed in Section 4. The subsequent sections describe how the induction strategy can be changed to avoid some of these problems and outline a proposal for a new algorithm, PRISM which, although based

KNOWLEDGE-BASED SYSTEMS Vol. 1
ISBN 0-12-273251-0

on techniques employed by ID3, produces its output as modular rules. The report concludes with an assessment of the performance of PRISM on a large training set.

2. The domain

The following example, taken from the world of ophthalmic optics, will be used throughout this report to illustrate the procedures involved in rule induction.

An adult spectacle wearer enters an ophthalmic practice with a view to purchasing her first pair of contact lenses. She has had her eyes examined recently elsewhere and has brought her prescription with her. She understands that there are different types of contact lenses available, and that it is the optician's decision as to whether or not she is suitable for contact lens wear, and if so, which type she should be fitted with.

From the optician's point of view, this is a three-category† classification problem. His decision will be one of:

δ_1: the patient should be fitted with hard contact lenses,
δ_2: the patient should be fitted with soft contact lenses,
δ_3: the patient should not be fitted with contact lenses.

In reaching his decision he must consider one or more of four† factors:

a: the age of the patient
  1. young,
  2. pre-presbyopic, or
  3. presbyopic

b: her spectacle prescription
  1. myope, or
  2. hypermetrope

c: whether she is astigmatic
  1. no, or
  2. yes

d: her tear production rate
  1. reduced, or
  2. normal

Table 1 shows the optician's decision for each combination of the four factors. However, the optician does not carry such a table around with him, either on his person or in his head. Instead, through his training and experience, he has learned to exercise his professional judgement in each individual case, and will make his decision almost instinctively. If questioned as to how he arrived at a particular decision, his answer is likely to be of the form:

> This patient is not suitable for contact lens wear because her tear production rate is reduced.

or

> This patient can only be fitted with hard contact lenses because she is astigmatic. As she is young and has a normal tear production rate, hard lenses are not contraindictated.

† It should be noted that this is a highly simplified example. In real life there are many types of contact lenses and many more factors affecting the decision as to which type, if any, to fit.

TABLE 1
Decision table for fitting contact lenses

	Value of attribute a	b	c	d	Decision† δ		Value of attribute a	b	c	d	Decision† δ		Value of attribute a	b	c	d	Decision† δ
1	1	1	1	1	3	9	2	1	1	1	3	17	3	1	1	1	3
2	1	1	1	2	2	10	2	1	1	2	2	18	3	1	1	2	3
3	1	1	2	1	3	11	2	1	2	1	3	19	3	1	2	1	3
4	1	1	2	2	1	12	2	1	2	2	1	20	3	1	2	2	1
5	1	2	1	1	3	13	2	2	1	1	3	21	3	2	1	1	3
6	1	2	1	2	2	14	2	2	1	2	2	22	3	2	1	2	2
7	1	2	2	1	3	15	2	2	2	1	3	23	3	2	2	1	3
8	1	2	2	2	1	16	2	2	2	2	3	24	3	2	2	2	3

† The reader is asked not to be tempted to use this decision table to determine whether or not (s)he is suitable for contact lenses as there are many factors, not mentioned here, which may radically influence the decision.

Each explanation is a justification of a decision in terms of the values of relevant attributes, and is based on one or more 'rules of thumb':

if tear production rate is reduced
then do not fit contact lenses,

if the patient is astigmatic, **and**
the patient is young, **and**
the tear production rate is normal
then fit hard contact lenses.

Although the optician is able to easily justify each individual decision, he would find it quite difficult to formalize his knowledge as a complete set of rules. ID3 seeks to establish this underlying set of rules, in the form of a decision tree, from examples of the optician's decisions. The algorithm is described in detail in Section 3. Table 1 is used as the training set of instances; δ_1, δ_2 and δ_3 are the decisions or classifications; *a, b, c* and *d* are the attributes. Attribute *a* has three possible values (1, 2 and 3) and attributes *b, c* and *d* each have two possible values (1 and 2). Each instance is a description of a classification in terms of values of the four attributes. The following assumptions have been made about the training set:

- the classifications are mutually exclusive
- there is no noise, i.e. each instance is complete and correct
- each instance can be classified uniquely
- no instance is duplicated
- the values of the attributes are discrete
- the training set is complete, i.e. all possible combinations of attribute–value pairs are represented

3. An information theoretic approach I

3.1. ENTROPY

The training set can be thought of as a discrete information system, i.e. it contains a number of discrete messages (values of attributes) which impart some information about an event (classification). The entropy of a set of events has been defined as a measure of the 'freedom of choice' involved in the selection of the event, or the 'uncertainty' associated with this selection (Edwards, 1964, Goldman, 1968, Shannon & Weaver, 1949). Given a training set, S, if the above assumptions hold, then each instance is classified correctly and uniquely, i.e. there is no uncertainty about the classification. The entropy of S is 0. The entropy of a decision tree or rule set, which fully describes S is also 0, but in most cases the decision tree is a generalization of S, which implies that some information offered by the training set is redundant. ID3 tries to reduce this redundant information as much as possible (and thus find the least complex decision tree which fully describes the training set) by partitioning S into the smallest possible number of subsets, each of which can be described by a set of features (attribute–value pairs) whose entropy is 0.

If all that is known about the classifications is their probabilities of occurrence, $p(\delta_i;\, i = 1, 2, 3)$, then the entropy of the set of classifications,

$$H = -\sum_i p(\delta_i) \log_2 p(\delta_i) \text{ bits.} \tag{1}$$

For the contact lens classification problem,

$$H = -p(\delta_1) \log_2 p(\delta_1) - p(\delta_2) \log_2 p(\delta_2) - p(\delta_3) \log_2 p(\delta_3) \text{ bits.}$$

The probabilities of occurrence of each of the classifications are

$$p(\delta_1) = 4/24,$$
$$p(\delta_2) = 5/24,$$
$$p(\delta_3) = 15/24.$$

Thus,

$$\begin{aligned} H &= -\frac{4}{24}\log_2\left(\frac{4}{24}\right) - \frac{5}{24}\log_2\left(\frac{5}{24}\right) - \frac{15}{24}\log_2\left(\frac{15}{24}\right) \\ &= 0{\cdot}4308 + 0{\cdot}4715 + 0{\cdot}4238 \\ &= 1{\cdot}3261 \text{ bits.} \end{aligned} \tag{2}$$

The induction algorithm partitions the training set into subsets in such a way as to reduce this entropy by the maximum amount, and continues doing so recursively until the entropy is 0.

3.2. REDUCING ENTROPY

If the training set, S, is divided according to the values of some attribute, α, then unless the classification, δ, is completely independent of α, the values will contain some information about δ. The total entropy of the subsets is known as the conditional entropy of S with known α, $H(S \mid \alpha)$. Let $p(\alpha_x)$ be the probability that attribute α has value x, and let $p(\delta_n \cap \alpha_x)$ be the probability that the classification is δ_n and the value of α is x. Then

$$H(S \mid \alpha) = H(S \cap \alpha) - H(\alpha), \tag{3}$$

where

$$H(S \cap \alpha) = -\sum_x \sum_n p(\delta_n \cap \alpha_x) \log_2 p(\delta_n \cap \alpha_x) \tag{4}$$

and

$$H(\alpha) = -\sum_x p(\alpha_x) \log_2 p(\alpha_x). \tag{5}$$

By performing this calculation for each attribute, it is possible to minimize the entropy of S by dividing it into subsets according to the values of that attribute for which $H(S \mid \alpha)$ is minimum.

The calculation can be simplified by using a frequency table, for example for attribute a:

No. of instances referencing	a_1	a_2	a_3	Total
δ_1	2	1	1	4
δ_2	2	2	1	5
δ_3	4	5	6	15
Total	8	8	8	24

$$\begin{aligned}
H(S \mid a) &= H(S \cap a) - H(a) \\
&= -\sum_x \sum_n p(\delta_n \cap a_x) \log_2 p(\delta_n \cap a_x) + \sum_x p(a_x) \log_2 p(a_x) \\
&= -3 \times \frac{2}{24} \log_2 \left(\frac{2}{24}\right) - 3 \times \frac{1}{24} \log_2 \left(\frac{1}{24}\right) - \frac{4}{24} \log_2 \left(\frac{4}{24}\right) \\
&\quad - \frac{5}{24} \log_2 \left(\frac{5}{24}\right) - \frac{6}{24} \log_2 \left(\frac{6}{24}\right) + 3 \times \frac{8}{24} \log_2 \left(\frac{8}{24}\right) \\
&= \frac{1}{24} (3 \times 8 \log_2 8 - 3 \times 2 \log_2 2 - 2 \times \log_2 1 - 4 \log_2 4 \\
&\quad - 5 \log_2 5 - 6 \log_2 6) \\
&= 1{\cdot}2867 \text{ bits.}
\end{aligned} \tag{6}$$

Similarly,

$$H(S \mid b) = 1{\cdot}2867 \text{ bits}, \tag{7}$$

$$H(S \mid c) = 0{\cdot}9491 \text{ bits}, \tag{8}$$

$$H(S \mid d) = 0{\cdot}7773 \text{ bits}. \tag{9}$$

Therefore, the entropy of S can be reduced by the greatest amount by dividing S according to the values of attribute d. Two subsets are formed, each of which is then further subdivided in the same way until the entropy of each subset is 0, i.e. all instances in the subset belong to the same classification. The final decision tree is

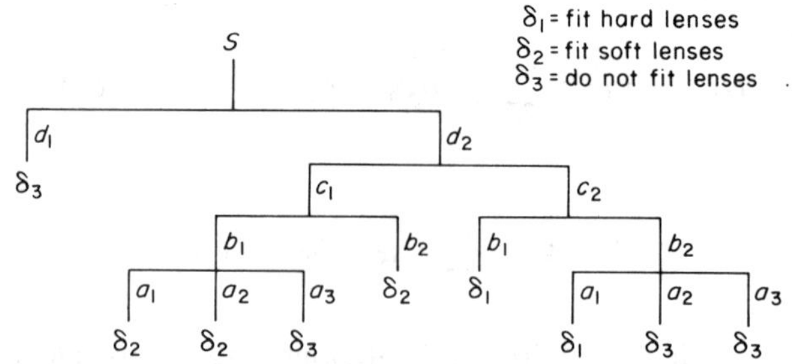

FIG. 1. Decision tree produced by ID3.

shown in Fig. 1. For convenience, this can be written as a set of individual rules:

1. $d_1 \rightarrow \delta_3$
2. $d_2 \wedge c_1 \wedge b_1 \wedge a_1 \rightarrow \delta_2$
3. $d_2 \wedge c_1 \wedge b_1 \wedge a_2 \rightarrow \delta_2$
4. $d_2 \wedge c_1 \wedge b_1 \wedge a_3 \rightarrow \delta_3$
5. $d_2 \wedge c_1 \wedge b_2 \rightarrow \delta_2$
6. $d_2 \wedge c_2 \wedge b_1 \rightarrow \delta_1$
7. $d_2 \wedge c_2 \wedge b_2 \wedge a_1 \rightarrow \delta_1$
8. $d_2 \wedge c_2 \wedge b_2 \wedge a_2 \rightarrow \delta_3$
9. $d_2 \wedge c_2 \wedge b_2 \wedge a_3 \rightarrow \delta_3$

4. Rule representation

One of the principal features of rule-based expert systems is that the modularity of the rules typically enables a knowledge base to be easily updated or modified. It also provides a means for explanation. There is a requirement, therefore, that rules should be both modular and comprehensible, whether they are elicited from experts or automatically induced from examples.

Although ID3 has been proved to be computationally efficient (Carbonell, Michalski & Mitchell, 1983; Michie, 1983; O'Rorke, 1982), it produces its output in the form of a decision tree (e.g. Fig. 1). This decision tree representation of rules has a number of disadvantages. Firstly, decision trees are extremely difficult to manipulate—to extract information about any single classification it is necessary to examine the complete tree. This problem is only partially resolved by trivially converting the tree into a set of individual rules, as the amount of information contained in some of these will often be more than an easily be assimilated. More importantly, there are rules that cannot easily be represented by trees.

Consider, for example, the following rule set:

$$\text{Rule 1: } a_1 \wedge b_1 \rightarrow \delta_1,$$
$$\text{Rule 2: } c_1 \wedge d_1 \rightarrow \delta_1.$$

Suppose that Rules 1 and 2 cover all instances of class δ_1 and all other instances are of class δ_2. These two rules cannot be represented by a single decision tree as the

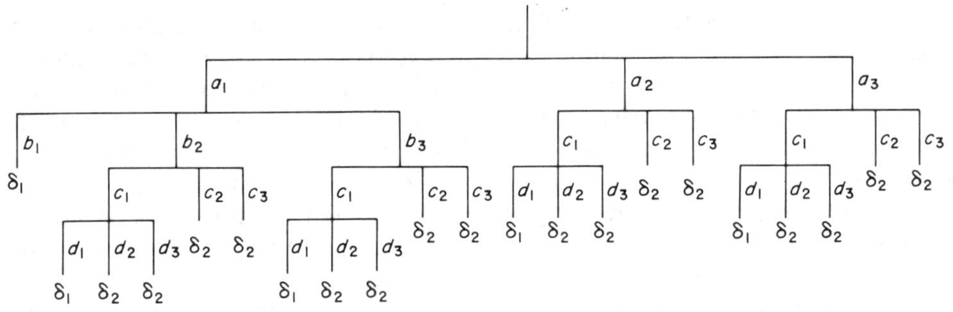

FIG. 2. Decision tree representation of Rules 1 and 2 (Section 4).

root node of the tree must split on a single attribute, and there is no attribute which is common to both rules. The simplest decision tree representation of the set of instances covered by these rules would necessarily add an extra term to one of the rules, which in turn would require at least one extra rule to cover instances excluded by the addition of that extra term. The complexity of the tree would depend on the number of possible values of the attributes selected for partitioning. For example, let the four attributes, *a, b, c* and *d* each have three possible values, 1, 2 and 3, and let attribute *a* be selected for partitioning at the root node. Then the simplest decision tree representation of Rules 1 and 2 above is shown in Fig. 2. The paths relating to class δ_1 can be listed as follows:

1. $a_1 \wedge b_1 \rightarrow \delta_1$,
2. $a_1 \wedge b_2 \wedge c_1 \wedge d_1 \rightarrow \delta_1$,
3. $a_1 \wedge b_3 \wedge c_1 \wedge d_1 \rightarrow \delta_1$,
4. $a_2 \wedge c_1 \wedge d_1 \rightarrow \delta_1$,
5. $a_3 \wedge c_1 \wedge d_1 \rightarrow \delta_1$.

Clearly, the consequence of forcing a simple rule set into a decision tree representation is that the individual rules, when extracted from the tree, are often too specific (i.e. they reference attributes which are irrelevant). This makes them highly unsuitable for use in many domains, as is illustrated by the following example.

Suppose the decision tree in Fig. 1 was used as the knowledge base for an expert system advising on contact lens suitability, and suppose the patient requiring contact lenses was a presbyope with high hypermetropia and astigmatism (attributes a_3 & b_2 & c_2). The optician would know immediately from the age of the patient and her prescription that she was not a suitable candidate for contact lens wear (a decision taking about 30 seconds to make and costing the patient nothing). The expert system, however, would be unable to make a decision without the result of a tear production rate test (attribute *d*). This test is normally carried out as part of a contact lens consultation requiring a lot of time and payment of a fee. Having spent all this time and money, it would be quite understandable if the patient became upset or angry on finding out that the consultation had been, after all, unnecessary. The consequences could be even more serious if the expert system was a medical one and attribute *d* involved surgery.

Clearly, a decision tree in its unmodified form is most unsuitable for some domains, not only because it an be incomprehensible, but because in many cases its use would demand irrelevant information to be supplied, information that could be costly to obtain. Attempts have been made at modifying the algorithm to avoid this problem by assigning a 'cost' to each attribute. Attempts have also been made at converting decision trees into simple rule sets by identifying and removing redundant nodes, or by incorporating extra information which enables the user to focus on only relevant parts of the tree, but the problem is not an easy one to solve, particularly for very large and complex decision trees.

Although simplification of the trees is possible by identifying common branches or parts of branches, the combinatorial explosion in the number of comparisons that have to be made as the complexity increases makes this method only feasible for small trees. Also, parts of a branch may be matched in different ways, and the question then arises as to which is the better generalization to make. This would involve either asking the expert, or using another rule induction program to induce new rules from the old ones.

5. An information theoretic approach II

5.1. ENTROPY VS. INFORMATION GAIN

The main cause of the problem described in the preceding section is either that an attribute is highly relevant to only one classification and irrelevant to the others, or that only one value of the attribute is relevant. For example, the attribute d in the contact lens problem is highly relevant to the classification δ_3, *if its value is* 1, and because of this, it is selected for partitioning the training set, for which all its values are used.

Figure 3 shows the decision tree after S has been partitioned according to the

S

d_1

	a	b	c	d	δ
1	1	1	1	1	3
3	1	1	2	1	3
5	1	2	1	1	3
7	1	2	2	1	3
9	2	1	1	1	3
11	2	1	2	1	3
13	2	2	1	1	3
15	2	2	2	1	3
17	3	1	1	1	3
19	3	1	2	1	3
21	3	2	1	1	3
23	3	2	2	1	3

$H(S|d_1)=0$ bits

d_2

	a	b	c	d	δ
2	1	1	1	2	2
4	1	1	2	2	1
6	1	2	1	2	2
8	1	2	2	2	1
10	2	1	1	2	2
12	2	1	2	2	1
14	2	2	1	2	2
16	2	2	2	2	3
18	3	1	1	2	3
20	3	1	2	2	1
22	3	2	1	2	2
24	3	2	2	2	3

$H(S|d_2)=1{\cdot}555$ bits

FIG. 3. S partitioned according to d.

values of attribute d. It can be seen that although the entropy of the branch d_1 has been reduced to 0, the entropy of the branch d_2 has actually increased to 1·555 bits. Attribute d was chosen because ID3 minimizes the *average entropy* of the training set, or alternatively, it maximizes the *average* amount of information contributed by an attribute to the determination of *any* classification.

In order to eliminate the use of irrelevant values of attributes and attributes which are irrelevant to a classification, the algorithm needs to maximize the *actual* amount of information contributed by knowing the value of the attribute to the determination of a *specific* classification.

5.2. INFORMATION CONTENT

As stated at the beginning of Section 3, the values of attributes can be thought of as discrete messages in a discrete information system. Now, the amount of information about an event in a message i,

$$I(i) = \log_2\left(\frac{\text{probability of event after the message is received}}{\text{probability of event before the message is received}}\right)\text{ bits.}$$

The training set, S, contains 4 instances belonging to class δ_1, 5 belonging to class δ_2 and 15 to class δ_3. Therefore, the probability of an instance belonging to class δ_1, $p(\delta_1)$ is 4/24 and thus if the message i was δ_1 (i.e. the class is δ_1) then the amount of information received in this message,

$$I(\delta_1) = \log_2\left(\frac{1}{p(\delta_1)}\right) = -\log_2\left(\frac{4}{24}\right) = 2{\cdot}585\text{ bits.} \tag{10}$$

Similarly, the amount of information received in the message δ_2,

$$I(\delta_2) = \log_2\left(\frac{1}{p(\delta_2)}\right) = -\log_2\left(\frac{5}{24}\right) = 2{\cdot}263\text{ bits.} \tag{11}$$

and in the message δ_3,

$$I(\delta_3) = \log_2\left(\frac{1}{p(\delta_3)}\right) = -\log_2\left(\frac{15}{24}\right) = 0{\cdot}678\text{ bits.} \tag{12}$$

Thus the lower the probability of occurrence of an event, the more information we receive if we are told that the event has occurred.

Now, if the message received was that attribute d has value 1, then the amount of information received in this message about δ_3,

$$I(\delta_3 \mid d_1) = \log_2\left(\frac{p(\delta_3 \mid d_1)}{p(\delta_3)}\right)\text{ bits.} \tag{13}$$

where $p(\delta_3 \mid d_1)$ is the probability of δ_3 given that the value of d is 1.

For S, $p(\delta_3 \mid d_1) = 1$, therefore

$$I(\delta_3 \mid d_1) = \log_2\left(\frac{1}{p(\delta_3)}\right) = 0{\cdot}678\text{ bits.} \tag{14}$$

Thus knowing that attribute d has value 1 contributes 0·678 bits of information to the belief that an instance belongs to class δ_3.

If, on the other hand, the message was that attribute d has value 2, then the amount of information received about δ_3,

$$I(\delta_3 \mid d_2) = \log_2\left(\frac{p(\delta_3 \mid d_2)}{p(\delta_3)}\right) = \log_2\left(\frac{3/12}{15/24}\right) = -1{\cdot}322 \text{ bits.} \tag{15}$$

The minus sign indicates that knowing that the value of d is 2 makes it less certain that an instance belongs to δ_3 than if the value of d was unknown. d_2 is therefore not a good choice for describing δ_3.

If an attribute-value pair, α_x, and a classification, δ_n, are completely independent, then $p(\delta_n \mid \alpha_x) = p(\delta_n)$ and $I(\delta_n \mid \alpha_x) = \log_2 1 = 0$, i.e. the fact α_x contributes no information to the belief that the class is δ_n.

5.3. MAXIMIZING INFORMATION GAIN

The task of an induction algorithm must be to find the attribute–value pair, α_x, which contributes the most information about a specified classification, δ_n, i.e. for which $I(\delta_n \mid \alpha_x)$ is maximum. Now,

$$I(\delta_n \mid \alpha_x) = \log_2\left(\frac{p(\delta_n \mid \alpha_x)}{p(\delta_n)}\right) \text{ bits.} \tag{16}$$

but $p(\delta_n)$ is the same for all α_x, and thus it is only necessary to find the α_x for which $p(\delta_n \mid \alpha_x)$ is maximum.

The values of $p(\delta_n \mid \alpha_x)$ for all α_x and $n = 1$ are listed in Table 2a. There are two candidates for 'best' α_x. These are c_2 and d_2. For c_2, chosen arbitrarily, the information gain,

$$I(\delta_1 \mid c_2) = \log_2\left(\frac{p(\delta_1 \mid c_2)}{p(\delta_1)}\right) = \log_2\left(\frac{4/12}{4/24}\right) = 1 \text{ bit.} \tag{17}$$

Had d_2 been chosen, the information gain would also have been 1 bit. Repeating the process now on a subset of S which contains only those instances which have value 2 for attribute c, it can be seen from Table 2b that $p(\delta_1 \mid \alpha_x)$ has the highest value for d_2. The information gain (for this subset),

$$I(\delta_1 \mid d_2) = \log_2\left(\frac{p(\delta_1 \mid d_2)}{p(\delta_1)}\right) = \log_2\left(\frac{4/6}{4/12}\right) = 1 \text{ bit.} \tag{18}$$

If the process is now repeated on the subset which contains only those instances which have value 2 for attribute c and value 2 for attribute d (Table 2c), there is again a choice for 'best' α_x. Suppose the second of these, b_1, is selected.† Then

$$I(\delta_1 \mid b_1) = \log_2\left(\frac{p(\delta_1 \mid b_1)}{p(\delta_1)}\right) = \log_2\left(\frac{1}{4/6}\right) = 0{\cdot}585 \text{ bits.} \tag{19}$$

From equation 10, the information provided by the message δ_1 before any attributes are known = 2·585 bits.

The information provided by c_2 = 1 bit.

† The reason for this choice is explained in Section 7.2.1.

TABLE 2a
Selecting the first term

α_x	$p(\delta_1 \mid \alpha_x)$
a_1	$2/8 = 0{\cdot}25$
a_2	$1/8 = 0{\cdot}125$
a_3	$1/8 = 0{\cdot}125$
b_1	$3/12 = 0{\cdot}25$
b_2	$1/12 = 0{\cdot}083$
c_1	$0 = 0$
c_2	$4/12 = 0{\cdot}333$
d_1	$0 = 0$
d_2	$4/12 = 0{\cdot}333$

TABLE 2b
Selecting the second term

α_x	$p(\delta_1 \mid \alpha_x)$
a_1	$2/4 = 0{\cdot}5$
a_2	$1/4 = 0{\cdot}25$
a_3	$1/4 = 0{\cdot}25$
b_1	$3/6 = 0{\cdot}5$
b_2	$1/6 = 0{\cdot}167$
d_1	$0 = 0$
d_2	$4/6 = 0{\cdot}667$

TABLE 2c
Selecting the third term

α_x	$p(\delta_1 \mid \alpha_x)$
a_1	$2/2 = 1$
a_2	$1/2 = 0{\cdot}5$
a_3	$1/2 = 0{\cdot}5$
b_1	$3/3 = 1$
b_2	$1/3 = 0{\cdot}333$

The information provided by d_2 when c_2 is known = 1 bit.

The information provided by b_1 when d_2 and c_2 are known = 0·585 bits.

Therefore, the information provided by $c_2 \wedge d_2 \wedge b_1 = 1 + 1 + 0{\cdot}585 = 2{\cdot}585$ bits.

i.e. the message $c_2 \wedge d_2 \wedge b_1$ provides the same amount of information as the message δ_1.

Specialization of (i.e. adding more attribute-value pairs to) $c_2 \wedge d_2 \wedge b_1$ does not increase the information gain. All other attributes are irrelevant in this description as all instances containing c_2 & d_2 & b_1 belong to class δ_1 ($p(\delta_1 \mid c_2 \wedge d_2 \wedge b_1) = 1$). The induced rule is therefore

$$c_2 \wedge d_2 \wedge b_1 \rightarrow \delta_1$$

and is known to be correct for S.

5.4. TRIMMING THE TREE

The decision tree at this stage of the induction process is shown in Fig. 4. The algorithm has concentrated on building the shortest branch possible for the class δ_1. The remaining branches are not yet labelled, and the next step in the induction process is to identify the best rule for the set of instances which are not examples of the first rule. This is done by removing from S all instances containing c_2 & d_2 & b_1

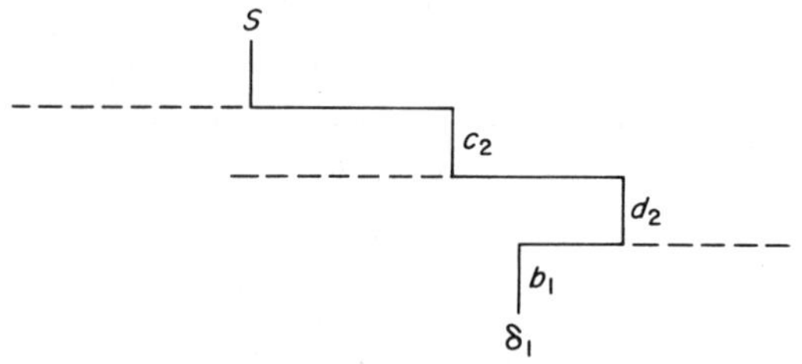

FIG. 4. 'Decision tree' after induction of the first rule.

and applying the algorithm to the remaining instances. If this is repeated until there are no instances of class δ_1 left in S, the result is not a decision tree but a collection of branches. The whole process can then be repeated for each classification in turn, starting with the complete training set, S, each time.

The final output is an unordered collection of modular rules, each rule being as general as possible (but see Section 7.2), thus ensuring that there are no redundant terms. The rule set for the optician's contact lens classification problem is as follows:

1. $c_2 \wedge d_2 \wedge b_1 \rightarrow \delta_1$,
2. $a_1 \wedge c_2 \wedge d_2 \rightarrow \delta_1$,
3. $c_1 \wedge d_2 \wedge b_2 \rightarrow \delta_2$,
4. $c_1 \wedge d_2 \wedge a_1 \rightarrow \delta_2$,
5. $c_1 \wedge d_2 \wedge a_2 \rightarrow \delta_2$,
6. $d_1 \rightarrow \delta_3$,
7. $a_3 \wedge b_1 \wedge c_1 \rightarrow \delta_3$,
8. $b_2 \wedge c_2 \wedge a_2 \rightarrow \delta_3$,
9. $b_2 \wedge c_2 \wedge a_3 \rightarrow \delta_3$.

Although the number of rules in this set is the same as the number of leaf nodes in the decision tree (Fig. 1), six of the rules have had redundant terms removed. The presbyopic patent with high hypermetropia and astigmatism no longer needs to undergo an examination to be told that she is not suitable for contact lens wear (Rule 9).

6. The 'correctness' of rules and predictability

Given that the assumptions listed at the end of Section 2 hold, the above algorithm produces a complete set of correct rules.† This section is devoted to explaining first the meaning, and then the importance of this statement.

6.1. A COMPLETE SET...

A set of rules is complete if for every possible example of a classification there is at least one rule which explains it. It is assumed that all examples can be adequately

† This statement applies to most training sets. For the remainder, the algorithm must first be modified as explained in Section 7.2.

described in terms of the attributes used for the training set. Such a set of rules can be used for predicting the classification of any instance, which is a basic requirement for any rule induction program. A set of rules *must* be complete if it is induced from a complete training set. Otherwise, a rule set can be either complete or incomplete.

6.2. ... OF CORRECT RULES

On the other hand, a rule which is not incorrect is not necessarily correct. There are different levels of 'correctness'. An incorrect rule is one which misclassifies instances. For example, the rule Rule 1: $a_1 \wedge b_1 \rightarrow \delta_1$ is incorrect if it is too general, because there will be some instances which have value 1 for attribute a and value 1 for attribute b, but which are of a class of other than δ_1. These instances will be misclassified as δ_1 by Rule 1. It is possible for a rule to be both too general and too specific; for example, if Rule 1 should have been $a_1 \wedge c_1 \rightarrow \delta_1$, then it is too general with respect to attribute c but too specific with respect to attribute b. However, this does not alter the fact that the rule is incorrect because it still misclassifies some instances. An incorrect rule is, therefore, one which does not reference all the relevant attributes.

A rule which is not too general is correct in the sense that it will not misclassify any instances. If it is too specific, however, it will fail to classify some instances which it should classify, although there may be other rules in the set which will cover these instances. A rule which is too specific is incorrect in the sense that it will not fire unless the value of an irrelevant attribute has been determined. The undesirability of this was discussed in Section 4.

A 'correct' rule, therefore, is one which references all the relevant attributes and no irrelevant ones. A complete set of correct rules classifies all possible instances correctly.

6.3. PREDICTABILITY

The algorithm described in Section 5 induces a complete set of correct rules, on the condition that the assumptions listed in Section 2 hold. However, these assumptions are extremely restrictive and unlikely to be applicable to 'real-life' classification problems. In particular, the last assumption—that the training set be complete—is most unrealistic. Relaxing any of the restrictions, even slightly, introduces into the set of induced rules the possibility of errors or uncertainty, thus reducing their predictability value. If the rule set cannot be guaranteed to be complete and correct (in the strict sense) when the training set does meet the assumptions then any errors or uncertainty introduced by relaxing the restrictions will be greatly increased. The importance of knowing that the rule set is complete and correct for a complete and noiseless training set cannot be over-emphasized.

7. Prism

The theory outlined in Section 5 has been embodied in a new rule induction program, PRISM. PRISM takes as input a training set entered as a file of ordered sets of attribute values, each set being terminated by a classification. Information about the attributes and classifications (e.g. name, number of possible values, list of possible values, etc.) is input from a separate file at the start of the program, and the results are output as individual rules for each of the classifications listed in terms of the described attributes.

7.1. THE BASIC ALGORITHM

The basic induction algorithm is essentially as described above, namely:

If the training set contains instances of more than one classification, then for each classification, δ_n, in turn:

Step 1: calculate the probability of occurrence, $p(\delta_n \mid \alpha_x)$, of the classification δ_n for each attribute–value pair α_x,

Step 2: select the α_x for which $p(\delta_n \mid \alpha_x)$ is a maximum and create a subset of the training set comprising all the instances which contain the selected α_x,

Step 3: repeat Steps 1 and 2 for this subset until it contains only instances of class δ_n. The induced rule is a conjunction of all the attribute–value pairs used in creating the homogeneous subset.

Step 4: remove all instances covered by this rule from the training set,

Step 5: repeat Steps 1–4 until all instances of class δ_n have been removed.

When the rules for one classification have been induced, the training set is restored to its initial state and the algorithm is applied again to induce a set of rules covering the next classification. As the classifications are considered separately, their order of presentation is immaterial. If all instances are of the same classification then that classification is returned as the rule, and the algorithm terminates.

Although the basic induction algorithm used by PRISM is based on techniques employed by ID3, it is quite unlike ID3 in many respects. The major difference is that PRISM concentrates on finding only relevant values of attributes, while ID3 is concerned with finding the attribute which is most relevant overall, even though some values of that attribute may be irrelevant. All other differences between the two algorithms stem from this. ID3 divides a training set into homogeneous subsets without reference to the class of this subset, whereas PRISM must identify subsets of a specific class. This has the disadvantage of slightly increased computational effort, but the advantage of an output in the form of modular rules rather than a decision tree.

7.2. THE USE OF HEURISTICS

The two algorithms are similar in that they both employ an information theoretic approach to discovering disjunctive rules by grouping together sets of instances with similar features. Consequently, they both encounter similar difficulties in certain circumstances. In particular, there is the problem of which attribute or attribute–value pair to choose when the results of the respective calculations indicate that there are two or more which are equal. In ID3, however, the choice is immaterial because the objective is to reduce entropy at the maximal rate and this is achieved equally well whichever attribute is chosen. On the other hand, if the wrong choice is made in PRISM, then the result is that an irrelevant attribute–value pair may be chosen. Fortunately, this most unwelcome feature can be avoided by incorporating some heuristics in the basic algorithm.

7.2.1. Opting for generality I

If there are two or more rules describing a classification, PRISM tries to induce the most general rule first. The rationale behind this is that the more general a rule is then the less likely it is to reference an irrelevant attribute. Thus where there is a choice of attribute–value pairs, PRISM selects that attribute–value pair which has the highest frequency of occurrence in the set of instances being considered. Referring back to Table 2c in Section 5 (selection of a third term for the first rule for class δ_1), it can be seen that the attribute–value pairs a_1 and b_1 both offer an equal information gain. PRISM selects b_1 because the resulting rule covers three instances, whereas the rule resulting from the selection of a_1 would only cover two instances. Thus the rule $c_2 \wedge d_2 \wedge b_1 \rightarrow \delta_1$ is more general than $c_2 \wedge d_2 \wedge a_1 \rightarrow \delta_1$. In this particular case, both rules are in fact equally correct, and so the order in which they are induced does not really matter, but opting for generality in this way has the advantage of reducing computational effort when there is a significant difference in the number of instances covered by each of the rules. Its true value, however, is realized when the training set is an incomplete one and there is a possibility that one potential rule is a specialization of another. In this situation PRISM *must* select the more general.

7.2.2. Opting for generality II

When both the information gain offered by two or more attribute–value pairs is the same and the numbers of instances referencing them is the same, PRISM selects the first. This is the only time that the order of input of the attributes affects the induction process, but in these cases it is still possible for an irrelevant attribute–value pair to be selected. To illustrate how PRISM copes with this situation, suppose there are four attributes, *a, b, c* and *d,* each having three possible values, 1, 2 and 3, and the rules to be induced for class δ_1 are:

$$\text{Rule 1: } c_1 \wedge d_1 \rightarrow \delta_1,$$
$$\text{Rule 2: } c_2 \wedge d_2 \rightarrow \delta_1,$$
$$\text{Rule 3: } c_3 \wedge d_3 \rightarrow \delta_1.$$

Thus, attributes *a* and *b* are irrelevant to δ_1, whereas all values of attributes *c* and *d* are equally relevant. If the training set is complete, then $p(\delta_1 \mid \alpha_x)$ is the same for all α_x and PRISM selects a_1. The subset containing only instances which have value 1 for attribute *a* also presents the same problem—$p(\delta_1 \mid \alpha_x)$ is equal for all α_x, so b_1 is selected, and so on. The result is the following set of rules:

$$\text{Rule 1: } a_1 \wedge b_1 \wedge c_1 \wedge d_1 \rightarrow \delta_1,$$
$$\text{Rule 2: } a_2 \wedge b_1 \wedge c_1 \wedge d_1 \rightarrow \delta_1,$$
$$\text{Rule 3: } a_3 \wedge b_1 \wedge c_1 \wedge d_1 \rightarrow \delta_1,$$
$$\text{Rule 4: } b_2 \wedge a_1 \wedge c_1 \wedge d_1 \rightarrow \delta_1,$$
$$\text{Rule 5: } b_3 \wedge a_1 \wedge c_1 \wedge d_1 \rightarrow \delta_1.$$

At this stage $p(\delta_1 \mid \alpha_x)$ is greater for c_2, c_3, d_2 and d_3 than for any other attribute–value pair, so the next two rules are induced correctly:

$$\text{Rule 6: } c_2 \wedge d_2 \rightarrow \delta_1,$$

$$\text{Rule 7: } c_3 \wedge d_3 \rightarrow \delta_1.$$

The remaining instances all have value 1 for attribute c and value 1 for attribute d, so the final rule is

$$\text{Rule 8: } c_1 \wedge d_1 \rightarrow \delta_1.$$

Rules 1–5 are all specializations of Rule 8. To avoid this happening, PRISM first induces all rules for a classification and then selects the most general of these on the basis of (i) the rule which covers the maximum number of instances, and (ii) the rule which references the fewest attributes. The instances covered by this rule are removed from the training set, and PRISM goes on to induce the remaining rules in the same way. For the above example, the result is that Rules 6 and 7 are induced first, and then Rule 8. These three rules account for all instances of class δ_1, so Rules 1–5 are discarded.

Although this iterative procedure is quite costly in terms of computational effort, it ensures (at least for a complete training set) that the induced rules are maximally general.

8. Induction from incomplete training sets

When PRISM is applied to a complete training set, the resulting set of rules can confidently be expected to be complete and correct. When the training set is incomplete, this confidence is reduced. The smaller the relative number of instances in the training set, the more likely it is that the rule set will contain errors. Errors in the induction process arise for a number of reasons and can be best explained using an (artificial) example. For this purpose, suppose there are four attributes, a, b, c and d. Attribute a has five possible values $(1, 2, 3, 4, 5)$, attributes b and c each have four possible values $(1, 2, 3, 4)$ and attribute d has three possible values $(1, 2, 3)$. Thus a complete training set would consist of $5 \times 4 \times 4 \times 3 = 240$ instances. Suppose that the rule set governing class δ_1 is

$$\text{Rule 1: } a_4 \wedge d_2 \rightarrow \delta_1,$$

$$\text{Rule 2: } c_1 \wedge d_1 \rightarrow \delta_1,$$

$$\text{Rule 3: } a_2 \wedge c_4 \wedge d_2 \rightarrow \delta_1,$$

$$\text{Rule 4: } a_5 \wedge c_4 \wedge d_2 \rightarrow \delta_1,$$

and that the 40 instances listed in Table 3 are the only ones available to the induction program.

The set of rules induced by PRISM for the class δ_1 is

$$\text{Rule A: } a_4 \wedge d_2 \rightarrow \delta_1,$$

$$\text{Rule B: } a_3 \wedge c_1 \wedge d_1 \rightarrow \delta_1,$$

$$\text{Rule C: } a_2 \wedge c_4 \rightarrow \delta_1,$$

$$\text{Rule D: } b_1 \wedge d_1 \wedge c_1 \rightarrow \delta_1.$$

TABLE 3
Example of incomplete training set

a	b	c	d	δ	a	b	c	d	δ	a	b	c	d	δ	a	b	c	d	δ
1	1	3	3	2	2	1	2	2	2	3	2	1	1	1	4	3	2	2	1
1	2	1	2	2	2	2	2	1	2	3	2	4	1	2	4	4	1	3	2
1	2	3	1	2	2	2	4	2	1	3	2	4	2	2	4	4	3	1	2
1	3	1	3	2	2	3	2	1	2	3	3	1	1	1	5	1	1	2	2
1	3	3	2	2	2	3	3	1	2	3	3	1	2	2	5	1	3	2	2
1	4	1	3	2	2	3	3	3	2	3	3	2	2	2	5	2	2	2	2
1	4	4	1	2	2	4	1	3	2	3	4	2	1	2	5	3	1	2	2
2	1	1	1	1	2	4	2	1	2	4	1	3	2	1	5	3	2	3	2
2	1	1	3	2	3	1	1	1	1	4	1	4	2	1	5	4	1	3	2
2	1	2	1	2	3	1	4	3	2	4	2	1	3	2	5	4	4	3	2

It can be seen that Rule 1 is induced correctly (Rule A), Rule 2 has been specialized in two ways (Rules B and D), Rule 3 has been generalized (Rule C) and Rule 4 has not been induced at all. The decision tree induced by ID3 from the same training set is shown in Fig. 5. The bold lines depict the branches for class δ_1.

8.1. FAILURE TO INDUCE A RULE

A rule will not be induced if there are no examples of it in the training set (e.g. Rule 4 above). This applies to all induction programs. Even human beings cannot be expected to induce rules from non-existent information.

8.2. OVER-GENERALIZATION

An induced rule may be too general if there are no counter-examples to it in the training set. For example, Rule C above ($a_2 \wedge c_4 \rightarrow \delta_1$) is a generalization of the correct rule, Rule 3 ($a_2 \wedge c_4 \wedge d_2 \rightarrow \delta_1$). As there are no instances containing a_2 & c_4 & d_1 or a_2 & c_4 & d_3 in the training set, then there are no counter-examples to $a_2 \wedge c_4 \rightarrow \delta_1$ and no reason to specialize. Any attempts to specialize automatically would have unwanted side-effects on rules which were not too general.

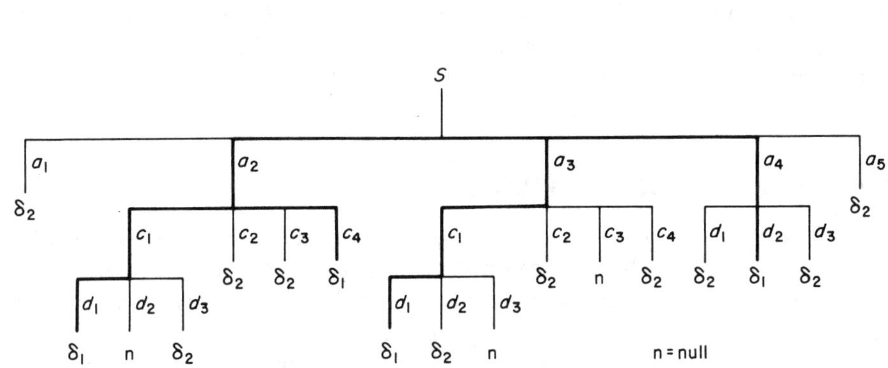

FIG. 5. Decision tree produced from the training set in Table 3.

TABLE 4
Relative frequency f vs. *probability p for a small training set*

α_x	$f(\delta_1 \mid \alpha_x)$	$p(\delta_1 \mid \alpha_x)$	α_x	$f(\delta_1 \mid \alpha_x)$	$p(\delta_1 \mid \alpha_x)$
a_1	0	0·083	b_4	0	0·107
a_2	0·182	0·167	c_1	0·267	0·357
a_3	0·333	0·083	c_2	0	0
a_4	0	0·125	c_3	0	0
a_5	0	0·083	c_4	0·167	0·071
b_1	0·222	0·107	d_1	0·286	0·25
b_2	0·222	0·107	d_2	0·091	0·063
b_3	0·1	0·107	d_3	0	0

8.3. OVER-SPECIALIZATION

Theoretically, the induction algorithm is based on finding the α_x for which $p(\delta_1 \mid \alpha_x)$ is a maximum. In practice, for an incomplete training set, the true probability of occurrence p is unknown, and is approximated by the relative frequency, $f(\delta_1 \mid \alpha_x)$. This approximation of p introduces errors in the estimation of information gain of each α_x, which become significant for small training sets, resulting in the selection of an irrelevant attribute–value pair as the best representative of δ_1. Rule B above ($a_3 \wedge c_1 \wedge d_1 \rightarrow \delta_1$) is an example of this type of error, in which a_3 is the unwanted term. The reason for the selection of a_3 becomes obvious when the values of p and f for each α_x are compared (see Table 4). It can be seen that $p(\delta_1 \mid a_3)$, is relatively small compared with $p(\delta_1 \mid c_1)$, but as the distribution of a_3 is inaccurately represented in the training set, $f(\delta_1 \mid a_3)$ is artifically high, thus leading to the selection of a_3 as 'best' attribute–value pair. This in turn leads to the induction of the second too-specific rule, Rule D.

However, this situation can frequently correct itself. Rule B is a specialization of Rule 2, induced incorrectly because of the inaccurate representation of a_3 in the training set. Once Rule B has been induced, the instances covered by it are removed from the training set, thus removing the offending bias towards a_3. At this stage it is possible that the training set still contains enough instances which are examples of the correct rule, Rule 2, so that Rule 2 can subsequently be correctly induced. As all instances covered by Rule B are also covered by Rule 2, Rule B becomes redundant and can be discarded in the manner described in Section 7.2.2.

These problems are inherent in many induction algorithms and successful solutions to them will be extremely difficult to find.

9. Comparison of ID3 and PRISM

This final section demonstrates the performance of PRISM on a training set containing a large number of examples. The training set is provided by the King–Knight–King–Rook chess end-game on which Quinlan performed his original experiments (Quinlan, 1979*a*). The problem is to find a rule set which will determine for each configuration of the four pieces, whether knight's side is lost two-ply in a black-to-move situation. Quinlan tackled the problem in stages, by first placing severe constraints on the number of allowable configurations of the pieces, and then gradually relaxing these constraints until he could apply his algorithm successfully to the original unrestricted problem. He identified a total of seven problems of increasing complexity. The training set described below is provided by the third of these problems.

There are seven attributes:

a: distance from black king to knight, values 1, 2 or 3,
b: distance from black king to rook, values 1, 2 or 3,
c: distance from white king to knight, values 1, 2 or 3,
d: distance from white king to rook, values, 1, 2 or 3,
e: black king, knight, rook in line, values t or f,
f: rook bears on black king, values t or f,
g: rook bears on knight, values t or f.

There are two possible classifications—lost and safe, and the training set consists of 647 instances†. The decision tree produced by ID3 is shown in Fig. 6. It has 52 branches, and if these are trivially converted into separate rules, there are a total of 337 terms. In contrast, the rule set produced by PRISM has 15 rules and 48 terms:

1. $e_f \rightarrow$ safe,
2. $f_f \rightarrow$ safe,
3. $g_f \rightarrow$ safe,
4. $b_1 \wedge d_2 \rightarrow$ safe,
5. $b_1 \wedge d_3 \rightarrow$ safe,
6. $a_1 \wedge c_2 \rightarrow$ safe,
7. $a_2 \wedge c_2 \rightarrow$ safe,
8. $a_1 \wedge c_3 \rightarrow$ safe,
9. $a_2 \wedge c_3 \rightarrow$ safe,
10. $a_3 \wedge b_2 \wedge e_t \wedge f_t \wedge g_t \rightarrow$ lost,
11. $b_3 \wedge c_1 \wedge e_t \wedge f_t \wedge g_t \rightarrow$ lost,
12. $a_3 \wedge b_3 \wedge e_t \wedge f_t \wedge g_t \rightarrow$ lost,
13. $b_2 \wedge c_1 \wedge e_t \wedge f_t \wedge g_t \rightarrow$ lost,
14. $a_3 \wedge b_1 \wedge d_1 \wedge e_t \wedge f_t \wedge g_t \rightarrow$ lost,
15. $a_2 \wedge b_1 \wedge c_1 \wedge d_1 \wedge e_t \wedge f_t \wedge g_t \rightarrow$ lost.

Both the decision tree and the above rule set classify all 647 instances correctly, but an expert system using the decision tree as its knowledge base would require significantly more tests to be performed.

There is also one less obvious difference between the outputs, which is that the decision tree would classify the illegal instance (a_1 & b_1 & c_1 & d_1 & e_t & f_t & g_t) as safe, whereas the rule set produced by PRISM is unable to classify it.

† There is one combination of the seven attributes (a_1 & b_1 & c_1 & d_1 & e_t & f_t & g_t) which is illegal and therefore not included in the training set.

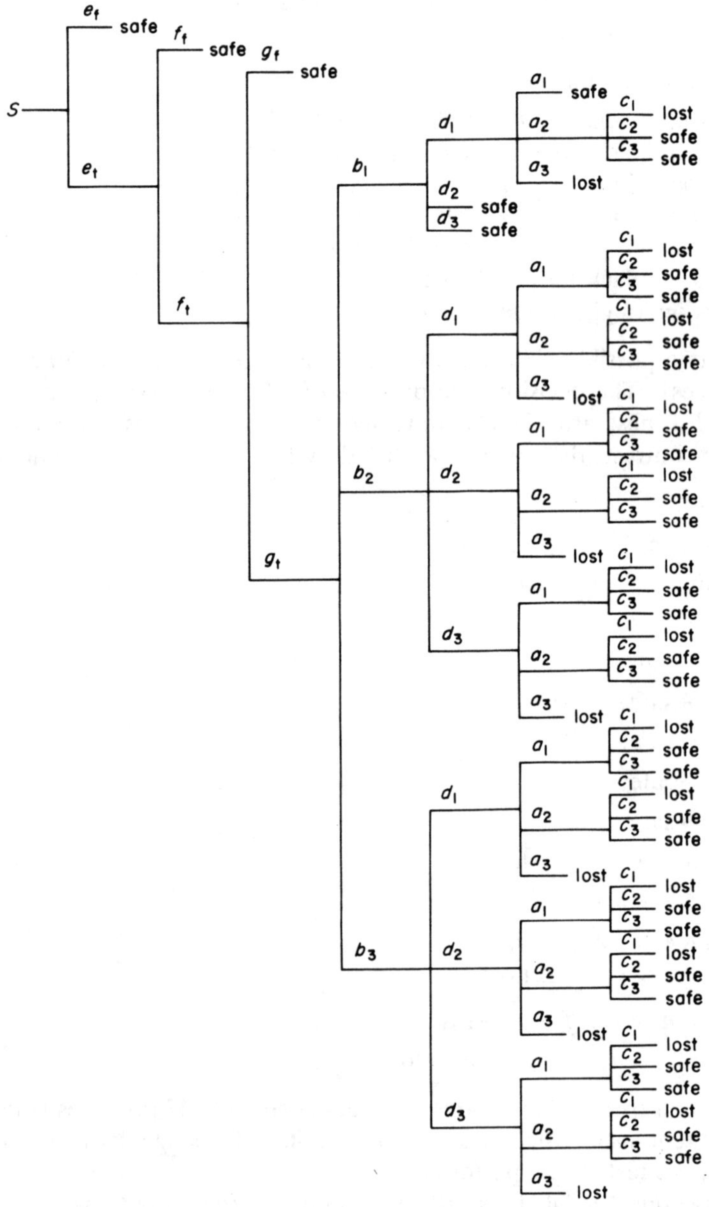

FIG. 6. Decision tree for Quinlan's third problem.

10. Summary and conclusions

One of the major criticisms of the ID3 algorithm is that its decision tree output is not suitable for use in expert systems whose control structure is based on the forward or backward chaining of modular rules, particularly if these rules are also used for explanation purposes. Attempts at converting decision trees into modular rules have had limited success because large and complex trees often contain a lot of redundancy, and simplification of these trees requires generalization techniques similar to those used in rule induction. It has been easier to implement expert systems whose control structure is designed to operate on decision trees.

However, the use of unmodified decision trees can have serious consequences in some domains, because the inherent redundancy requires that the results of irrelevant tests be known before a decision can be made. In medicine, these tests may require surgery, or alternatively may take up valuable time; in other domains, they may be extremely costly to perform. An expert system which uses such a decision tree must know the result of a requested test before it can decide on the next test to perform.

Redundancy is clearly an undesirable feature of a decision tree, but as this report points out, it is an inherent weakness in the strategy employed for induction, and can only be remedied by radically altering this strategy. By minimizing the average entropy of a set of instances, ID3 does not pay any attention to the fact that some attributes or attribute values may be irrelevant to a particular classification. This report suggests that a better strategy would be to maximize the information contributed by an attribute–value pair to knowing a particular classification. The report outlines a new induction algorithm, PRISM, which is based on this strategy, and describes some of the results obtained by applying it to different training sets.

PRISM produces its results as a set of modular rules which are maximally general when the training set is a complete one. The accuracy of rules induced from an incomplete training set depends on the size of that training set (as with all induction algorithms) but is comparable to the accuracy of a decision tree induced by ID3 from the same training set, despite the gross reduction in number and length of the rules.

References

A-Razzak, M., Hassan, T. & Pettipher, R. (1985). EX-TRAN7 (expert translator); a FORTRAN-based software package for building expert systems. In Bramer, R. A. Ed., *Research and Development in Expert Systems: Proceedings of the Fourth Technical Conference of the British Computer Society Specialist Group on Expert Systems.* Cambridge: Cambridge University Press. pp. 23–30

Bundy, A., Silver, B. & Plummer, D. (1984). An analytical comparison of some rule learning programs. *Technical Report 215,* Department of Artificial Intelligence, University of Edinburgh.

Carbonell, J. G., Michalski, R. S. & Mitchell, T. M. (1983). An overview of machine learning. In Michalski, R. S., Carbonell, J. G. & Mitchell, T. M. Eds, *Machine Learning: An Artificial Intelligence Approach.* Palo Alto: Tioga. pp. 3–23

Cendrowska, J. (1984). Practical requirements for rule induction: a critical analysis of current methodologies. *Technical Report,* The Faculty of Mathematics, The Open University, Milton Keynes.

EDWARDS, E. (1964). *Information Transmission.* London: Chapman and Hall.

GOLDMAN, S. (1968). *Information Theory.* New York: Dover Publications.

HART, A. E. (1985). Experience in the use of an inductive system in knowledge engineering. In BRAMER, M. A., Ed., *Research and Development in Expert Systems: Proceedings of the Fourth Technical Conference of the British Computer Society Specialist Group on Expert Systems.* Cambridge: Cambridge University Press. pp. 117–126

LAVRAC, N., VARSEK, A., GAMS, M., KONONENKO, I. & BRATKO, I. (1986). Automatic construction of the knowledge base for a steel classification expert system. In *Proceedings of the Sixth International Workshop on Expert Systems and their Applications.* Avignon, France, pp. 727–740.

MICHIE, D. (1983). Inductive rule generation in the context of the fifth generation. In MICHALSKI, R. S. Ed., *Proceedings of the International Machine Learning Workshop.* University of Illinois. pp. 65–70

O'RORKE, P. (1982). A comparative study of inductive learning systems AQ11P and ID-3 using a chess endgame problem. *Technical Report UIUCDCS-F-82-899,* Department of Computer Science, University of Illinois.

QUINLAN, J. R. (1979*a*). Discovering rules from large collections of examples: a case study. In MICHIE, D., Ed., *Expert Systems in the Micro-Electronic Age.* Edinburgh: Edinburgh University Press. pp. 168–201

QUINLAN, J. R. (1979*b*). Induction over large databases. *Technical Report HPP-79-14,* Heuristic Programming Project, Stanford University.

QUINLAN, J. R. (1983*a*). Learning efficient classification procedures and their application to chess endgames. In MICHALSKI, R. S., CARBONELL, J. G. & MITCHELL, T. M., Eds, *Machine Learning: An Artificial Intelligence Approach.* Palo Alto: Tioga. pp. 463–482

QUINLAN, J. R. (1983*b*). Learning from noisy data. In MICHALSKI, R. S., Ed. *Proceedings of the International Machine Learning Workshop.* University of Illinois.

SHANNON, C. E. & WEAVER, W. (1949). *The Mathematical Theory of Communication.* Urbana: University of Illinois Press. (Published in 1964).

Induction of Horn clauses: methods and the plausible generalization algorithm

WRAY BUNTINE

New South Wales Institute of Technology and Macquarie University, Computing Science, N.S.W.I.T., P.O. Box 123, Broadway, 2007 Australia

We are considering the problem of induction of uncertainty-free descriptions for concepts when arbitrary background knowledge is available, to perform constructive induction, for instance. As an idealised context, we consider that descriptions and rules in the knowledge-base are in the form of definite (Horn) clauses. Using a recently developed model of generality for definite clauses, we argue that some induction techniques are inadequate for the problem. We propose a framework where induction is viewed as a process of model-directed discovery of consistent patterns (constraints and rules) in data, and describe a new algorithm, the Plausible Generalization Algorithm, that has been used to investigate the sub-problem of discovering rules. The algorithm raises a number of interesting questions: How can we identify *irrelevance* during the generalization process? How can our knowledge-base answer queries of the form "What do (objects) X and Y have in common that is relevant to (situation) S?"

1. Introduction

Concept induction is a key technique for use in any semi-automatic acquisition system for knowledge-based systems (Michalski, 1983; Michie, 1986; Quinlan, Compton, Horn & Lazarus, 1986). By *concept induction,* we mean the discovery of a description for a concept when only positive and negative examples of the concept are known (Dietterich, London, Clarkson & Dromey, 1982; Rendell, 1986).

In practice, such induction may proceed in the presence of a rich body of background knowledge, for instance, to allow new descriptors to be derived about examples. So *constructive induction,* induction incorporating the derivation of new descriptors (Michalski, 1983), may well be required.

Suppose concept descriptions are to be in the form of a set of rules. How should search for these descriptions proceed? If a suitable set of rules are to be created, could the induction of other kinds of knowledge assist? And, what role does background knowledge play? For instance, how does it affect the construction of common generalizations of examples? What further issues are important here?

In this paper we consider these questions afresh† by framing the problem in the (idealized) language of Horn clauses; we are considering the induction of logical, uncertainty-free descriptions, called here *logic-based induction,* in the presence of arbitrary, logical background knowledge. Both the background knowledge and

† They have, of course, been considered by Michalski (1983), Stepp & Michalski (1986) and Sammut & Banerji (1986), among others.

KNOWLEDGE-BASED SYSTEMS Vol. 1
ISBN 0-12-273251-0

concepts are to be described in the rule form for Horn clauses, definite clauses, the language on which PROLOG is based (Robinson, 1979; Lloyd, 1984).

Arguments supporting the use of Horn clauses to pursue logic-based induction are given in section 1.1. We believe that by having a good understanding of inductive learning and its applications in the context of Horn clauses, we can improve our understanding of comparable problems in more extensive knowledge representations, for instance, ones incorporating uncertainty or using a different subset of first-order logic—as often found in knowledge-based systems.

Our treatment is incomplete, however, as it makes only passing mention of the roles that *credibility of hypotheses* and *noise* have to play in logic-based induction, or of *heuristic search* of a space of hypotheses (Quinlan, 1986, 1987; Cheeseman, 1984, unpubl. Rendell, 1986).

In section 4 we review a number of existing AI logic-based induction methods and consider which approaches to induction are required for the context considered here, that is, induction of definite clauses in the context of background knowledge or constructive induction. This analysis is made in the light of a recently devised model of generality for definite clauses (Buntine, 1986, 1987), reviewed in section 2.

Concurrently, we develop a framework where induction is viewed as a process of model-directed discovery of consistent patterns in data as observed from the standpoint of our relevant background knowledge. By noting patterns in the form of constraints and rules, such a process would gradually converge to a hypothesis (or more precisely, a class of equivalent hypotheses). The constraints and rules framework, as well as providing an equivalent but more intuitive formulation to the logic-based induction problem than Version Spaces, also allows a cleaner formulation of the factoring of a space of rules ordered by generality (Subramanian & Feigenbaum, 1986).

In section 5 we present a new algorithm for constructing common generalizations that addresses the problem of discovering rules from data. The Plausible Generalization Algorithm is based on a method of constructing most specific generalizations in the presence of arbitrary definite clause background knowledge (Buntine, unpubl.). It highlights the need for constraining irrelevance when generalizations are made utilizing the background knowledge. The algorithm can be easily tailored to perform a search of a class of models applicable to a particular sequence prediction induction problem, for instance.

1.1. HORN CLAUSES AS A LANGUAGE FOR INDUCTION

We believe that Horn clause logic provides an ideal framework for the investigation of logic-based induction. A good deal of previous work can be rephrased into the language of Horn clauses. For example, Bundy, Silver & Plummer (1985) compare a number of very different induction techniques by describing them in this unifying context. Mooney (1986) and DeJong & Mooney (1986) use a similar language when analysing a related learning technique, Explanation-Based Learning.

Both *disjunctive* and *conjunctive* concepts can be described in the language. Background knowledge provides the means for *additional descriptors* to be derived about any example. *Recursion* is an important feature in any language used for solving sequence prediction problems and is essential to conveniently express concepts about recursively defined structures such as trees, lists and language grammars. The language also has a well-understood means of *performing inferences* (Lloyd, 1984), essential to derive relevant descriptors from background knowledge during the induction process. This is critical to the development of a computational model of generality.

The techniques for incorporating *negation* into this language have not been considered here (see for example Lloyd, 1984). Negation should be a fundamental component of any general concept description language. One method of induction that deals with negation has been suggested by Vere (1980). However, to our current knowledge, a complete analysis of the effects of negation on the induction process has not yet been made.

2. A model of generality

To perform logic-based induction, certain tools are required. Induction is usually considered a process of searching for descriptions consistent with the background knowledge and known examples (Michalski, 1983). The search space for descriptions can be partially ordered by the generality of possible descriptions (Mitchell, 1982; Dietterich *et al.*, 1982). So to practice induction of definite (Horn) clauses, we need tools that can manipulate the generality of definite clauses and move around in the resultant search space.

Only recently has a thorough analysis been made of this search space for definite clauses based on a model of generality incorporating arbitrary definite clause background knowledge (Buntine, 1986, 1987). This was inspired by the work of Sammut & Banerji (1986). In this section, we introduce this model; it is used throughout the paper.

Previous models of generality have considered the case where: (1) background knowledge is in the form of facts (Plotkin, 1971; Vere, 1977); (2) in the form of equalities on predicates, type hierarchies and some additional forms (Kodratoff & Ganascia, 1986); and (3) in the form of concept hierarchies (Mitchell, Utgoff, Nudel & Banerji, 1981). Background knowledge of a similar kind to that considered here has been employed in conceptual clustering (Stepp & Michalski, 1986). Michalski (1983) stresses the importance of background knowledge for practical induction and mentions a number of generalization and specialization rules.

Before describing the notion of generality considered here, we introduce the terminology used.

2.1. PRELIMINARIES

Corresponding to a pure PROLOG rule, a *definite clause,* usually abbreviated to clause or called a *rule,* is in the form

$$\forall_X(A_0 \leftarrow A_1 \wedge \cdots A_n)$$

where $n \geq 0$, the A_is are *atoms* (*predicate symbols* applied to some arguments, see the examples below), and X is the set of variables occurring in the A_is. The universal quantification is usually implicit. The atoms on the right-hand side of a rule are referred to as the *conditions* for the rule, and their conjunction, the *body.*

For $n = 0$, a clause is called a *fact* and written simply as A_0.

The *description* of a concept takes the form of a *predicate definition,* a set of clauses with the same predicate symbol on the left-hand side of the rule. This predicate symbol is used to give examples of the concept. For instance, the concept "friends", denoting who is friends with whom, could be specified by the following definition for the two place predicate *friends.*

$$friends\,(bob, mary)$$

$$friends\,(mary, bill)$$

$$friends\,(X, Y) \leftarrow friends\,(Y, X)$$

We note for later use that a predicate definition, with the aid of the equality relation and trivial logical manipulation, can be shown to be equivalent to a single rule whose body consists of a disjunction of conjunctions.

Informally, the *task* of induction is to construct a predicate definition such that all and only positive examples of the concept will be proven true by a series of successful applications of the clauses in the predicate definition and in the current definite clause background knowlege.

A second kind of knowledge construct used here is a *constraint.* Constraints are introduced because they are a useful kind of knowledge about concepts that can feasibly be induced from data and, as we shall see, can also be used to simplify the induction of rules. The model of generality in its current state is not able to account for constraints computationally, however.

In this paper, we consider constraints of the form

$$\forall_X(A_0 \rightarrow \exists_{Y_1}(A_{1,1} \wedge \cdots A_{1,n_1}) \vee \cdots \exists_{Y_m}(A_{m,1} \wedge \cdots A_{m,n_m})) \quad \text{or} \quad \forall_X(\sim A_0)$$

where: A_0 and the $A_{i,j}$s are atoms; X is the set of variables occurring in A_0; the number of conjuncts in the disjunction on the right-hand side of the constraint, $m \geq 1$; and for the ith conjunct, Y_i is the set of remaining variables and $n_i \geq 1$. The quantification is usually implicit. In the database literature constraints would be referred to as *tuple generating dependencies* (Sagiv, 1986) when $m = 1$. These constraints could, for instance, be checked at each step of a computation.

Each constraint is actually the *converse* of a particular set of rules. To see this, swap the direction of the arrow in the constraint form given above then distribute " $\vee$ " over " $\leftarrow$ ". For instance, consider the following predicate definition which may be suggested to explain instances of the *sort* predicate. (*sort* (X, Y) is true if the list Y is a sorted version of the non-empty list X).

$$sort\,(X, Y) \leftarrow X = Y = [U]$$

$$sort\,(X, Y) \leftarrow Y = [A, B \mid L] \wedge A \leq B$$

Giving "sort" its usual interpretation, this description is incorrect as the second rule says *sort* ([3, 4], [1, 2]) is true. The description is logically equivalent to:

$$sort\,(X, Y) \leftarrow (X = Y = [U]) \vee (Y = [A, B \mid L] \wedge A \leq B)$$

and taking the converse we get:

$$sort\,(X, Y) \rightarrow \exists_U(X = Y = [U]) \vee \exists_{A,B,L}(Y = [A, B \mid L] \wedge A \leq B)$$

which is a useful constraint on the *sort* predicate.

2.2. THE MODEL

We adapt Mitchell's view of "more general" (1982) as follows. A fuller treatment of the model is presented by Buntine (1987).

Informally, a first description is *more general* than a second if it is possible to explain using the current background knowledge that the first description will apply to some example whenever the second description does. We give three different but equivalent views of this notion below.

A predicate definition, *R1*, is *more general* than another, *R2*, with respect to given background knowledge, if, in any possible world (sometimes referred to formally as an *extension* or *interpretation*) consistent with the background knowledge, whenever a clause from *R2* can be successfully applied to show, in the possible world, an example of the concept should be a positive example, then a clause from *R1* must be able to be successfully applied to show the same. We say a clause *C1* is more general than another clause *C2* when the equivalent condition for clauses occurs.

Alternatively, if *R1* is represented in the form $\forall_X(c(X) \leftarrow B1(X))$ and *R2* the form $\forall_X(c(X) \leftarrow B2(X))$ were "*c*" is the concept being described, *X* is some parameterization of the concept, and *B1* and *B2* represent the bodies of the descriptions (in the case of predicate definitions, a disjunction of conjunctions), then *R1* is more general than *R2* with respect to the background knowledge if and only if:

$$\forall_X(B2(X) \rightarrow B1(X))$$

is a logical consequence of the background knowledge.

If a first clause or predicate definition is more general than a second, then the second is also said to be *more specific* than the first.

To illustrate these views of generality and later ideas, we have concocted the following background knowledge, called the *social-knowledge-base*:

likes (bill, mary)
likes (bob, jane)
right-handed (bill)
right-handed (bob)
friends (bill, paul)
friends (kate, bob)
friends (paul, sally)
friends (paul, mary)
friends (jane, kate)
friends (sally, kate)
friends $(X, Y) \leftarrow$ *friends* (Y, X)
plays-tennis (mary)
member-tennis-social-club (jane)
member-tennis-social-club $(X) \leftarrow$ *plays-tennis* (X)
plays-social-tennis $(X) \leftarrow$ *member-tennis-social-club* (X).

Under this background knowledge, the description {*likes (bill, mary), likes (bob, jane)*} is more specific than the description consisting of the single rule:

$$likes\ (X, Y) \leftarrow member\text{-}tennis\text{-}social\text{-}club\ (Y) \wedge friends\ (X, Z) \wedge friends\ (Z, Y) \quad (1)$$

To justify this, we must show that rule 1 confirms *likes* (*bill, mary*) and *likes* (*bob, jane*) in any possible world consistent with the *social-knowledge-base*. That is, we must show that for (*X, Y*) set to (*bill, mary*) or (*bob, jane*), the right-hand side of rule 1 will be true in any possible world consistent with the *social-knowledge-base*. In the language of logic, this is equivalent to showing that rule 1 together with the *social-knowledge-base* logically implies *likes* (*bill, mary*) and *likes* (*bob, jane*).

An operational view of generality follows. In this view, background knowledge can only consist of definite clauses.

Buntine (1987) shows that, with respect to given definite clause background knowledge, a predicate definition *R1* is more general than another, *R2*, if and only if for every clause in *R2*, there exists a clause in *R1* more general than it. That is, to test the comparative generality of two predicate definitions, we only need compare the generality of their component clauses.

Furthermore, a test of the comparative generality of two clauses consists of a single query to the logic program representing the background knowledge (Buntine, 1987). Effectively, a clause is more general than another with respect to given definite clause background knowledge if the more general clause can be converted to the other by repeatedly turning variables to constants, adding conditions, or partially evaluating by applying some clause in the background knowledge to the body. When the background knowledge is absent the test degenerates to a test of *θ-subsumption* (Plotkin, 1970). Clause *C θ-subsumes D* if there exists a set of replacements for the variables in *C* (a *substitution*) making the atoms in *C* a subset of those in *D*. That is, clause *C* can be converted to clause *D* by turning some variables to constants and adding conditions.

For instance, the rule:

$$likes\ (X, Y) \leftarrow plays\text{-}social\text{-}tennis\ (Y) \wedge friends\ (X, Z)$$

is more general with respect to the *social-knowledge-base* than the rule 1, the rule:

$$likes\ (X, Y) \leftarrow plays\text{-}tennis\ (Y) \wedge friends\ (X, sally) \wedge friends\ (Z, sally)$$

is more specific, and finally, the rule:

$$likes\ (X, Y) \leftarrow member\text{-}tennis\text{-}social\text{-}club\ (Y) \wedge friends\ (X, Z) \wedge friends\ (Z, Y)$$
$$plays\text{-}social\text{-}tennis\ (Y) \wedge friends\ (Z, X)$$

is of equal generality. The last two conditions in this clause are *redundant* as, by the *social-knowledge-base,* they follow on from previous conditions in the clause.

This last clause above is just one of many clauses of equal generality to clause 1. Clause 1, however, is the shortest of all such clauses. Any other clause will have additional atoms in its body, *redundant conditions,* and consequently generate larger proofs when used. A kind of canonical form (called *reduced* after Plotkin, 1970, see Buntine, 1987) exists for a set of clauses of equal generality, the form being the shortest possible clause. Constructing this form is denoted reducing a clause and is a method of removing redundancy from the clause.

This model of generality formalises and considerably extends the work of Plotkin (1970, 1971), Vere (1977) and to some extent Kodratoff and Ganascia (1986). We denote this model of generality *generalized subsumption.*

3. Applications to knowledge maintenance

Redundancy detection is one of a number of knowledge maintenance tasks. Kitakami, Kunifuji, Miyachi and Furukawa (1984) argue that knowledge maintenance should be a task performed by an integrated knowledge acquisition system. Recent experiences with XCON indicate (Pundit, 1985) that knowledge maintenance will become a key task in the future.

Buntine (1987) shows that the model of generality given in section 2 has applications to the detection of redundancy in knowledge-bases. We are currently developing techniques for detecting *unused* rules and *redundant* rules for removing *redundant conditions* from a rule and for generally cleaning-up the form of a rule. These techniques are based on a stronger model of generality than generalised subsumption, a model incorporating background knowledge in the form of constraints as well as rules. Related results derived using a different technique based on the notion of *equivalence* rather than *generality* appear in the deductive database literature (Sagiv, 1986).

What does the model have to say about induction?

4. Methods for induction

A number of techniques have been devised by the AI community for the induction task of searching a space of descriptions ordered on generality. Some of these techniques allow a description to consist of only a single conjunctive rule; others allow disjunction as well. In this section, we consider the search strategies embodied in these techniques in the light of the generalised subsumption model and practical experience.

We begin with an overview of some current techniques for logic-based induction.

A number of these are based on a specific-to-general search, for instance, *climbing the generalization hierarchy* (Sammut & Banerji, 1986), *constructing maximally specific generalizations* of individual pairs of rules in the current description (Vere, 1977) and hybrid techniques (Vrain, 1986; Kodratoff & Ganascia, 1986).

The second technique above is based on the notion of a *most specific generalization* (MSG). The MSG of two clauses *C1* and *C2* with respect to background knowledge *P* is a clause that is a common generalization of *C1* and *C2* with respect to *P* but is more specific than any other such generalization. In the context of a number of different knowledge representations, MSGs have been employed in various ways as a useful induction tool (Vere 1977, 1980; Mitchell *et al.*, 1981; Fu & Buchanan, 1985).

Other search strategies are based on a general-to-specific search (Langley, 1981; Shapiro, 1981, 1983). The process of specializing a rule found to be inconsistent with new negative examples, performed during general-to-specific search, is called *discrimination* (Langley, 1981; Bundy *et al.*, 1985). A number of different search styles have been suggested. Mitchell (1982), Langley (1981) and Shapiro (1981) use a breadth-first search, Michalski (1983) a beam search, and Bundy *et al.* (1985) suggest a depth-first search.

As specific-to-general search has the effect of constructing plausible rules suggested by the data whereas general-to-specific search constrains potential rules from being overly general, it is now well accepted that these complementary search strategies be coupled together. The two most widely known techniques are *Version Spaces* (Mitchell, 1982) and *Focusing* (Bundy *et al.*, 1985). A recent improvement to the Version Spaces algorithm, justifying that the space can be split into independent components, is the technique of *factoring*† (Subramanian & Feigenbaum, 1986). This splitting is already found in the Focusing algorithm.

A further kind of search is *model-directed* search, discussed and applied to the induction problem, for instance, by Dietterich & Michalski (1986).

In the following subsections we consider the two search strategies, general-to-specific and specific-to-general, in more detail, and discuss what is appropriate for the context we are considering. For this, it is sufficient to view Version Spaces and Focusing in their decoupled form.

4.1. PERFORMING SPECIFIC-TO-GENERAL SEARCH

We first consider the problem of performing a specific-to-general search of the space of predicate definitions.

Given background knowledge expressed as definite clauses there exists a method of climbing the corresponding generalization hierarchy built around this background knowledge (Buntine 1986). The Marvin system (Sammut & Banerji, 1986) demonstrates a practical induction method based on a similar technique.

4.1.1. Using most specific generalizations

However, many of the techniques reviewed use a potentially more powerful method of performing specific-to-general search: constructing MSGs. This allows much larger jumps up the generalization hierarchy to be made, jumps that are well justified assuming—we explore this assumption in more detail in section 4.1.2—that the search space contains a correct hypothesis. This is done, for instance, in the specific-to-general component of the Version Spaces algorithm.

In its usual sense, the MSG of two predicate definitions is, simply, the union of the sets of clauses they are constructed from. For example, the MSG of the two descriptions:

$$\{member\ (1, [1])\};$$

$$\{member\ (2, [2, 3]),\ member\ (2, [3, 2]\},$$

is:

$$\{member\ (1, [1]),\ member\ (2, [2, 3]),\ member\ (2, [3, 2])\}.$$

As this achieves no induction gain, we can instead form a generalization by additionally replacing two clauses in the union by their MSG. For instance, the original two descriptions above could be generalized to:

$$\{member\ (X, X.Y),\ member\ (2, [3, 2])\}$$

The first atom in this set is the MSG of *member* (1, [1]) and *member* (2, [2, 3]) with respect to no background knowledge.

This is the idea behind Vere's Thoth (1977). In this way, MSGs or common generalizations of clauses can be usefully employed when performing specific-to-

† Though, they use this to increase the efficiency of "experiment generation", the asking of questions. We have not considered this here.

general search of the space of predicate definitions. We initialise the description of a concept to the set of positive examples and then repeatedly apply an MSG algorithm to pairs of rules in the description, backtracking when overgeneralization occurs. For instance, overgeneralization can be identified by counter-examples or by querying the user, done by Sammut & Banerji (1986).

4.1.2. Sufficiency of the knowledge

A common generalization consistent with the intended concept will only be constructed, according to the notion of generality considered here, if sufficient knowledge is available from which the generalization can be logically justified. After all, it is only these generalizations which have been given some degree of plausibility by the examples. This is in contrast with the induction framework developed by Delgrande (1987) where any hypothesis consistent with the current knowledge can be proposed.

This restriction can be a problem with recursively defined concepts. For example, the MSG of *member* (4, [3, 4]) and *member* (2, [5, 1, 2]) with respect to the background knowledge:

$$\begin{array}{l} \mathit{list}\ ([\ ]) \\ \mathit{list}\ (X.Y) \leftarrow \mathit{list}\ (Y) \\ \mathit{member}\ (X, X.Y) \leftarrow \mathit{list}\ (Y) \\ \mathit{member}\ (4, [3, 4]) \\ \mathit{member}\ (2, [5, 1, 2]), \end{array}$$

is $\mathit{member}\ (X, Y.\,Z.\,W) \leftarrow \mathit{list}\ (W)$ which would be inconsistent with most negative examples as it says anything is a member of a list with more than one element. The problem here is that the fact *member* (2, [1, 2]) is not available so the valid generalization:

$$\mathit{member}\ (X, Y.Z.W) \leftarrow \mathit{member}\ (X, Z.W) \wedge \mathit{list}\ (W).$$

cannot be justified from the existing knowledge.

This kind of problem occurs whenever the concepts on which a predicate definition is to be based are themselves not fully known.

Specific-to-general search, then, must proceed under the *justifiability assumption* (Buntine, 1987). That is, it is assumed that the instructor/user has supplied sufficient knowledge from which valid generalizations of at least some seed facts can be justified and hence constructed. Alternatively, general-to-specific search together with a querying strategy (for example, as done by Shapiro, 1983) can serve to obtain the extra facts needed.

4.1.3. Problems with most specific generalizations

Recently, an algorithm has been developed (Buntine, unpubl.) for constructing the shortest MSG of two clauses in the presence of arbitrary definite clause background knowledge. This was based on some ideas presented by Buntine (1987). Experiences with the algorithm brought to light the problems shown below.

One problem occurring is that an MSG can, in fairly simple circumstances, be too overspecific to be of any practical use as a plausible lower bound on inductive hypotheses. MSGs may not even exist in some circumstances (this is an open problem). If an MSG does not exist, it would be because the clause is in fact infinite.

For example, an MSG of the example facts *member* (4, [3, 4]) and *member* (2, [5, 1, 2]) with respect to the background knowledge:

$$\begin{array}{l} list\ ([\ ]) \\ list\ (X.Y) \leftarrow list\ (Y) \\ member\ (X, X.Y) \leftarrow list\ (Y) \\ member\ (4, [3, 4]) \\ member\ (2, [1, 2]) \\ member\ (3, [5, 3]) \\ member\ (2, [5, 1, 2]), \end{array}$$

cannot be computed by the MSG algorithm, is known to be at least several hundred atoms long, and may actually be an infinite clause. The trouble here is that an infinite number of facts can be inferred from the background knowledge which are able to be associated directly or indirectly with the two example facts.

A variation on this theme is that an abundance of *irrelevant* knowledge, as considered by Michalski (1983), can be inferable from the background knowledge. This knowledge contributes to the body of an MSG conditions that have arisen out of mere coincidence. Borgida, Mitchell and Williamson (1986) report a similar problem. For example, an MSG of the known facts *likes* (*bill, mary*) and *likes* (*bob, jane*) with respect to the *social-knowledge-base* is:

$$\begin{array}{l} likes\ (A, B) \leftarrow friends\ (A, C) \wedge friends\ (C, B) \wedge friends\ (C, sally) \\ \qquad \wedge\ member\text{-}tennis\text{-}social\text{-}club\ (B) \wedge right\text{-}handed\ (A). \end{array}$$

Right-handedness is not usually a property causing one person to be attracted to another. Commonsense tells us it is *irrelevant* in this situation. So this rule could be made more general but equally plausible by removing the condition about right-handedness. When a larger, less selective knowledge-base is used the small amount of irrelevance in the above clause can become considerably greater. There is a need to *constrain irrelevance* when constructing generalisations in the context of arbitrary background knowledge. Which of the multitude of facts that can be derived from the background knowledge about given examples should *not* be used? And, how do we identify them?

4.1.4. Implications

These problems with MSGs and the related issue of relevance mean that current logic-based induction techniques such as Version Spaces or Focusing will encounter problems when used in the more extensive context where arbitrary background knowledge is available. They would need to approximate MSGs or work with a limited class of background knowledge in which MSGs are well behaved, for instance, restricted background knowledge using non-recursive rules or composed of relevant knowledge only.

To retain the power of the approach of using MSGs, gained through constructing common generalisations likely to be true, we can modify an MSG algorithm to constrain some of the possible conditions that could be included in the body of the constructed clause. The aim is to make generalisations which are as specific as can be consistent with an acceptably terse solution and knowledge about irrelevance. The *Plausible Generalization Algorithm* (PGA), discussed in some detail in section 5, has been designed with this aim in mind. The algorithm could still be used in the manner suggested in section 4.1.1.

4.2. PERFORMING GENERAL-TO-SPECIFIC SEARCH

We next consider the problem of general-to-specific search of the space of predicate definitions.

4.2.1. Model-directed heuristic search

Heuristic and model-directed search are underlying themes in artificial intelligence and in induction. Successful demonstrations of their use are the Meta-DENDRAL system (Buchanan & Mitchell, 1978) and the SPARC/E system for predicting sequences (Dietterich *et al.*, 1986). Further evidence supporting their use is presented below, based on a variety of systems.

Experiences with Shapiro's Model Inference System indicate that a breadth-first search in the case of definite clauses is highly impractical. Shapiro was actually working with a model of generality incorporating no background knowledge (called θ-subsumption, mentioned in section 2). But in this case of general-to-specific search it has sufficiently similar characteristics to the stronger model used here. The *generalization hierarchy,* the general-to-specific search space, still has a huge branching factor. A relevant point is that the search spaces considered by Bundy *et al.* (1985), Langley (1981) and Mitchell *et al.* (1981) are smaller by comparison due to the restricted languages or generality models used.

Shapiro effectively employed a model-directed technique to attempt to overcome the problem of a huge branching factor. He states (1981, p. 27):

> In practice, some information on the structure of the set of hypotheses to be inferred is usually known. In such a case we always prefer to [use a particular model], to guarantee more efficient [induction] of the [concepts] of interest.

For instance, many of the examples given by him used a model where no constants appear in the conditions of clauses (Shapiro, 1983, Appendix II, PDSREF). Instantiating constants is one of several rule specialization steps, so this had the effect of reducing the branching factor of a generalization hierarchy considerably. Another model he used is definite clause grammars (*ibid.*, DCGREF).

This model-directed technique is useful in the situation where we have *a priori* belief that the description should belong to a particular class of hypotheses.

Another technique possible is the heuristic search of a general-to-specific space. It is with this that the *credibility of hypotheses* plays an important role (Rendell, 1986). A key factor for success in Quinlan's ID3 (1986) and its more recent versions (Quinlan *et al.*, 1986; Quinlan, 1987) is their techniques for determining how to expand the decision tree. They tend to but not always develop a shorter tree, sometimes trading off accuracy for a decrease in size (Quinlan, 1987). There are strong arguments to suggest that without any other source of prior information this produces a more credible hypothesis (Georgeff & Wallace, 1985; Cheeseman, 1984).

As it turns out, the clausal search space considered here is usually infinite. No amount of examples will, consequently, enable a single hypothesis to be identified as the correct one. To identify a reasonably likely hypothesis from the set of all possible hypotheses, a heuristic is essential. Shapiro used a similar heuristic to Quinlan in that the shorter clauses are preferred. This is a weak heuristic for the more extensive context of clauses where some background knowledge is available. The development of more powerful heuristics is an important research goal. We suggest below that this will be an easier task if we reformulate general-to-specific search somewhat.

The two techniques, model-directed and heuristic, are not at odds with each other. We believe they should work in collaboration. Dietterich *et al.* (1986) suggest a number of options.

4.2.2. Searching the space of constraints

General-to-specific search of the space of rules can also be thought of as a weak-to-strong search of the space of constraints. (A *stronger* constraint constrains more examples from being positive.) This alternative view allows a much cleaner formulation of the search problem.

This change in formulation is allowed because of the following property. We say a predicate definition is *overly general* if it can be used to prove true all possible positive examples of the concept, and in addition some negative ones as well. All members of the G-set for the Version Spaces algorithm, for instance, are overly general. A predicate definition for a concept is overly general if and only if its *converse* represents a constraint that is true of the concept. A proof of this is quite straight forward. Furthermore, a more specific predicate definition has a stronger constraint as its converse.

Therefore, the technique of coupled specific-to-general and general-to-specific search can also be thought of as a specific-to-general search of the space of descriptions/rules together with a weak-to-strong search of the space of constraints. Constraints balance rules and vice versa, as in the coupled paradigm. Induction, then, becomes a process of searching for and generalizing consistent *patterns* in the data, regardless of whether those patterns are expressed as rules or constraints.

One advantage of phrasing the search in terms of constraints is that the question "What is the strength of your belief that constraint C applies to concept S?" is more natural to consider than the question "What is the strength of your belief that rule C is overly general for concept S?" Heuristic search of the general-to-specific search space of descriptions essentially involves repeatedly asking this question. The mathematician Polya, whose work was a prime inspiration behind the AM program (Lenat & Davis, 1982), demonstrates that reasonable heuristics may exist on a search for constraints (Polya, 1954, Ch. IV). He gives a number of examples with suitable heuristics in the domain of number theory.

Another advantage is that search of the space of constraints can be made more efficient as search can proceed *independently* for different constraints. Any number of constraints induced can later be combined to produce a single more powerful constraint. This result rests on the following property. Given two valid constraints about concept "c"

$$\forall_X\big(c(X)\rightarrow \exists_{Y_1}(A_{1,1}\wedge\cdots A_{1,n_1})\vee\cdots\exists_{Y_m}(A_{m,1}\wedge\cdots A_{m,n_m})\big)$$
$$\forall_X\big(c(X)\rightarrow \exists_{Y_1}(B_{1,1}\wedge\cdots B_{1,n_1})\vee\cdots\exists_{Y_p}(B_{p,1}\wedge\cdots B_{p,n_p})\big)$$

then, after reformulating the right-hand side of the implication sign, the following is also a valid constraint:

$$\forall_X\Big(c(X)\rightarrow\Big(\big(\exists_{Y_1}(A_{1,1}\wedge\cdots A_{1,n_1})\vee\cdots\exists_{Y_m}(A_{m,1}\wedge\cdots A_{m,n_m})\big)\wedge\big(\exists_{Y_1}(B_{1,1}\wedge\cdots B_{1,n_1})\vee\cdots\exists_{Y_p}(B_{p,1}\wedge\cdots B_{p,n_p})\big)\Big)\Big)$$

This new constraint is stronger than the previous two. Considering that the converse of any one of these constraints represent a set of rules about concept "c", this means that general-to-specific search can be decomposed into independent components. In the case where constraints do not have a disjunction, the resulting independent components effectively factor the search space (Subramanian & Feigenbaum, 1986).

4.3. SUMMARY

Induction of predicate descriptions can be achieved by generalizing rules from examples and by constructing stronger and stronger constraints. Constraints serve the same purpose as the G-set in Version Spaces; they provide an upper bound on potential rules. Their use allows a natural factoring of the search space and is more conducive to the use of heuristics.

A weak-to-strong search of the space of constraints, by virtue of its equivalence with general-to-specific search of the space of rules, should require model-directed, heuristic search. A model-directed approach is required because *a priori* knowledge can restrict search initially to some subset of hypotheses, or dictate that certain models for hypotheses should be explored before others. We need to make best use of this *a priori* knowledge to attain efficiency. A heuristic approach is required because the remaining search space may still be huge and so require a best-first search. This would be based on our judgement of the *a posteriori* credibility of constraints in view of the current examples. Search for several constraints can proceed independently.

A method of generalizing rules from examples has been proposed based around a technique for constructing common generalizations. MSGs have been shown to be sometimes unsuitable for this in the context of arbitrary background knowledge as they do not constrain irrelevance sufficiently. These issues are explored more in section 5.3.

When performing induction myself, I typically proceed with (1) an inductive phase, a model-directed, heuristic discovery of patterns (constraints and rules), and then complete the process with (2) a deductive phase, to patch together the induced patterns to construct a concept description that is, hopefully, both operational and complete. A critical part of my approach is reasoning about relevance and the likelihood of certain constraints applying.

This discussion shows that there are strong theoretical arguments to support the use of this commonly used intuitive approach, in particular, how the induction of constraints can assist the task of inducing rules.

5. The plausible generalization algorithm

Having discussed the induction problem in general, we now consider a particular subproblem: generalizing plausible rules from examples. Our intention is to explore the framework proposed and issues raised in section 4.1.

Our vehicle for doing this is the *Plausible Generalization Algorithm* (PGA), an algorithm for constructing useful inductive hypotheses by forming common generalizations of pairs of rules or examples. The context in which this algorithm would be used was described in sections 4.1.1 and 4.1.4. With it, the useful generalization:

$$member\ (X,\ Y.Z.W) \leftarrow member\ (X,\ Z.W) \wedge list\ (W)$$

for the member example introduced in section 4.1.2 can be quickly constructed using a notion of relevance such as allowing in the body of a rule only sub-lists of lists in the head of the rule.

The algorithm is based on a theory of most specific generalizations and redundancy in the context of generalized subsumption (Buntine, 1987). When used without relevance constraints, it is guaranteed to find a most specific generalization should one exist.

5.1. AN INFORMAL INTRODUCTION

Below, we describe the algorithm informally, applied to the common situation where the clauses being generalized are merely positive examples of the concept. We are to find a common generalization of the positive examples *likes* (*bill, mary*) and *likes* (*bob, jane*) with respect to the *social-knowledge-base.*

We will have to consider the following kinds of arguments. These have a certain intuitive appeal about them suggesting they may be useful for a broader class of induction problems coping with background knowledge.

To find some common reason for Bill liking Mary and Bob liking Jane we must find potentially relevant statements common about both Bill and Bob and/or Mary and Jane. To do this we need to query the knowledge base to find out, for instance: "What do Bill and Bob have in common?"

Of course not all answers to this query will be relevant to the situation that Bill likes Mary and Bob likes Jane so we will need some *relevance constraints* to prune obviously irrelevant details. Such a constraint would prevent such details as Bill and Bob's right-handedness from being considered. These constraints are the form:

Only ground facts satisfying condition X are relevant to ground fact Y

In our example, we assume that only facts concerning social-life of people are relevant to someone liking someone else. Operationally, a fact is irrelevant to Bill liking Mary if it would never need to appear on the right-hand side of an instance of a rule that proves Bill likes Mary. This check is made in step 2a of the algorithm in Fig. 1.

Furthermore, some answers will be *a joint consequence* of existing facts we have already found and so be redundant. We notice that Mary and Jane are *both* members of the tennis social club and so add this to the conditions in the body of the rule. But because it is a consequence of this that they also *both* play social tennis, we do not need to consider their playing social tennis in the constructed rule. This check is made in step 2b of the algorithm.

Finally, as we are concerned with constraining the size of the final rule constructed, certain controls are provided in the algorithm. One bound specifies the depth that inference can go when searching for relevant facts (mentioned in step 2a). Another specifies the maximum depth of nesting of existential quantification allowed to occur in the body of the constructed rule (mentioned in step 2d).

5.2. THE ALGORITHM

The actual algorithm proceeds as follows.

We first construct the left-hand side of the final generalization (step 1 of the algorithm). This is done by finding the most specific atom which has *likes* (*bill, mary*) and *likes* (*bob, jane*) as instances of it. This is called the *least common anti-instance* of the two facts. For instance, a least common anti-instance of the atoms $a(f(g(c), d), h(d))$ and $a(f(b, c), h(c))$ is $a(f(X, y), h(Y))$, or any variant where the variables have been renamed to other variables. We construct these using Plotkin's algorithm to find the least generalization of terms (Theorem 1, 1970) or similar; it is a purely syntactic operation. In the case being considered here, it is the atom *likes* (A, B) where the variables A and B could take the values *bill* and *mary* or *bob* and *jane* respectively. The ordered pair (*bill, bob*) is called the *conflict pair* for A and (*mary, jane*) likewise for B.

Our task of induction is to find potentially relevant statements common to both Bill and Bob and/or Mary and Jane, common to corresponding members of a conflict pair. These are to become conditions in the rule constructed. So far, our rule must look as follows:

$$\textit{likes}\ (A, B) \leftarrow \{\text{some conditions on } A \text{ and } B\}.$$

The conditions on the right-hand side must currently be known to hold true when A and B take the values *bill* and *mary* or *bob* and *jane*, so that the rule can be shown to be a generalization of the initial examples. Our initial set of conflict pairs is {(*bill, mary*), (*bob, jane*)}.

From the background knowledge it can be seen that:

member-tennis-social-club (*mary*), *member-tennis-social-club* (*jane*)

and *friends* (*bill, paul*), *friends* (*bob, kate*)

are two pairs of statements concerning Mary and Jane, and Bill and Bob. Before using these, we must check to see if they are relevant (step 2a) and that they will not be redundant in the final clause (step 2b). These checks both pass; both pairs of facts are about "social life" and for each pair of facts, each member has been derived using a different rule/fact from the *social-knowledge-base*.

We then construct the atoms generalizing the pairs of facts found and link them to the rule to produce

$$\textit{likes}\ (A, B) \leftarrow \textit{member-tennis-social-club}\ (B) \wedge \textit{friends}\ (A, C)$$
$$\wedge\ \{\text{further conditions on } A, B \text{ and now possibly } C\}$$

Notice that when generalizing the atom we create a new conflict pair (*paul, kate*) associated with the variable C. Common statements about Paul and Kate could indirectly effect Bill and Bob so these must also be considered. The set of conflict pairs subsequently becomes $\{(bill, mary), (bob, jane), (paul, kate)\}$. The variable C is said to have an indirection of depth 1. Indirection of a variable is related to the depth of existential quantification for the variable Restricting indirection (step 2d) is a useful means of constraining the size of the constructed clause.

Proceeding in this fashion, we gradually build up potentially relevant conditions on the right-hand side of the rule, discarding irrelevant or redundant conditions as we go. The final rule would look as follows:

$$likes\ (A, B) \leftarrow friends\ (A, C) \wedge friends\ (C, B) \wedge member\text{-}tennis\text{-}social\text{-}club\ (B)$$

In Fig. 1, we summarize the version of the algorithm that finds a common generalization of two facts. A more exact account of the full algorithm and the theory behind it occurs in (Buntine, unpubl.).

(1) Construct the left-hand side by finding the least common anti-instance of the two facts, and find the initial set of conflict pairs.
e.g. *likes* (A, B) with conflict pairs (*bill, bob*) and (*mary, jane*) for variables A and B respectively.
(2) Construct the right-hand side:
Repeat
(2a) Look for a pair of possibly relevant facts (i.e. they pass the relevance constraints) inferable from the background knowledge (searching to some depth n) concerning corresponding members of a conflict pair.
e.g. the pair *friends* (*bill, paul*) and *friends* (*bob, kate*) are possibly relevant and concern Bill and Bob.
e.g. the pair *friends* (*bill, paul*) and *friends* (*kate, bob*) do not concern Bill and Bob as *bill* and *bob* do not occur in corresponding positions.
(2b) Disregard the pair (so back to 2a) if both facts have been derived from other relevant facts using exactly the same rule. This is a sufficient condition for them to be redundant in the final rule constructed (Buntine, unpubl. Lemma 3).
(2c) Construct the atom generalizing the pair of facts.
e.g. *friends* (A, C).
(2d) Add the atom to the conditions accumulated so far. Add any new conflict pairs found to the set of conflict pairs if their indirection is not greater than some depth m.
e.g. (*paul, kate*) is a new conflict pair for C.
Until step 2a fails to find a pair of facts that has not already been considered. (A particular fact can be paired several times but only with different other facts. For instance *friends* (*bill, paul*) could be paired with *friends* (*bob*, X) for each of the different friends X of Bob).
(3) Strip away any redundant conditions from the right-hand side of the constructed rule, for instance, using a reduction algorithm (Buntine, 1987, Theorem 6.1).

FIG. 1. The plausible generalization algorithm.

5.3. FEATURES OF THE ALGORITHM

The algorithm and its use raises several noteworthy issues discussed below.

Using relevance constraints

Some relevance constraints we have found effective for inducing list related concepts say that the facts relevant to the example fact can only be those about lists of smaller length, sub-lists or lists containing only constants occurring in the given example. Relevance constraints for sequence prediction problems can, for instance, specify that the current member in the sequence is dependent on only the previous L elements (termed *lookback* by Dietterich *et al.*, 1986) or is periodic with some specified period.

Thus, relevance constraints can be used as a means of specifying a particular model in which to perform induction. In our system, written in Prolog, a new model can be specified simply by redefining the *relevance* predicate. This provides great flexibility for performing model-directed search using classes of pre-defined and, indeed, user-defined models.

Relevance constraints can also be viewed as an alternative to Utgoff's approach for specifying bias (1986).

Constraining redundancy

In the context of certain kinds of (arbitrary) background knowledge, the detection of redundant conditions in clauses can become a computationally expensive factor when constructing common generalizations (Buntine, unpubl.). This and relevance are the main factors that influenced the design of the algorithm.

Commonality queries

Critical to the algorithms performance are queries to the knowledge-base of the form: "What statements are known common to (objects) X and Y?" occurring in step 2a. In our system we currently use a brute force approach. We refer to these as *commonality queries*. Knowledge-base systems can feasibly be tuned to answer this variant of object-centred querying.

Of course, potential answers can be further constrained by the need for relevance. So one could also ask: "What statements are known common to (objects) X and Y that are potentially relevant to the truth of S and T?" This is asking: "What is of interest to the current induction problem?" We believe efficient answering of such queries will be an important prerequisite for practical inductive inference systems sourcing knowledge from a knowledge-base.

The method of constructing common generalizations presented by Vrain (1986) demonstrates an approach for the answering of commonality queries.

Determining relevance

A prerequisite for the use of the algorithm is the necessary knowledge about relevance, relevance constraints. Below we list a number of options for their acquisition.

Relevance is one kind of knowledge that people can often articulate even when they are not able to articulate rules. This is the principle on which successful applications of decision-tree induction systems rely (Michie, 1986; Hart, 1985; Quinlan *et al.*, 1986). Using the ID3 paradigm, it has been found that experts usually have the ability to articulate knowledge in the form of tutorial examples and relevant attributes (usually, features about examples that are true or false). If their same ability scales up to the induction problem as it is phrased in more extensive rule languages, then experts should also be able to provide at least some knowledge about relevance of the kind suggested here.

Furthermore, relevance is related to the notion of statistical dependence, but is also strongly tied to the notion of causality (Pearl & Paz, 1986). If all events in the world have some cause, then the occurrence of one event can only be relevant to the occurrence of another if they share, directly or indirectly, a common cause.

A strong model of causality usually forms the basis of Explanation-Based

Learning (Mitchell, Keller & Kedar-Cabelli, 1986). Pazzani, Dyer and Flowers (1986) explore the role of a theory of causality in generalization; to their analysis should be added that a partial theory of causality gives a means of tying down relevance to some degree. If the partial theory says that two events cannot possibly share a common cause, they cannot be relevant to one another. This kind of reasoning, in conjunction with induction yields a form of learning halfway between induction and explanation-based learning.

Finally, Borgida *et al.* (1986) argue that heuristics exist for determining relevance from data itself—certainly the case if we consider that relevance is related to statistical independence. The problem of determining relevance is beginning to look similar to that of determining which of the applicable constraints on a concept are likely to hold (see section 4.2.2). Using the approach of section 4.2.2, the construction of common generalizations from examples could also be viewed as a data-driven, strong-to-weak generation of constraints.

Sufficiency of the knowledge

As discussed in section 4.1.2, generalizations will only be found by PGA if sufficient knowledge is available (can be derived in step 2a of the algorithm) from which the generalization can be justified. Of course, this can be used to advantage to constrain irrelevance.

6. Conclusion

We have based the paper on a theory of generality for definite clauses where generality is set in the context of background knowledge in the form of facts and other definite clauses.

Guided by this, we have shown that existing induction techniques such as Version Spaces and Focusing are inadequate for the induction of logical descriptions in the extended context where arbitrary background knowledge is available. This is due to a number of factors. When generality takes into account background knowledge most specific generalizations may be infinite or weighted down with irrelevant conditions. Furthermore, the full definite clause search spaces, for the general-to-specific search particularly, have a huge branching factor and search needs to be model-directed and/or heuristic. At least a weak heuristic is essential when search spaces are infinite—the usual case. These problems can only be magnified when more extensive knowledge representations are used.

Though we have not presented a completed replacement technique, we have rephrased and redirected the existing techniques. We view induction as the model-directed discovery of patterns (constraints and rules) in data. The problem of heuristic search of the space of constraints can be factored into independent searches for constraints and necessarily raises the kind of question: "How strongly do you believe constraint C is valid?" The problem of heuristic search of these inductive search spaces remains an open problem.

We have also presented an algorithm, the Plausible Generalization Algorithm that is intended to form the basis of a technique for discovering rules in data. The algorithm has been implemented in PROLOG and highlights issues in the discovery of rules.

The algorithm generalizes examples to form rules. The algorithm accepts additional input in the form of relevance constraints that allow a trainer to focus the algorithm on relevant information rather than all possible information inferable from the knowledge base. These constraints also allow simple and flexible specification of a model in which a model-directed induction could proceed.

Finally, we have identified problems that must be considered if induction is to be performed drawing on a body of background knowledge, for instance constructive induction. The first is the problem of determining irrelevance. Partial models of causality, statistical analyses and *a priori* beliefs of the trainer can be used here. The second problem is the answering of commonality queries, queries of the form "What do (objects) X and Y have in common that is relevant to (situation) Z?"

Thanks are due to Ross Quinlan and Donald Michie for their suggestions about an earlier draft of this paper and to Jenny Edwards, John Hughes and Graham Wrightson for their support. Attendance at the Knowledge Acquisition for Knowledge-Based Systems Workshop was sponsored by the Australian Department of Science and the N.S.W.I.T. Key Centre for Computing Sciences.

References

BORGIDA, A., MITCHELL, T. M. & WILLIAMSON, K. (1986). Learning improved integrity constraints and schemas from exceptions in databases and knowledge bases. In BRODIE, M. L. & MYLOPOULOS J. Eds, *On Knowledge Base Management Systems,* pp. 259–286. New York: Springer–Verlag.

BUCHANAN, B. G. & MITCHELL, T. M. (1978). Model-directed learning of production rules. In WATERMAN, D. A. & HAYES-ROTH, F. Eds, *Pattern-directed Inference Systems.* New York: Academic Press.

BUNDY, A., SILVER, B. & PLUMMER, D. (1985). An analytical comparison of some rule learning programs. *Artificial Intelligence,* **27**.

BUNTINE, W. L. (1986). Towards a practical theory of Horn clause induction. *Proceedings Ninth Annual Australian Computer Science Conference,* Canberra.

BUNTINE, W. L. (1987). Generalised subsumption and its applications to induction and redundancy. *Artificial Intelligence.* In press. Extended version of a paper with the same title in *Proceedings of the European Conference on Artificial Intelligence* (1986), Brighton, U.K.

CHEESEMAN, P. (1984). Learning of expert systems from data. *Proceedings of the IEEE Workshop on Principles of Knowledge-Based Systems,* Denver.

DEJONG, G. & MOONEY, R. (1986). Explanation-based learning: an alternative view. *Machine Learning,* **1,** 145–176.

DELGRANDE, J. P. (1987). A foundational approach to autonomous knowledge acquisition in knowledge-based systems. *International Journal of Man–Machine Studies,* **26,** 123–141.

DIETTERICH, T. G., LONDON, B., CLARKSON, K. & DROMEY, G. (1982). Learning and inductive inference. In COHEN, P. R. & FEIGENBAUM, E. A. Eds, *The Handbook of Artificial Intelligence, Vol. III,* pp. 323–512. Los Altos: Kaufmann.

DIETTERICH, T. G. & MICHALSKI, R. S. (1986). Learning to predict sequences. In MICHALSKI, R. S., CARBONELL, J. & MITCHELL, T. M. Eds, *Machine Learning–an Artificial Intelligence Approach, Vol. II.* Los Altos: Morgan Kaufmann.

FU, L.-M. & BUCHANAN, B. G. (1985). Learning intermediate concepts in constructing a hierarchical knowledge base. *Proceedings of the Ninth International Joint Conference on Artificial Intelligence.* Los Angeles, California.

GEORGEFF, M. P. & WALLACE, C. S. (1985). A general selection criterion for inductive inference. *SRI Technical Note 372,* Stanford Research Institute.

HART, A. (1985). The role of induction in knowledge elicitation. *Expert Systems,* **2,** 24–28.

KITAKAMI, H., KUNIFUJI, S., MIYACHI, T. & FURUKAWA, K. (1984). A methodology for implementation of a knowledge acquisition system. *Proceedings of the IEEE International Symposium on Logic Programming,* Atlanta City, New Jersey.

KODRATOFF, Y. & GANASCIA, J.-G. (1986). Improving the generalization step of learning. In MICHALSKI, R. S., CARBONELL, J. & MITCHELL, T. M. Eds, *Machine Learning—An Artificial Intelligence Approach, Vol. II.* Los Altos: Morgan Kaufmann.

LANGLEY, P. (1981). Language acquisition through error recovery. *CIP working paper 43,* Carnegie-Mellon University.

LENAT, D. B. & DAVIS, R. (1982). *Knowledge-based Systems for Artificial Intelligence.* McGraw–Hill.

LLOYD, J. W. (1984). *Foundations of Logic Programming.* Springer–Verlag.

MICHALSKI, R. (1983). A theory and methodology of inductive learning. *Artificial Intelligence,* **20,** 111–161.

MICHIE, D. (1986). Current developments in expert systems. In QUINLAN, J. R. Ed., *Applications of Expert Systems.* London: Addison Wesley.

MITCHELL, T. M. (1982). Generalisation as search. *Artificial Intelligence,* **18,** 203–226.

MITCHELL, T. M., KELLER, R. M. & KEDAR-CABELLI, S. T. (1986). Explanation-based generalisation: a unifying view. *Machine Learning,* **1,** 47–80.

MITCHELL, T. M., UTGOFF, P. E., NUDEL, B. & BANERJI, R. (1981). Learning problem solving heuristics through practice. *Proceedings Seventh International Joint Conference on Artificial Intelligence,* Vancouver, Canada, pp. 127–134.

MOONEY, R. (1986). A domain independent explanation-based generaliser. *AAAI-86, Proceedings of the Fifth National Conference on Artificial Intelligence,* Philadelphia.

PAZZANI, M., DYER, M. & FLOWERS, M. (1986). The role of prior causal theories in generalization. *AAAI-86, Proceedings of the Fifth National Conference on Artificial Intelligence,* Philadelphia, pp. 545–550.

PEARL, J. & PAZ, A. (1986). On the logic of representing dependencies by graphs. In *Proceedings of Canadian Artificial Intelligence,* Montreal, Canada.

PLOTKIN, G. D. (1970). A note on inductive generalisation. In MELTZER, B. & MICHIE, D. Eds, *Machine Intelligence* **5,** pp. 153–163. New York: Elsevier North-Holland.

PLOTKIN, G. D. (1971). A further note on inductive generalisation. In MELTZER, B. & MICHIE, D. Eds, *Machine Intelligence* **6,** pp. 101–124. New York: Elsevier North-Holland.

POLYA, G. (1954). *Mathematics and Plausible Reasoning,* Vol. I. *Induction and Analogy in Mathematics.* Princeton, New Jersey: Princeton University Pres.

PUNDIT, N. (1985). On AI Applications. Invited talk, *Conference on Commercial Applications of Expert Systems.,* Sydney.

QUINLAN, J. R., (1986). Induction of decision trees. *Machine Learning,* **1,** 81–106.

QUINLAN, J. R., (1987). Simplifying decision trees. *International Journal of Man–Machine Studies,* **26,** 000–000.

QUINLAN, J. R., COMPTON, P. J., HORN, K. A. & LAZARUS, L. (1986). Inductive knowledge acquisition: a case study. In QUINLAN, J. R. Ed, *Applications of Expert Systems.* London: Addison Wesley.

RENDELL, L. (1986). A general framework for induction and a study of selective induction. *Machine Learning,* **1,** 177–226.

ROBINSON, J. A. (1979). *Logic Form and Function—the Mechanisation of Deductive Reasoning.* New York: North-Holland.

SAGIV, Y. (1986). Optimizing datalog programs. In MINKER J. Ed., *Workshop on Foundations of Deductive Databases and Logic Programming,* Washington.

SAMMUT, C. A. & BANERJI, R. B. (1986). Learning concepts by asking questions. In MICHALSKI, R. S., CARBONELL, J. & MITCHELL, T. M. Eds, *Machine Learning—an Artificial Intelligence Approach,* Vol. II. Los Altos: Morgan Kaufmann.

SHAPIRO, E. Y. (1981). Inductive inference of theories from facts. *TR 192, DCS,* Yale University.

SHAPIRO, E. Y. (1983). *Algorithmic Program Debugging.* MIT Press.

STEPP, R. E. & MICHALSKI, R. (1986). Conceptual clustering: inventing goal-oriented classifications of structured objects. In MICHALSKI, R. S., CARBONELL, J. & MITCHELL, T. M. Eds, *Machine Learning—An Artificial Intelligence Approach, Vol. II.* Los Altos: Morgan Kaufmann.

SUBRAMANIAN, D. & FEIGENBAUM, J. (1986). Factorization in experiment generation. *AAAI-86, Proceedings of the Fifth National Conference on Artificial Intelligence,* Philadelphia, pp. 518–522.

UTGOFF, P. (1986). Shift of bias for inductive concept learning. In MICHALSKI, R. S., CARBONELL, J. & MITCHELL, T. M. Eds, *Machine Learning—an Artificial Intelligence Approach,* Vol. II. Los Altos: Morgan Kaufmann.

VERE, S. A. (1977). Induction of relational productions in the presence of background information. *Proceedings of the Fifth International Joint Conference on Artificial Intelligence,* Cambridge, Massachusetts.

VERE, S. A. (1980). Multilevel counterfactuals for generalisations of relational concepts and productions. *Artificial Intelligence,* **14,** 139–164.

VRAIN, C. (1986). The use of domain properties expressed as theorems in machine learning. *Proceedings of the International Meeting on Advances in Learning,* Les Arcs, France.

Learning Under Uncertainty

Generalization and noise†

YVES KODRATOFF, MICHEL MANAGO

Inference and Learning Group, Laboratoire de Recherche en Informatique, A. 410 du CNRS, Université de Paris-Sud, Bât. 490, 91405 Orsay, France

AND

JIM BLYTHE

GEC Research, Marconi Research Laboratories, West Hanningfield Road, Great Baddow, Chelmsford, Essex CM2 8HN, U.K.

This paper describes a research project which aims at applying Machine Learning (ML) techniques to ease Knowledge Acquisition (KA) for Knowledge Based systems. Since noise in real life data has a drastic effect on ML, we examine in detail problems connected with noise. The learning system integrates two apparently distinct approaches: the numeric approach and the symbolic approach. It uses a filtering mechanism that is driven by statistical information and by comparison between several sources of knowledge (multi-expertise and experts-users "cross-examination" of input). The system also attempts to generate concepts which are resilient to noise and to improve the language of description. While it is usually thought that noise prevents using ML techniques in real applications, we attempt to show that on the contrary existing techniques can be stretched to cope with noise and to obtain better results than traditional KA techniques.

1. Introduction

Formalizing the knowledge needed to solve a real world problem is far from being a trivial task. As noted in Clancey (1986) and McDermott (1986), Knowledge Acquisition (KA) is not the process of transferring a mental model that lies somewhere within the brain of a human expert, but the familiar scientific and engineering problem of formalizing a domain for the first time.

The classical methods to achieve this result are based on the domain expert's ability to explain his (or her) behavior:

(a) a knowledge engineer interviews experts and attempts to formalize their knowledge,
(b) the expert himself is trained to construct a computable model that is extentionally equivalent to his own model.

This is usually a cumbersome process because human experts are trained to solve a task and not to explain how they obtained the solution. Furthermore, the final model frequently contain bugs and is only representative of that specific expert (two different experts almost never agree on what should go in the knowledge base). This

† This research has been partially supported by the European Economic Community (ESPRIT contract P1063, the INSTIL project, in collaboration with the French company Cognitech). LRI's research has also been supported by the GRECO-PRC "Intelligence Artificielle'.

KNOWLEDGE-BASED SYSTEMS Vol. 1
ISBN 0-12-273251-0

causes several problems, in particular those connected with the system's maintenance.

An alternative to the traditional methods is to use Machine Learning (ML) tools. From a set of examples of expertise, the learning system automatically constructs the rules. Several experiments have shown that very good results can be obtained this way (for instance, see Michalski & Chilauski, 1980; Quinlan, 1986*b*). These techniques solve partially the problem of getting the expert to formalize his rules. However, the expert still plays a critical role since he must provide the vocabulary needed to describe the events (the language of description), the domain specific properties (valid axioms and constraints, default and common sense knowledge and so on), and that he must validate (or even provide) the set of training examples.

In real applications, the data provided to the system by the expert contains noise. Noise can originate from incorrect information as well as from a lack of information or information that is not reliable or which contains some uncertainty. We view noise as a critical problem that must be treated in order to apply ML techniques on a larger range of applications.

In this paper, we look at different types of noise, which affect KA at different steps and which yield different procedures of detection and treatment. Some methods developed here are common to both traditional and automatic KA techniques (for example, detecting and filtering noisy information by cross-examination of experts) but we actually studied noise from a ML perspective.

The material presented here is based on a research project which aims at automatically building a Knowledge Base to diagnose tomato plant diseases and to compare the results with an existing Knowledge Base that contains over 350 rules (This work started in October 1985 and is done in collaboration with the French company Cognitech and the French "Institut National de Recherche en Agronomie"). The example base contains observed examples and examples generated by several different experts. The observed cases are obtained through questionnaires that have been completed by students in agriculture and validated by domain experts.

At LRI, we are currently studying another real scale application, air traffic control. A case library is currently being built (work started in October 1986) in collaboration with the "Centre d'Etudes de la Navigation Aérienne": by looking at a radar picture, the human expert describes what he sees and explains his behavior. Some of the material presented here was and is still being tested on a smaller application in law (work started in early 1987). The examples are judgements in a court of law that describe the legal actions of the mayor of a city. We will compare the rules generated with existing laws.

At GEC Research, we are also using the techniques presented here to learn rules predicting the location of an address on a parcel. The domain combines structural information with a large proportion of numerical data.

2. Learning in noisy domains

The traditional approach to handling noise consists in attaching numerical coefficients such as certainty factors (Buchanan & Shortliffe, 1984) to the rules. These techniques are well adapted in some cases, but this approach does not solve all the problems relating to noise and it generates some new ones. For example:

(a) computing numericial CFs when the case library does not contain valid statistical information.

If the system is used in Rennes (north-west of France) and the case library was built in Avignon (south of France), the diseases will affect the plants differently.

(b) Updating CFs when statistical information varies over time.

If there is an epidemic of one disease one year, the statistics will be different the next year. Furthermore several diseases will not appear in a given year (in tomato plant pathology, according to the domain experts it takes roughly seven years before all the diseases appear)

(c) Combining uncertain rules when the independance assumption does not hold.

Consider the case where the expert sees one out of two symptoms, and reaches a certain conclusion. If he sees both of the symptoms he derives the same conclusion with the same confidence. Therefore firing the second rule after the first one does not increase the certainty factor.

(d) The resulting system lacks explanatory capabilities.

(e) It is difficult for a human expert to relate to a large set of such rules and to generate and/or validate these.

(f) Always handling noise through numerical methods sometime masks the problem instead of solving it.

Our goal is to use numerical and statistical methods to handle noise only when it is appropriate. We feel that some of the problems connected with noise must be handled at the symbolic level by using general knowledge about the domain, by improving the language of description, by finding good intermediary knowledge and so on. By gathering knowledge from several sources (several experts, the end users of the system) we increment our background knowledge on the domain learning the relative reliability of the features that can be tested. This can be used to generate rules that are robust with respect to noise or to filter noisy training data before initiating the learning process. Our goal is to stretch and make better use of some known ML techniques in order to handle noise (learn from noisy data as well as generate concepts which are more resilient to noise).

Several applications have experimentally shown that model-driven search performs better in a noisy environment than data-driven search. This is true for consultation systems as well as concept learning systems [as noted in Mitchell (1985), generalization can be viewed as a search problem] in various domains. See for example the following applications in signal processing DENDRAL (Lindsay *et al.*, 1980) and HASP (Nii *et al.*, 1982), medical diagnosis (Kuntz *et al.*, 1978; Aiello, 1983], numerical analysis (Diday & Simon, 1979) and concept learning (Fu & Buchanan, 1985; Quinlan, 1986*b*; Michalski *et al.*, 1986). Using some domain specific knowledge about the features (their relative weights, the constraints between the features and so on) or some broad universal knowledge (such as information theory), these systems are able to limit the search space and only examine paths which are likely to yield accurate results.

In the case of ID_3 (Quinlan, 1983) (or its multi-class extention IP) the system uses information theory and optimizes a numerical evaluation function (information gain, based on the entropy measure) to grow a search tree and separate the high level concepts most efficiently. It uses a heuristic (hill-climbing) search procedure, heading toward the highest information gain at each level. Various numeric methods to simplify the decision trees by pruning [pre-puning at construction time or post-pruning at the end (Niblett, 1987; Quinlan, 1986*b*] and thus allow some inconsistency in the final classification (some clusters containing both positive and negative examples) have also been developed as a means to treat erroneous data. There is no need to explicitly represent the uncertainty introduced by pruning since the error generated in the classification has been shown to be significantly lower than than the one introduced by a decision tree that outputs a total classification (Quinlan, 1986*b*).

In the case of RL (Fu, 1985), the system begins to do a full graph search and prune the branches when the significance of the clusters (according to some statistical function that takes into account the number of positive examples covered and the number of negative examples rejected), falls below a given threshold. The basic idea is similar to the pre-pruning mechanism mentioned above but, since the goal is different, the numeric test for pruning is performed differently and the threshold is set at a much higher value. An explicit representation of the uncertainty introduced is provided by computing a numerical certainty factor that is attached to each path of the Directed Acyclic Graph that leads to a terminal node to generate a recognition rule. Another depth bound of 7 is also set (for efficiency reasons as well as understandability of the final rule) and the rules must chain to derive a conclusion.

Several characteristics make these two systems behave well in presence of noisy training data. First, using a model to guide the search favors attributes which are more "important". Note that although these systems rely very much on the data to build the model using information theory, the overall systems are model-driven since they consider the data as whole as opposed to considering each individual piece of the data. Furthermore, RL also incorporates in its model background domain knowledge (a half-order theory of the domain). Due to the top-down, generate and test clustering method, individual examples have a limited influence on the algorithm. The higher up in the tree (or the DAG) you are, the smaller is the influence of the training example. Thus, individual examples that are too far from the others will be naturally isolated. Terminating the algorithm before finding a total classification will produce good results when the example was incorrectly classified since its statistical weight is low.

However, during consultation, when only one individual feature of the example is incorrect and the system happens to test for that feature, the result will be drastic. Attaching numerical uncertainty to the terminal nodes only manages to hide the problem. Note that in this case, RL performs better than ID_3 because of its ability to learn rules with negative CF's which allows it to get back on the right track. Another way to treat this problem symbolically, is to gather knowledge on the reliability of a test (either by asking for it or by comparing knowledge from different sources) and use more reliable tests higher up in the tree or the DAG. This simple idea is one way to really treat noise and generate robust rules. Since there are several possible ways to reach a conclusion, the system will be able to find the most reliable one. This minimizes the amount of unreliable tests in the decision tree and also generates concepts which are more robust since the S generalization of the clusters (see next section) will often generalize away these unreliable tests.

In this paper, we present a learning algorithm which is able to learn disjunctive concepts and show how to use it to handle noisy training data and generate robust rules. It integrates two learning systems, NEDDIE (a descendant of ID_3) used to build clusters and MAGGY which generalizes these clusters.

3. Background software

3.1 MAGGY

MAGGY is an object-oriented generalization algorithm (see next section). It outputs the Most Specific Conjunctive Generalizations (MSCG) of the positive examples (a characteristic description of the high level concept). In Version Space terminology, MAGGY finds the near (S) set.

MAGGY can generalize characteristic rules that are given by the experts (training examples with un-instantiated variables) or observed cases whose classes are known (instantiated training examples). When generalizing rules, the positive examples are the Left-Hand Sides (LHS) of the rules concluding to the concept studied, while the premises of all the other rules are the negative examples. Note that all the variables of a rule are universally quantified while the variables of an example are free. One could also view the LHS as being lambda-bound (we are generalizing concepts or boolean functions that partition the space into the instances that are mapped onto the value TRUE and other instances).

For example, the rule: "$\forall x, \forall y. SPOT(x)$ & $YELLOW(x)$ & $ON(x\,y)$ & $FRUIT(y) \Rightarrow SUNBURN$ *generates the recognition function for the concept* *SUNBURN*: $\lambda x, \lambda. SPOT(x)$ & $YELLOW(x)$ & $ON(x\,y)$ & $FRUIT(y)$.

Intuitively, another function is more general if it recognizes the same sets (or subsets of these) and more (Vere, 1980), (Sammut, 1981). Note that the format of the output is identical to the format of the input which allows incremental learning capabilities (we can generalize rules with variables and fully instantiated training examples).

3.1.1. Pattern matching algorithm

MAGGY uses a modified version of the *structural matching* algorithm of AGAPE (Kodratofff & Ganascia, 1986; Vrain *et al.*, 1986). There are four steps in the algorithm:

(a) selection of conceptual objects or units (belonging to different examples) to match;
(b) transfer of the differences between the objects into variables of generalization;
(c) adjunction of hidden links by applying background knowledge on the difference set;
(d) dropping of everything that differs throughout the examples [A more general than A & B (Michalski, 1983)] and simplification of the resulting expression.

The last step is the only inductive one of the algorithm which allows complete control over the information that is lost during generalization. Most of the effort is spent on the first three deductive steps (introducing a variable is deductive as long as the binding of the variable is kept, an axiom may be used deductively since $A \Rightarrow B$ is logically equivalent to $A \Leftrightarrow A$ & B and so on). Hence, this information is lost as a side-effect since nothing is given up until the last step of the algorithm. Thus, during the third step (adjunction of links) the system can focus its attention on the difference set and apply the background knowledge only on the information that will be voluntarily abandoned. For example:

E_1: INSTANCE-OF(x SPOT) & COLOR(YELLOW x) & INSTANCE-OF(y MOULD) & COLOR(YELLOW y)

E_2: INSTANCE-OF(x_1 SPOT) & COLOR(WHITE x_1) & INSTANCE-OF(y_1 MOULD) & COLOR(WHITE y_1)

The matcher will choose to unify units x and x_1, and y and y_1 (the other possibility will be discarded as it loses more information and yields an expression that is more general). After structural matching we have:

E_1: INSTANCE-OF(ob_1 SPOT) & COLOR-OF(u obj_1) & INSTANCE-OF(obj_2 MOULD) & COLOR-OF(v obj_2) & EQ(u YELLOW) & EQ(v YELLOW)

E_2: INSTANCE-OF(obj_1 SPOT) & COLOR-OF(U obj_1) & INSTANCE-OF(obj_2 MOULD) & COLOR-OF(v obj_2) & EQ(u WHITE) & EQ(v WHITE)

Notice that the left portion of both expressions is identical. Thus, if it is possible to discover other similaritites that are hidden [constructive generalization (Michalski, 1983)], it is sufficient to focus the efforts on the difference set (on the right of the &). This is important since constructive generalization is a computationally costly mechanism (it requires theorem proving capabilities). In the example above, if in background knowledge the system knows that EQ is a commutative and transitive predicate, it will add EQ(u v) to both expressions. After using the dropping rule, the generalization will become:

INSTANCE-OF(obj_1 SPOT) & COLOR-OF(u obj_1) & INSTANCE-OF(obj_2 MOULD) & COLOR(v obj_2) & EQ(u v)

The semantics of this are: there is a SPOT and a MOULD with the same COLOR. Notice that an inductive bottom-up algorithm performing induction at each step would have either lost the fact that the two objects were of the same color, or it would have added the redundant information that MOULD and SPOT were both SYMPTOM before generalizing.

An important aspect of the algorithm is the way the variables are bound. In MAGGY, variables represent distinct conceptual objects or units that may not be instantiated by the same object. This is one of the differences between MAGGY's structural matching and AGAPE's (in AGAPE there can be a many-to-one variable binding). Consider the following example of Fig. 1.

E_1 and E_2 are generalized by MAGGY into two expressions:

G_1: ISA(x CUBE) & SIZE (x SMALL) & ISA(y CUBE) & SIZE(y BIG) & ON(x y)

G_2: ISA(x_1 CUBE) & COLOR(x_1 BLACK) & ISA(x_2 CUBE) & COLOR(x_2 STRIPES)

G1 and G2 are both elements of the S set (they are both MSCG whose generality cannot be compared). However, AGAPE would find the following generalization:

G: CUBE(x) & SMALL(x) & CUBE(y) & BIG(y) & ON(xy) & CUBE(x_1) & BLACK(x_1) & CUBE(y_1) & STRIPES(y_1) & DIFFERENT(x y) & DIFFERENT(x_1 y_1)

The generalization G would then cover both generalizations by allowing variables (x, x_1) and (y, y_1) to be bound to the same objects (there is an implicit May-Be-the-Same (F. Hayes-Roth & J. McDermott, 1978) link between these variables which are pure logical variables). We define this as a feature-oriented generalization. In

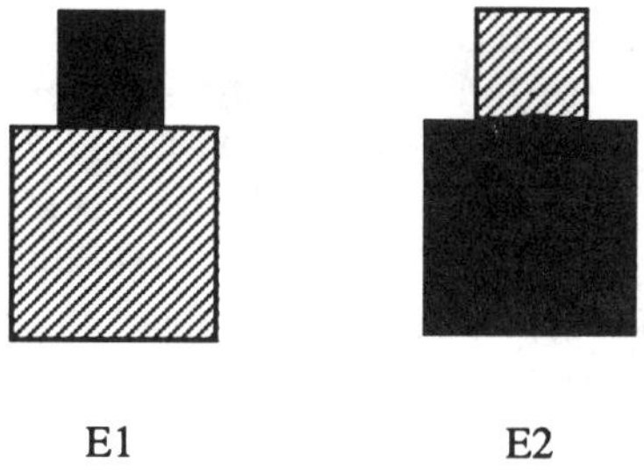

FIG. 1. Two examples from the world of blocks.

object-oriented generalization we add the constraint that variables cannot be instantiated by the same objects (this property is part of the initial theory).

We will not discuss here in detail why in our applications object-oriented generalization is more appropriate than feature-oriented generalization. As noted in Dietterich & Michalski (1983, page 65) about THOTHP (Vere, 1980), "A difficulty resulting from the relaxation of semantic constraints is that the algorithm creates generalizations involving a many-to-one binding of variables. While such generalizations may be desirable in some situations, they are usually meaningless, and their uncontrolled generation is computationally expensive."

Thus, even if it appears to be more powerful, in our large scale application we could not afford the combinatorics inolved with feature-oriented generalization. Furthermore, as will see later, the system uses a model which tells it the relative importance of the attributes in order to pick the right generalization. If these are equivalent, the system can output the disjunctive expression G1 ∨ G2 as it is done in version space.

3.1.2. Knowledge representation

MAGGY uses first order logic to represent the examples [a representation centered on units (Nilsson, 1984, chapter 9.1)]. Predicates can be valued to TRUE, FALSE and UNKNOWN.

For instance, there is a REPULSIVE GREEN MOULD ON A RED FRUIT WITH NO YELLOW SPOTS is represented as:
[x: ⟨INSTANCE-OF MOULD⟩ ⟨IS REPULSIVE⟩ ⟨COLOR GREEN⟩ ⟨IS-ON y⟩] & [y: ⟨INSTANCE-OF FRUIT⟩ ⟨COLOR RED⟩] & ¬[z: ⟨INSTANCE-OF SPOT⟩ ⟨COLOR YELLOW⟩ ⟨IS-ON y⟩]

Notice the negation. It can be used to avoid introducing quantifiers (for example "all the elephants are grey" can be converted to "the elephants are grey and no elephant is not grey"). MAGGY generalizes a negation by relaxing the constraint that it conveys. In other words, it generalizes (NOT(A) by specializing A or by dropping the negation (the latter seldom occurs because when an expert specifies negative information it is usually important information which allows the discrimination of a negative example and thus which cannot be generalized in such a drastic manner).

MAGGY uses frames (Minsky, 1975), (Charniak, 1977) to represent background knowledge (and background knowledge only). This is an improvement over the

classical taxonomic representation for the following reasons:
(a) Frames can represent more than one relation of generality. We use the AKO relationship (*for instance MOULD AKO SYMPTOM*) AND THE PART-OF relationship (*LEAVES PART-OF PLANT*) to represent composite objects. While the first one is found in taxonomies, the second is not. Information may be inherited over these links during generalization. Inheritance and default reasoning can be useful for filling missing information (a specific kind of noise as discussed in section 7). Note that the user has the ability to easily redefine the generalization/inheritance relationships by changing the values of the appropraite variables.
(b) We can represent several properties that are always true of a given object and that act as a constraint which must be verified (reducing the combinatorics when a specific match violates the constraint).
For instance, CONCAVE is a legal value for the FORM of a SPOT in general. However, it is not a legal value for a SPOT on a LEAF which can only be FLAT.
(c) A unit may have more than one father and belong to several different frames (*Although biologists view a TOMATO as being a FRUIT, the user might view it as a VEGETABLE*). Frame representation allows representing multiple taxonomies.[2] However, these must be ordered and cannot contain loops (these are Directed Acyclic Graphs).
(d) Frames represent prototypical situations and allow (easy) default reasoning and common sense reasoning. *For example, if the COLOR slot of the object LEAVES is filled with GREEN, we will not need to write down that a specific instance of LEAVES is GREEN* (*it is inherited*). This is an elegant (and efficient) method to represent axioms such as "every object has a color" to avoid using the dropping condition rule (Kodratoff & Ganascia, 1986). Note that a range may be imposed on the value of an attribute (*for example,* (*tomato-fruit* (*color* (*$RANGE green red*)))).
(e) Axioms (deamons) can be attached to slots. This is similar to KRL's (Bobrow & Winograd 1977) procedural attachment. *There are $IF-NEEDED deamons such as* $MATURE(x) \Rightarrow RED(x)$ *and* $YOUNG(x) \Rightarrow GREEN(x)$ *which are attached to the color slot of the unit FRUIT and are fired when we query the value of the slot* (*by default, a mature tomato fruit is red and a young tomato fruit is green*). *When the color of the fruit is unknown and is needed, MAGGY attempts to prove the Right-Hand-Side of the rule in backward chaining.*
(f) Finally, a limit on taxonomies (often theoretical) is that the sons must form a partition of their father. They must exhaust the possibilities of their father (*If RED and BLUE are the only sons of COLOR in a given taxonomy, then there cannot be any other unknown colors*) and must be mutually exlusive (an object is either classified as RED or BLUE but cannot be both RED and BLUE). While this may hold in an ideal environment, this is unlikely to be the case in practice. Polymorphy of the low level concepts (concepts with a non empty intersection) must be taken into account (*for example, when does dark brown stop being brown to become black*?).

3.2 NEDDIE

NEDDIE (Corlett, 1983) is a learning system similar to ID_3 (Quinlan, 1983). Given a set of examples that may belong to a number of different classes, it produces a decision tree that discriminates between the classes. NEDDIE has several improvements over ID_3 which will not all be detailed here.

The system has been designed to use, at will, a number of preference criteria for building the tree, including entropy and a chi-squared test of the reliability of an attribute. Many such criteria are still under experiment. Apart from this, the tree can be terminated at any point if its overall reliability, based on a chi-squared test, falls below a certain value called *min-branch-safety*. Thus the conclusions of a tree can be "fuzzy" since it may terminate before an exact classification has been reached (in such cases, the method for choosing a class for an unseen object, for instance the probability or majority methods (Quinlan, 1986*a*) is left to the user).

It has been argued that logical rules are easier to understand than decision trees (O'Rorke, 1982), (Quinlan, 1986*b*). Thus NEDDIE takes the decision tree that it generates, and separates it into its component rules. These are conjunctions of tests along with their conclusions. Each rule is then simplified as follows (Corlett, 1983):

Each test of the conjunction is experimentally dropped. If the new conclusion based on the examples that satisfy the new rule is weaker than the original one, the test is put back in. Otherwise it is dropped permanently. Thus all tests in a rule that were originally in the tree but not directly associated with that rule's conclusion are dropped. Thus, the path in the tree is generalized against the negative examples by using the dropping condition rule (Michalski, 1983). We are currently upgrading this module of the program so that it will be able to use other generalization rules that MAGGY uses. These are well documented through the literature (Michalski, 1983). *For example, one of the selectors in the path could be generalized by climbing the generalization tree.*

These rules can be given a "flat" representation, understanding that all the descriptors of their Left-Hand Side must simultaneously take the value TRUE, to allow the recognition of a high level concept. On the other hand, one might consider that some descriptors are more important or better defined or less noisy, or more efficient than others. One should then take advantage of this knowledge by checking the better ones first (meta-level knowledge). This choice will clearly influence the content of the final clusters. In our approach, we generate a partial ordering of the descriptors that will be used by MAGGY during generalization.

When the classification is carried to its end, this method finds the known necessary conditions to belong to a concept [it finds generalizations against the negative examples (Michalski, 1983)]. However, when learning in an uncertain domain from noisy data, it is preferable (during the learning stage) to find a maximally specific characterization of the clusters (these become characteristic descriptions of high-level or intermediary level concepts whose positive examples are in the clusters).

4. The need for "good" disjunctive generalizations

MAGGY is a system which outputs conjunctive generalizations. Nevertheless, conjunctive generalizations (even minimal ones) are sometimes inconsistent (i.e. overgeneral). In this case, a disjunctive generalization rejecting every counter-example is needed. Consider the following:

Let E_1, E_2 and E_3 be 3 positive examples, and CE a negative example:

E_1: [x: ⟨INSTANCE-OF FRUIT⟩ ⟨COLOR RED⟩ ⟨SIZE VERY-BIG⟩ ⟨TEXTURE SOFT⟩]

E_2: [x: ⟨INSTANCE-OF FRUIT⟩ ⟨COLOR GREEN⟩ ⟨SIZE BIG⟩ ⟨TEXTURE HARD⟩]
E_3: [x: ⟨INSTANCE-OF FRUIT⟩ ⟨COLOR GREEN⟩ ⟨SIZE VERY-BIG⟩ ⟨TEXTURE HARD⟩]
CE: [x: ⟨INSTANCE-OF FRUIT⟩ ⟨COLOR RED⟩ ⟨SIZE BIG⟩ ⟨TEXTURE HARD⟩]
A conjunctive generalization of the positive examples is:
G: [x: ⟨INSTANCE-OF FRUIT⟩ ⟨COLOR ANY⟩ ⟨SIZE LARGE⟩ ⟨TEXTURE ANY⟩]

It is intuitively clear that it is minimal but covers CE as well. A less general expression (which necessarily contains one or several disjunctions) must be found. There are three possibilities [Gen is the generalization operator (Michalski & Stepp, 1983)]:

$Gen(E_1, E_2) \vee E_3$
$Gen(E_1, E_3) \vee E_2$
$Gen(E_2, E_3) \vee E_1$

The first solution is inconsistent with the training data ($Gen(E_1\ E_2) = G$) and to lift the inconsistency we would have to specialize $Gen(E_1\ E_2)$ finding "$E_1 \vee E_2 \vee E_3$" which is the most trivial characteristic generalization. We are interested in a generalization that contains a minimal number of disjunctions, thus minimizing the numbers of clusters of positive examples which in turn minimize the number of rules that will be finally used by the knowledge-based system.

The other two generalizations are consistent disjunctive generalizations. In solution no. 2, the discriminant features are:
$Gen(E_1, E_3)$ (first cluster): ⟨SIZE VERY-BIG⟩
E_2 (second cluster): ⟨COLOR GREEN⟩
In solution no. 3, the discriminant features are:
$Gen(E_2, E_3)$ (first cluster): ⟨COLOR GREEN⟩
E_1 (second cluster): ⟨SIZE VERY-BIG⟩ and ⟨TEXTURE SOFT⟩

Depending on the heuristics of generalization/specialization, one of the two will be found. In the present case, since symbolic noise is lower for GREEN and RED than for BIG and VERY-BIG (the cost parameter is higher for the latter) the system will favor solution 3. This is achieved by merging the two learning systems, the clustering algorithm described in section 3.2. which finds the disjunctions that are compulsory and the generalizer described in section 3.1. which outputs characteristic descriptions of the concepts.

We can now see how knowledge about the environment (the relative reliability of the descriptors) can be used to generate clusters that are resilient to noise and thus rules which are more reliable. This is one way to symbolically deal with noise without using any numerical uncertainty. Note that we have encountered in the original rule base (that was built manually), rules where the numerical coefficient was voluntarily lowered by the expert because he would not trust the answer given by the user. Thus, the answer to a question such as "Is it BIG or is it very BIG" was considered suspect. Although some experts view this as a good knowledge-engineering practice, numerical coefficients are usually supposed to represent statistical information (Buchanan & Shortliffe, 1984).

5. Merging MAGGY and NEDDIE

The reason for the success of NEDDIE and similar systems in the presence of noise is strongly linked to the top-down numeric approach of building a decision tree. As we have seen earlier, it can be used to globally explain the input by setting thresholds for an early termination of the tree. For the sake of simplicity, in this section, we assume that the classification is taken to its end. We are currently working on the generalization of clusters which contain some uncertainty.

NEDDIE does not use symbolic connections between concepts (*for example, it cannot use the fact that WHITE and YELLOW are sons of LIGHT-COLOR*) and it can generate rules which are not easily understandable by a human being. By considering only the "best" discriminant features of the examples according to some numerical evaluation to efficiently split the search space, it eliminates relevant characteristic features of the examples.

Imagine you are trying to find rules to recognise vehicles such as cars, motorcyles and bicycles. Depending on the outcome of the numerical evaluation, NEDDIE can produce rules such as "a car is a vehicle with a door handle" (*see Fig.* 2). *Although this is true, a human expert is not likely to relate to this description. For him, a car is a vehicle with* 4 *wheels, an engine, doors etc. . . . The fact that all cars have a door-handle is a detail which is not always meaningful and there are alternative "conceptually better" explanations for being a car.*

When dealing with unknown information and features that are not independant, it is important not to abandon any explanations (*at consultation time, the user can answer that he does not know if there was a door handle or not and we must be able to find dynamically and for the characteristic generalizations another path to reach the*

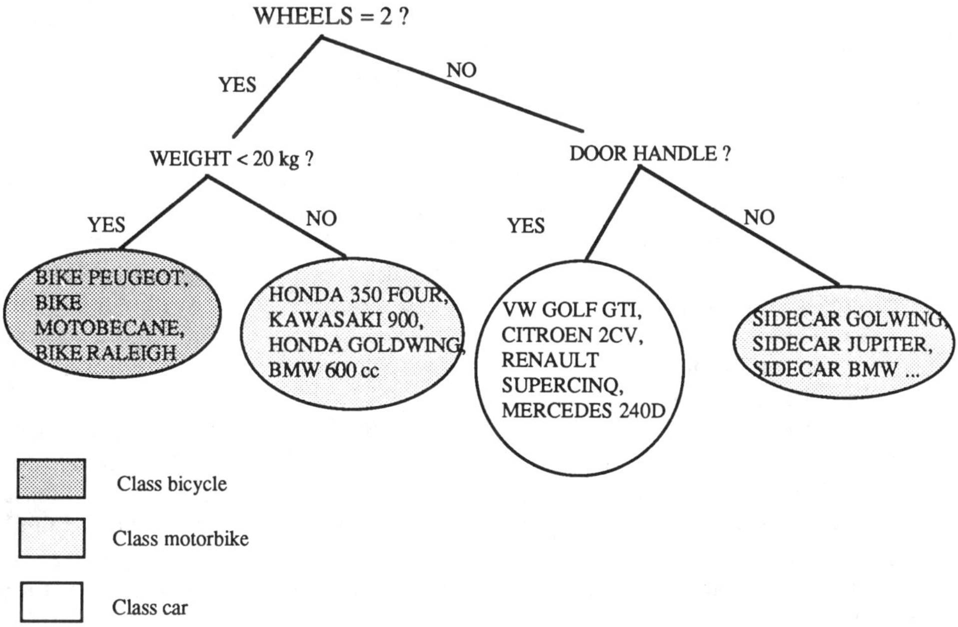

FIG. 2. A decision tree to classify vehicles.

conclusion). We must handle this kind of noise in our applications. For instance, the user can try to identify a disease by looking at a picture of a fruit. If the system generated a decision tree (and rules) about the state of the leaves, the system would not recognize the disease since the user can only answer questions about the state of the fruits. However, by looking only at the state of the leaves, one could find the correct diagnosis.

One way to improve could be to generate rules from several decision trees. Nevertheless non-discriminant characteristic features of the classes would be lost. *For example, we could forget the fact that a car has an engine, a windshield, a rear view mirror and so on because a motorcycle also does.* This information is important to improve the conceptual cohesiveness of the clusters and to generate characteristic rules that the experts can understand. This information is what we call the *explicability* of the concepts.

Thus, the system generates a characteristic description of the classes. This is achieved by growing a decision tree with NEDDIE and applying MAGGY on the individual clusters of positive examples.

The generalization (or cover) is the logical disjunction of the conjunctive generalizations of the individual clusters [the complexes (Michalski, 1983)]. By using MAGGY, the system finds the disjunction of the characteristic descriptions of the complexes while the method described in section 3.1.1 finds a star against the negative examples (Michalski, 1983, page 112). It benefits from NEDDIE (the efficient top-down model-driven noise-robust method of building a decision tree) and from MAGGY (the rich conceptual description of the examples which is well adapted for interaction with human beings).

Drawing analogies with the version space method is difficult since the positive examples may represent truly disjunctive concepts. Nevertheless, in the simple case where there is only one single cluster of positive examples in the decision tree, using the method described in section 3.1.2 generates an element in the G set while generalizing the cluster with MAGGY produces the S set. Combining MAGGY and NEDDIE has some consequences for each algorithm.

NEDDIE can use a different evaluation function by computing a *generalization gain*. It not only uses information gain to grow a discriminant decision tree on the basis of the differences among the classes (i.e. a tree which is maximally efficient in terms of separating the classes) but it uses the generalizer to find a tree which maximizes the explicability of the individual clusters (i.e. minimizes the loss of information in the description of the positive examples that is obtained with MAGGY). This is useful for selecting a test when several tests are more or less equivalent in terms of information gain and when the statistical information provided by the examples is not relevant.

MAGGY uses the information brought by NEDDIE (the model) to select units to match. It matches units so that the attributes in the path will not be generalized. Since NEDDIE has built the clusters in this manner, we know that there is at least one unification so that this will happen. Thus, the conjunctive generalization found by MAGGY will be consistent (it will be less general than the generalization computed from the path in the decision tree that is known to be consistent). In other words, MAGGY can decide to match units both in terms of the quantity and the quality (how discriminant it is, how noisy it is, how important it is for the model and so on) of the information which is lost during generalization.

6. Getting the initial knowledge from the experts

Knowledge located at different levels must be obtained from the human experts (Alexander *et al.*, 1986) or learned automatically. For example:

(a) the language of description;
(b) object level knowledge (stereotypical objects and default values, relations and links between objects, domain specific axioms and constraints between objects, example base, rules etc . . .);
(c) meta level knowledge (intermediary concepts, strategies, heuristics etc . . .).

These levels are not separated or modular. For example, as the learning system attempts to generalize the example base (and obtain object-level knowledge), it might try to refine the language of description by introducing new intermediary terms [shift of bias (Utgoff, 1986, page 125), symbolization of taxonomy points (Fu & Buchanan, 1985), creation of new frames (Manago, 1986)]. In this paper, we have divided the KA process into the following three steps:

(1) Expert Interaction + Literature on the domain (obtaining background knowledge);
(2) Concept formation from examples (learning full descriptions of high level and intermediary level concepts);
(3) Rule learning (learning diagnostic rules and meta-level knowledge).

Some of the methods that we will describe for treating noise such as filtering rely heavily on the ability to go back and forth between the first and second steps.

Note that we use the term *concept* for a function f that partitions the space of instances into the instances that verify the function (i.e. the set of instances whose image is TRUE) and those that do not. Thus, we equate a concept with its description instead of considering that it is some abstract entity. A low level concept is the basic building block (for instance the symptoms), a high level concept is what we are trying to learn (for example the disease) and an intermediary level concept is something in between that is "interesting" (for example, the diseases caused by a fungus). It is important to emphasize that there is an important intentional component (its purpose) in a concept.

In the next sections we describe in detail how to obtain the initial data from the expert, the different kinds of noise that we have encountered, and propose methods to detect and treat some of it.

6.1 GETTING THE INITIAL VOCABULARY

There is no universal method to obtain the initial vocabulary. The descriptors (low level concepts) can be gathered by searching the literature, by interviewing an expert and translating natural language in the proper formalism [see for example (Regoczei & Plantinga, 1986)]. In our application (plant pathology) we started with an existing expert system which was built by an expert and we asked another expert to filter off the irrelevant (noisy) descriptors and to add some which he thought would be more universal. *For example, a descriptor like BIRD-EYE-SPOT was transformed into "two round spots, the darker one in the center of the other". This generated the unary object descriptors ROUND, SPOT and the binary relational descriptors DARKER-THAN and IN-CENTER-OF. This information was then encoded as part of the background knowledge in a frame-based language (see next section).*

This idea of comparison or *cross-examination* between several sources of knowledge for filtering noise or learning which descriptors are unreliable will come up several times in this paper. This is similar to a technique used in signal processing (computer vision and so on) to filter noisy pixels introduced by the recording instruments. Two pictures are taken and AND-ed together to remove false pixels or OR-ed together to fill missing pixels. We take "a picture" of the domain with an expert and compare it with "pictures" taken by other experts and/or by the end-users of the expert system.

The choice of descriptors depends on the underlying logic used by the system. For example, with propositional calculus, the descriptors BLUE, SQUARE and BLUE-SQUARE are needed to describe the world. Using a feature–value pair language, the system must learn valid attributes and attribute values. Using first order logic the system needs valid attribute names (predicates), attribute values (constants) and links between objects (relational predicates). The choice of the logic to be used depends on practical considerations such as the need to represent high level concepts involving several objects, variables and so on (there is often a trade-off between generality and efficiency of the representation).

To represent the examples and the rules we use the unit-based first order logic language used by MAGGY (see section 3.1.2). The format of the examples is similar to the format of the generalized examples which allows incremental learning. Thus it is feasible to maintain the system automatically by providing other examples as the domain expands or changes over time.

Using first order logic enables using low level concepts such as BLUE and SQUARE. Intermediate concepts such as SQUARE(x) & BLUE(x) will be represented as BLUE-SQUARE(x) only when they are "interesting" [switchover points (Fu & Buchanan, 1985) which we call *intermediary subgoals* and expert's input]. Nevertheless, we emphasize the use of low-level concepts to allow communication and cross-examination between different experts. It also removes some of the initial bias introduced by the expert who provided the language of description (Utgoff, 1986) so that the ML tool can discover the most efficient ones (Fu & Bunchanan, 1985; Utogoff, 1986).

Note that we do not advocate totally ignoring the expert's intermediary knowledge, but we use it as a guide instead of relying totally on it. We try to discover new frames and subgoals. As noted in (Clancey, 1983) "intermediate knowledge provides better explanation capabilities for the system and thereby increases the understandability for the user" (this is true for intermediate goals as well as intermediate descriptors). There can also be some redundancy in the vocabulary which must be preserved (*for instance, a local tomato expert from Avignon will use different terms than an expert from Paris to describe the same concept*). This background information is encoded in the frames (see next section).

The choice of what is a constant and what is a predicate is usually independent of the application. When there is only one person named MARY in the universe, MARY can be a constant. If there are several persons with this name, MARY must be a unary predicate. The atomic formula MARY(x) takes the value TRUE when x is instantiated by one of the possible MARYs. However, the choice of the low-level concepts (or primitives) varies from one application to the other even in the same domain. *For example, in the world of blocks, if the application is to move blocks around as in SHRDLU* (*Winograd,* 1972), *CUBE is a valid primitive. On the other hand, if the system is to perform analysis of visual scenes in that same world of blocks, the concept of CUBE will be described in terms of concave/convex interior/boundary lines* (*Waltz,* 1975) *etc*. . . . Identifying low-level concepts can be fairly easy (this is often the case in applications with strong domain theory) or extremely difficult (like in our air traffic control application where the expert is having a hard time describing in natural language the knowledge contained in the radar picture).

The initial choice of primitives to represent knowledge is necessarily ad hoc (Schank & Carbonell, 1979). Thus the choice of descriptors is never considered as final and is modified depending on the results. It is clear that asking the experts to fill up the questionnaires several times as the language of description evolves is a cumbersome process. Some Explanation Based Learning (Dejong, 1981; Mitchell *et al*., 1983) could be useful in order to bring modifications to the information already learned without the need for filling again the questionnaires.

We are very sensitive to the importance of a good set of descriptors and its relation to noisy data. A good choice of canonical descriptors enables all things to be said cleanly (Hayes, 1984). A lot of problems connected with noise can be solved by improving the language of description. In these cases, using numeric uncertainty is not appropriate: it only hides noise instead of solving it.

6.2. OBJECT LEVEL KNOWLEDGE

We also obtain some domain specific knowledge from the experts (expert's common sense and specialized knowledge). This is:

(a) generic units (*all fruits have a color, a size, a texture etc* . . .);
(b) default properties (*leaves are normally green*);
(c) relations between generic units that always hold (*a leaf is part of the plant, yellow is a light color*);
(d) axioms which deduce low level concepts from other low level concepts (*if there are a lot of spots on a leaf, then the size of the spots must be small*).

We call *unit* a class of physical objects (*for example the fruits*), or a conceptual object (*for example yellow*). No distinction is made between generic objects and feature values and the system represents structures (relations) between these the same way.

In this paper we consider the two relations ISA and PART-OF (the reverse links are SUBSUME and ONE-PART-IS). A ISA B Means "A is a member of class B" or "any object belonging to class A also belongs to class B" (in other work ISA denotes set membership and set inclusion). *For example, SPOT ISA SYMPTOM and SPOT087/ISA SPOT*. A PART-OF B means that B is a composite object and that A is one of its components.

There are various ways to represent background knowledge. We have chosen to encode it using a frame-based language. *For example, the frame-unit LEAVES has a slot COLOR filled with* ($*DEFAULT* (*GREEN*)). *This property may be inherited by a specific instance of that frame* (*or a SUBSUME descendant*) *when information is* *missing. Slot fillers may be simple attributes or procedural calls (deamons). For example, the COLOR slot of the TOMATO-FRUIT frame-unit is filled with an $IF-NEEDED deamon: If the fruit is young, this procedure returns green otherwise if it is mature it returns red.* We have implemented all axioms as deamons to deduce missing low-level concepts. Axioms could also be implemented as constraints (*if the color of the fruit is not red, then the fruit is not mature*) to detect wrong information when a constraint is violated. Nevertheless, this would cause problems related to non-monotonic reasoning and theorem proving which would go beyond the scope of our application.

Last, we obtain the training example base. Our approach was to design a set of questionnaires which contains all the information [a similar approach has been taken in [Michalski & Chilauski, 1980)]. We asked some students of INRA to fill them up and we asked the experts to provide the diagnosis for the disease. We asked several experts and several students to do this task in order to cross-examine their knowledge, cross-examine user-given knowledge and expert-given knowledge, filter the information and/or detect noisy unreliable low level concepts (see next section). Thus, the experts are asked to do what they know (i.e. provide a diagnosis) and are not asked to reconstruct a set of rules. From these, the system is able to automatically learn the most specific concepts consistent with the training data (MAGGY's generalizations of the clusters, which we call the S-rules) and diagnostic rules (the G-rules computed from the paths in the decision tree). The experts were also asked to provide stereotypical examples of diseases using pictures because several diseases could not be observed (it takes roughly seven years to observe all the tomato diseases). Thus, we are learning from a set of existing examples as well as from experts given characteristic descriptions/rules.

7. Noise in a knowledge base

Noise is present in a knowledge base when it does not truly reflect the environment we want to learn from. In our application, we are interested in learning from noisy data and setting up rules that are less sensitive to noise. We define noise as being:

(a) Wrong information (false positive/negative examples of a concept);
(b) incompleteness of the data/language of description;
(c) unreliable data.

There are different sources of noise, different effects of noise and different kinds of noise. Hence, it is rather difficult to speak about noise as if it were an entity. We present here an attempt at classifying noise in a knowledge base and at detecting/treating each kind of noise.

7.1 UNRELIABLE INFORMATION

In this section (and only in this section), we assume that the rules are correct and that the language of description is adequate. Unreliable data is noise that naturally originates from the concepts themselves and that is likely to introduce errors during consultation. We stress that this kind of noise cannot be handled by a filtering mechanism and that it will always be present inside the training data. Thus, there is a limit beyond which noise cannot be eliminated. Note that some people might view some of this as being uncertainty in the domain as opposed to noise. We have identified three kinds:

(a) Symbolic noise

This can be because the concept is **hard to see**.

For instance, a discriminating feature of "Colletotrichum coccodes" is the presence of tiny black marks on the roots, less than one millimeter in size which are hard to see and often missed during consultation.

Natural noise can also be due to *concept polymorphy*.

For example, when does dark grey stop being grey to become black? If the system asks "is the color grey?" and the user answers "no" because he thinks it is black while the expert intended that he answered yes, the system will mis-diagnose the disease (or it will be unable to conclude).

In multi-expert knowledge acquisition, symbolic noise can be detected by the fact that two experts disagree on the value of a certain descriptor.

(b) Difference between the skill of the expert and the skill of the user

While an expert is not likely to confuse a brown spot with a rotting tissue, an unskilled user is. This will also produce mis-diagnosis

Currently, we ask the experts to guess how noisy a test is (in the questionnaire, the expert can specify the reliabilty of a test). This knowledge is used when growing a decision tree to discourage the use of noisy concepts. The cost parameter is set higher for noisy attributes and, as we have seen at the end of section 4, avoiding noisy descriptors when building the tree can generate clusters which are more resilient to noise (the noisy attributes being generalized away). In this case, some good results can also be achieved by using binarization of attributes when growing the tree.

At any rate, we emphasize representing this knowledge explicitly as opposed to simply lowering Certainty Factors when a test is not reliable. This is one example where treating noise symbolically gives much better results than to treat it numerically.

(c) Uncertainty in the measure of an attribute

When testing numerical parameters, the measuring instrument might not be 100% accurate. The system should not rely on the outcome of a test like "Is $A < B$" when A is in the neighborhood of B (Zucker, 1978).

For example the instrument which measures the acidity of the soil might not be 100% accurate.

Clearly, when A gets close to B, we do not want to rely heavily on the outcome of the test. A solution is to replace the test "$A < B$" by two tests "$A < B - \Delta$" and "$A > B + \Delta$" where Δ is given by the expert and ignore the result of the test when A gets within Δ of B.

Currently, handling unreliable data relies on information given by the expert. Improvements to this could be made by performing any analysis of a wrong diagnosis and adjusting the parameters. This is still a research topic.

7.2. RANDOM ERRORS IN THE TRAINING DATA

These are human introduced mistakes. These can be:

(a) *Giving the wrong value to a descriptor*
(b) *Describing a class and attributing the description to another*
(c) *Giving too many descriptors*

(this can happen when our expert uses by mistake a plant which has several diseases as a reference for a specific disease).

As noted in (Fu, 1985), a false positive can drive the system into producing over-general expressions which sometimes introduce inconsistencies in the knowledge base. When growing a total classification decision tree, a false positive example could be isolated in a cluster of its own because it is too "far" from the other positive examples. Since we are not learning only from a case library (and thus, we cannot rely fully on the statistical information), the example cannot simply be discarded [this technique has been used for example in METAXA3 (Emde, 1986) or AQ15 (Michalski *et al.*, 1986)]. Indeed, it could be a perfectly good example thought of by a single expert who happens to be more specialized than the others (as noted in Cestnik (1987) this method removes "good" exceptions as well as random errors].

Another method to handle errors in the training examples is to allow some inconsistency in the leaves of the decision tree. This method has been used in Corlett (1983), Fu & Buchanan (1985), Quinlan (1986*b*). One way to achieve this result is to use the pre-pruning feature of the algorithm described in section 3.2. We can then focus our attention on the "uncertain" clusters to detect potentially noisy training examples which are filtered by the experts. If the examples are indeed correct, then we are faced with two concepts that are close to one another and, as noted (Winston, 1975) the learning will be very rich (we could define these as *statistical near misses*).

Note that when random errors do not introduce some inconsistency in the leaves of the decision tree, the noisy information is generalized away by the algorithm (yielding an expression that is overly-general, but that is consistent with the training data).

Briefly, in the case of random errors in the training data, we have chosen to use statistical methods as a means of detecting noise rather than as a means of treatment. We feel that there cannot be any regularities with respect to random phenomena (up to this day, we have never encountered anyone who can walk into a casino with a $10 bill and walk out with $1 000 000 without taking any risks). Thus, assuming that random errors occur according to a certain statistical pattern that is represented using numerical uncertainty is, in our opinion, the wrong approach. This is one case where we use numerical methods to drive the symbolic methods (filtering, enriching the language of description, learning with low-level features are very important etc · · ·).

7.3 INCOMPLETE INFORMATION

Although some people might not strictly consider this as a form noise, when using Machine Learning tools a lack of information is just as bad (and is sometimes worse) as errors in the data. It can drive the system into finding overly general expressions which will generate consultation errors. This can be:

(a) Forgetting an example

This is a problem only when we are asking the experts for prototypical examples and not when we are generating rules from an adequate case library. By asking several experts for prototypical examples, this problem will probably vanish since it is unlikely that all the experts forget the same example.

(b) Forgetting a descriptor

This can be due to the fact that people forget to mention negative features (*example: the leaves must not be dry*) or because the descriptor is a default value. It also happens that sometimes people simply forget to mention relevant features for no apparent reason (Pazzani 1986).

When needed, we use the frames to inherit default values and avoid using the dropping condition rule during generalization [A is more general than A & B (Michalski, 1983)]. Hence, if the expert does not fill up the value of all the descriptors which corresponds to the norm, the system does (if it needs to do so during generalization).

When an expert forgets to mention a descriptor because it is naturally noisy to see, this appears during cross-examination with other expert rules. In fact, this seldom occurs because experts usually know what to look for and tend to look carefully for these noisy descriptors. When a single expert forgets a descriptor that has been seen by all the others and the descriptor is noisy, the example is marked as being suspect.

Often people forget to mention relevant features such as "*the leaves must not be wet* (*they must be dry which is not a default value*)". When the error introduces an inconsistency, if there are near misses the system can automatically detect where the problem comes from Manago (1986). If the faulty positive example has several differences with the negative examples, then we rely on the expert to fix the problem (note that the system can make suggestions of what the problem might be).

(c) Giving a value for a descriptor that is too general

This can be treated in a similar manner as when the expert forgets the descriptor

(d) Providing a language of description that is not rich enough

This kind of incomplete information comes from the fact that there are not enough terms in the language of description to discriminate diseases (the original example base is inconsistent) or discriminate generalizations of diseases (yielding overly-specific rules). In the first case, this introduces an inconsistency in the knowledge base and we call the expert. The generalization tool we use (MAGGY (Manago, 1986)] has the capacity of suggesting improvements to the language of description in the second case. Consider the following examples of sunburn:

E_1: [x: ⟨ISA PLANT⟩ ⟨AGE MATURE⟩] & [spots1: ⟨ISA SPOT⟩ ⟨NUMBER SEVERAL⟩ ⟨COLOR WHITE⟩ ⟨FORM CONCAVE⟩] & [fruit] ⟨ISA FRUIT⟩] & [face1: ⟨PART-OF fruit1⟩ ⟨IS EXPOSED-TO-SUN⟩]⇒SUNBURN

E_2: [X: ⟨ISA PLANT⟩ ⟨AGE MATURE⟩] & [spots1: ⟨ISA SPOT⟩ ⟨NUMBER

SEVERAL⟩ ⟨COLOR YELLOW⟩ ⟨FORM CONCAVE⟩] & [fruit1 ⟨ISA FRUIT⟩] & face1: ⟨PART-OF fruit1⟩ ⟨IS EXPOSED-TO-SUN⟩]⇒SUNBURN

By climbing the generalization tree (lattice), one could conclude that a characteristic description of SUNBURN is: there are light colored spots on the face of the fruit exposed to the sun. However, consider the following negative example of sunburn.

CE: [x: ⟨ISA PLANT⟩ ⟨AGE MATURE⟩] & [spots1: ⟨ISA SPOT⟩ ⟨NUMBER SEVERAL⟩ ⟨COLOR GREY-BEIGE⟩ ⟨EVOLUTE-INTO spots2⟩ ⟨FORM CONCAVE⟩] & [spots2: ⟨ISA SPOTS⟩ ⟨COLOR BROWN⟩] & [fruit1 ⟨ISA FURIT⟩] & [face1: ⟨PART-OF fruit1⟩ ⟨IS EXPOSED-TO-SUN⟩ & [leaves1: ⟨ISA LEAVES⟩] & [symptom1: ⟨ISA SYMPTOM⟩1⇒NARCOSIS-OF-FRUIT-EXTREMITY

Since grey-beige is also a light color, the negative example is covered by the generalization. Let us assume that the origin of the problem is the generalization of the color attribute. One solution is to add to the generalization: COLOR(LIGHT-COLOR,spots1) & → COLOR(GREY-BEIGE spots1) as done for example in (Clavieras, 1984). Nevertheless, one does not want to collect these negations, because on the conceptual level this is a hint that WHITE and YELLOW share some conceptual characteristics that GREY-BEIGE does not have. A better way to solve this problem is to introduce an intermediary level node G0001 in the frame such that WHITE and YELLOW are sons of G0001, G0001 is a son of LIGHT-COLOR and GREY-BEIGE is not a son of G0001. The generalization then becomes COLOR(G0001,spots1) which rejects the negative example. This is similar to RL symbolization of taxonomic points (Fu, 1985) or Utgoff's shift of bias (Utgoff, 1986). Nevertheless, while RL introduces intermediary concepts in a systematic manner, MAGGY only does this when there is a need for them (to avoid generating expressions such as A and B and not (C) where A B and C have the same father).

On the other hand, this is still not the optimal solution (try to find a name for G0001 that a human expert can relate to). The optimal solution obtained through interaction with the experts is what WHITE and YELLOW belong to a different frame which had been omitted, the TRANSLUCENT frame. Indeed, WHITE and YELLOW are not only colors but they also correspond to a color loss (when exposed too long to the sun, the colors fade away and become translucent). While in some cases one can clearly see that the white or yellow is a color and not a color loss (this means that the preceding solution does not work), in some other cases it is more complicated (in other words, we cannot intoduce two new colors such as TRANSLUCENT-WHITE and TRANSLUCENT-YELLOW).

Note that in the present case (the SUNBURN disease), the fact there are no symptoms on leaves is also relevant while the fact that the color does not evolve into brown is not (this information is obtained from the expert). The correct generalization will then be: there are translucent spots on the part of the fruits exposed to the sun and no symptoms on the leaves.

7.4. MISCELLANEOUS

(a) Uncertainty in the domain

This can be caused by several diseases that can be distinguished at certain stages of the development, but not at some others. This can also be caused by the fact that the outcome of a test which is needed in order to discriminate between two diseases is never available to the user.

For example, finding which virus causes a disease requires some lab tests. The results of these lab tests will never be known by the person who uses the system · · · . Hence, as far as the system is concerned, it will never be able to discriminate between different viral diseases.

Currently, our solution to the problem is to merge all the uncertain classes into a single class (like VIRUS).

(b) Randomness of natural phenomena

Sometimes, not all the symptoms that are supposed to be there appear. The idea is then to be able to diagnose a disease when n out of $m(n < m)$ of the symptons on the left hand side of the rule are verified. We consider that each combination of n elements is an example and automatically generate these (we then have to ask the expert to filter impossible combinations). In fact, when learning from a case library that represents reality, one does not need to be concerned at all with this "classical" type of noise.

(c) Noise in the background knowledge

We are currently totally empty-handed with respect to this kind of noise. The effect of noise in the model when the system is model-driven mechanism is clearly drastic (Nii, 1986). Detecting and solving errors in the background knowledge requires extensive analysis of consultation errors. Missing background knowledge is already a little bit easier and involves using some Explanation-Based learning techniques (see for instance ODYSSEUS (Wilkins, 1986)].

8. Conclusion

We have presented here several problems and solutions in relation to noise. There is a fundamental distinction that ought to be made between learning from noisy training data and learning concepts resilient to noise. To solve the first one we use a filtering mechanism driven by statistical methods and by cross-examination of experts. We still heavily rely on the experts to fix the problems but we are able to help them focus their attention on potential problems. The data can be "fixed" by removing errors, improving the language of description or by using some other symbolic methods.

Cross-examination between experts, or between experts and end-users of the system is also used to learn the relative reliability of the low-level concepts. This information is used to learn robust concepts. We stress that this kind of noise will always remain and cannot be fixed by a filtering mecanism or by improving the language of description. An explicit representation of the relative reliability of the tests can be used to influence the way the clusters are built. Often, the unreliable tests are generalized away which produces concepts that are more resilient to noise and therefore leads to rules that give more accurate results. When failing to achieve this result, using numerical uncertainty is appropriate since there is a certain regularity in the errors. Note that in this case, the numerical coefficients ought to be computed from the data given by the end-user of the system and not from the experts.

We wish to thank Ed Pacello at GEC-research for the great job he has done correcting our English. Any errors remaining in here are probably due to the last minute changes we have made before sending in the paper. We also thank the people at Cognitech who are involved with the INSTIL project for providing "real life" examples for this paper. Thank you to Bruce Buchanan and to the Helix group at Stanford's Knowledge Systems Laboratory, for all the useful and relevant comments they have made on this work. The learning algorithms described in this paper are implemented in COMMON LISP on EXPLORER LISP machines and SUN 3 workstations.

References

AIELLO, N. (1983). A Compative Study of Control Strategies for Expert Systems: AGE Implementation of Thee Variations of PUFF. *Proceedings of the AAAI Conference*. Los Altos, CA: Morgan Kaufmann.

ALEXANDER *et al.* (1986).

BOBROW, D. G. & WINOGRAD, T. (1977). An overview of KRL, a knowledge representation language. *Congnitive Science* **1**(1), 3–46.

BUCHANAN, B. G. & SHORTLIFFE, E. H. (1984). *Rule-Based Expert Systems, the MYCIN Experiments of the Stanford Heuristic Programming Project*. Reading, MA: Addison-Wesley.

CHARNIAK, E. (1977). A framed painting: The representation of a common sense knowledge fragment. *Congnitive Science* **1**(1), 355–394.

CLANCEY, W. J. (1983). The epistemology of rule-based expert system: a framework for explanation. *Artificial Intelligence* **20**.

CLANCEY, W. J. (1986). Transcript of plenary session: congnition and expertise. In: *1st AAAI Workshop on Knowledge Acquisition for Knowledge Based Systems,* Banff Canada.

CORLETT, R. (1983). Explaining induced decision trees. *Proceeding Expert Systems*. pp. 136–142.

DEJONG, G. (1981). Generalizations based on explanation. Los Altos, CA: Morgan-Kaufmann *Proceedings, 7th IJCAI*. pp. 67–69.

DIDAY, E. & SIMON J. C. (1979). Clustering Analysis In: FU, K. S. Ed. *Communication and Cybernetics Digital Pattern Recognition*. Berlin: Springer Verlag.

DIETTERICH, T. G. & MICHALSKI, R. S. (1987). A comparative review of selected methods for learning from examples. In: MICHALSKI R. S., CARBONELL J. G. & MITCHELL T. M. Eds. *Machine Learning*: *An Artifical Intelligence Approach, Volume 1*. Los Altos, CA: Morgan Kaufmann.

EMDE, W. (1986). Great flood in the block world *Proceedings, EWSL*, Orsay.

FU L. M. (1985). Learning object-level and meta-level knowledge in expert systems. PhD thesis, Knowledge Sytems Laboratory, Computer Science Department, Stanford University.

FU, L. M. & BUCHANAN, B. G. (1985). Inductive knowledge acquisition for rule-based expert systems. *Technical Report*. Knowledge Systems Laboratory, Computer Science Department, Stanford University.

HAYES, P. (1984), Liquids. In: HOBBS, J. Ed. *Formal Theories of the Commense World.* Hillsdale: Ablex.

HAYES-ROTH, F. & MCDERMOTT, J. (1978). An interference matching technique for inducing abstractions *ACM Journal,* **21,** 401–411.

KODRATOFF, Y. & GANASCIA J. G. (1986). Improving the generalization step in learning. In: MICHALSKI R. S., CARBONELL J. G. & MITCHELL T. M. Eds. *Machine Learning: An Artificial Intelligence Approach*, *Volume 2*. Los Altos, CA: Morgan Kaufmann.

KODRATOFF, Y., MANAGO, M., BLYTHE, J., SMALLMAN, C. & ANDRO, T. (1986). Generalization in a noisy environment: the need to integrate symbolic and numeric techniques in learning. *Proceedings of the AAAI Workshop on Knowledge Acquisition for Knowledge Based Systems*. Banff, Canada.

KUNTZ, J., FALLAT, R., MCCLUNG, D., OSBORN, J., VOTTERI, B., NII, H. P., AIKINS, J. S.,

FAGEN, L. & FEIGENBAUM, E. A. (1978). A physiological rule-based system for interpreting pulmonary function test results. *Technical Report HPP-78-19*, Knowledge Systems Laboratory, Stanford University.

LINDSAY, R., BUCHANAN, B. G., FEIGENBAUM, E., & LEDERBERG, J. (1980). *Application of Artificial Intelligence for Organic Chemistry: The DENDRAL Project*. New York: McGraw-Hill.

MANAGO, M. (1986). Object oriented generalization: a tool for improving knowledge based systems. *Proceedings of the First International Meeting on Advances in Learning*, Les Arcs, France.

MICHALSKI, R. S., CHILAUSKI, R. L. (1980). Learning by being told and learning from examples: an experimental comparison of two methods of knowledge acquisition in the context of developing an expert system for soybean disease diagnosis *Policy Analysis and Information Systems*, **4**(2).

MICHALSKI, R. S., CARBONELL, J. G. & MITCHELL, T. M. (1983). An overview of machine learning. In: MICHALSKI R. S., CARBONELL J. G. & MITCHELL T. M. Eds. *Machine Learning: an Artificial Intelligence Approach, Volume 1.* Los Altos, CA: Morgan Kaufmann.

MICHALSKI, R. S. (1983). Theory and methodology of inductive learning. In: MICHALSKI R. S., CARBONELL J. G., & MITCHELL T. M. Eds. *Machine Learning: an Artificial Intelligence Approach, Volume 1.* Los Altos, CA: Morgan Kaufmann,

MICHALSKI, R. S., MOZETIC, I., HONG, J. & LAVRAC N. (1986). The AQ15 inductive learning system: An overview and experiments. *Proceedings of AAAI Conference.* Los Altos, CA Los Altos,

MINSKY, M. (1975). A framework for representing knowledge. In: WINSTON, P. H., Ed. *The Psychology of Computer Vision.* New York: McGraw-Hill.

MITCHELL, T. M. (1985). Generalization as Search. In: Webber B. L., Nilsson W. J. Eds. *Readings in Artificial Intelligence*, Los Altos, CA: Morgan Kaufmann.

MITCHELL, T. M., UTGOFF, P. E. & BANERJI, R. (1983). Learning by experimentation: acquiring and refining problem-solving heuristics. In: MICHALSKI R. S., CARBONELL J. G., & MITCHELL T. M. Eds. *Machine Learning, An Artifiical Intelligence Approach, Volume 1.*

MCDERMOTT, J. (1986). Transcripts of plenary session: interactive tools 1. *1st AAAI Workshop on Knowledge Acquisition for Knowledge Based Systems*, Banff, Canada.

NIBLETT, T. (1987). In: BRATKO, I. & LAVRACS, N., Eds. *Progress in Machine Learning. I.* Sigma Press.

NII, H. P., FEIGENBAUM, E. A., ANTON, J. J. & ROCKMORE, A. J. (1982). Signal to symbol transformation: HASP/SIAP case study. *The AI Magazine* **3**(2): 23–35.

NILSSON, N. J. (1984). *Principles of Artificial Intelligence*. Los Altos, CA: Morgan Kaufmann.

O'RORKE, P. (1982). A comparative study of inductive learnings systems AQ11 and ID3 using a chess end-game test problem. *Report ISG 82-2, UIUCDCS-F-82-899,* Department of Computer Science, University of Illinois in Urbana-Champaign.

PAZZANI, M. J. (1986). Explanation-based learning for knowledge-based systems. *Proceedings of the 1st AAAI Workshop on Knowledge Acquisition for Knowledge Based Systems*, Banff, Canada.

QUINLAN, J. R. (1983). Learning efficient classification procedures and their application to chess end games. In: MICHALSKI R. S., CARBONELL J. G. & MITCHELL T. M. Eds. *Machine Learning, An Artificial Intelligence Approach, Volume 1.* Los Altos, CA: Morgan Kaufmann.

QUINLAN, J. R. (1986*a*). The effect of noise on concept learning. In: MICHALSKI R. S., CARBONELL J. G. & MITCHELL T. M. Eds. *Intelligence Approach, Volume 2.* Los Altos, CA: Morgan Kaufmann.

QUINLAN, J. R. (1986*b*). Symplifying decision trees. *Proceedings of the AAAI workshop on Knowledge Acquisition for Knowledge Based Systems*, Banff, Canada.

REGOCZEI, S. & PLANTINGA, E. O. (1986). Ontology and inventory: a foundation for a knowledge acquisition methodology *Proceedings of the 1st AAAI Workshop on Knowledge Acquisition for Knowledge Based Systems*, Banff, Canada.

SAMMUT, C. (1981). Learning concepts by performing experiments. PhD thesis, Department of Computer Science, University of New South Wales, Australia.

SHANK, R. & CARBONELL, J. G. (1979). Re: the Gettysburg Address: representing social and political acts. In: FINDLER, N. V. Ed. *Associative Networks: Representation and Use of Knowledge by Computers*. New York: Academic Press.

TERRY, A. *The CRYSALIS project: Hierarchical Control of Production Systems,* Technical Report HPP-83-19, Knowledge Systems Laboratory, Stanford University, 1980.

UTGOFF, P. E. (1986). Shift of Bias for Inductive Concept Learning. In: MICHALSKI R. S., CARBONELL J. G. & MITCHELL T. M. Eds. *Machine Learning 2: An Artificial Intelligence Approach*. Los Altos, CA: Morgan Kaufmann.

VERE, S. A. (1980). Multilevel counterfactuals for generalization of relational concepts and productions. *Artificial Intelligence Journal*, **14,** 139–164.

VRAIN, C., MANAGO, M., GANASCIA, J. G. & KODRATOFF Y. (1986). AGAPE: an algorithm that learns from similarities. In: *Proceedings, European Working Session on Learning 1986*, Université d'Orsay, Paris-sud, France.

WALTZ, D. L. (1975). Understanding line drawing of scenes with shadows In: WINSTON P. H. Ed. *The Psychology of Computer Vision*. New York: McGraw-Hill.

WINOGRAD, T. (1972). *Understanding Natural Language*. New York: Academic Press.

WINSTON, P. H. (1975). Learning structural descriptions from examples. In: WINSTON, P. H. Ed. *The Psychology of Computer Vision*. New York: McGraw-Hill.

ZUCKER, S. Production systems with feedback. In: WATERMAN D. & HAYES-ROTH F. Eds. *Pattern Directed Inference Systems*. New York: Academic Press.

Acquisition of uncertain rules in a probabilistic logic

JOHN G. CLEARY

Department of Computer Science, University of Calgary, 2500 University Drive, Alberta T2N 1N4, Canada

The problem of acquiring uncertain rules from examples is considered. The uncertain rules are expressed using a simple probabilistic logic which obeys all the axioms of propositional logic. By using three truth values (true, false, undefined) a consistent expression of contradictory evidence is obtained. As well the logic is able to express the correlations between rules and to deal with uncertain rules where the probabilities of correlations between the rules can be directly computed from examples.

Uncertain rules

Uncertainty is an important part of many ruled based expert systems. For example, applications such as medical diagnosis do not allow anything but the weakest inferences to be made from available evidence. To report certainty in conclusions is both incorrect and misleading. A number of schemes for expressing and computing such uncertainties have been developed. For example, in MYCIN and EMYCIN "certainty factors" are used (Shortliffe, 1976). A certainty factor is a number between −1 and +1, −1 is intended to express sure knowledge that something is false and +1 sure knowledge that it is true. Intermediate values express varying degrees of ambivalence about the truth. For example 0 expresses a complete lack of knowledge about truth or falsity. Such certainty factors can be used in a number of ways. For example a rule such as:

will-rain ← dark-cloud and falling-presure with certainty 0·6;

says that it is almost certainly true that if there is dark cloud around and the barometric pressure is falling then it will rain, although, there will be some cases where this is not true. Certainty factors can be propagated through the system to evaluate the certainty of conclusions. For example the conclusion "will-rain" would be given a certainty factor based on the certainty of the rule above and the certainties of "dark-cloud" and "falling-pressure". Unfortunately there are grave problems with using certainty factors in this way. Consider the additional rule:

will-rain ← lightning with certainty 0·4;

If both these rules fire then "will-rain" is given a higher certainty factor then if only one of them fires. Unfortunately the second rule is merely another way of saying that storm clouds are present and the conclusion is not much more true as a result.

These problems can be seen more starkly by considering the expression "(dark-cloud or not dark-cloud)" or the expression "(dark-cloud and not dark-cloud)" which are respectively always true or false. However, certainty factors

KNOWLEDGE-BASED SYSTEMS Vol. 1
ISBN 0-12-273251-0

ignore the fact that the two parts of the expression are correlated (the same) and report intermediate values for both expressions.

Another way to approach this is to use the probability that an expression is true rather than certainty factors. Again problems arise that are similar to those above. For example one scheme used to compose such probabilities is obtained by assuming the various parts of an expression are uncorrelated:

$$p(x \text{ and } y) = p(x) \times p(y)$$
$$p(x \text{ or } y) = p(x) + p(y) - p(x) \times p(y)$$
$$p(\text{not } x) = 1 - p(x)$$

Let $p(x) = \frac{1}{2}$ then $p(x \text{ and not } x) = \frac{1}{4}$ and $p(x \text{ or not } x) = \frac{3}{4}$, neither of which is correct.

A second evaluation scheme (Zadeh, 1965) assumes that some correlation can occur between expressions and lets:

$$p(x \text{ and } y) = \min\,(p(x), p(y))$$
$$p(x \text{ or } y) = \max\,(p(x), p(y))$$
$$p(\text{not } x) = 1 - p(x)$$

Again let $p(x) = \frac{1}{2}$ then $p(x \text{ and not } x) = \frac{1}{2}$ and $p(x \text{ or not } x) = \frac{1}{2}$ is still wrong. This scheme does no overweight rules which are similar but does underweight rules which are independent.

Another weaker scheme (Quinlan, 1983) is available which uses intervals of probabilities. It is based on the principle that the following inequalities always hold:

$$\max\,(0, 1 - p(x) - p(y)) \le p(x \text{ and } y) \le \min\,(p(x), p(y))$$
$$\max\,(p(x), p(y)) \le p(x \text{ or } y) \le \min\,(1, p(x) + p(y))$$
$$p(\text{not } x) = 1 - p(x)$$

By only asserting that the probability for an expression lies in some range it never asserts anything which is false. For example if $p(x) = \frac{1}{2}$ it deduces that:

$$0 \le p(x \text{ and not } x) \le \tfrac{1}{2}$$

and

$$\tfrac{1}{2} \le p(x \text{ or not } x) \le 1.$$

While true these inequalites are too weak to be generally useful.

A related problem is that some conclusions may have evidence both indicating that they are true and evidence indicating that they are false. It seems, intuitively, that the situation where there is no evidence about something is different from the one where there is a known 0·5 probability that it is true and a known 0·5 probability that it is false. The schemes mentioned above also have problems in consistently accommodating both positive and negative evidence.

Considerations such as these show that it does not seem to be possible to provide a quantitative theory of truth and falsity using just probability values for expressions. In the next section an alternative is introduced which circumvents this by using (potentially infinite) sequences of bits to represent the truth or falsity of a statement. Probabilities can be extracted after reasoning is complete but the calculations cannot take place just using the probabilities. Other attempts to provide a logical basis for uncertain reasoning have been made (Shapiro, 1983), (van Emden, 1986) but because they are based on the probability values of expressions they also are heir to the ills described above.

A probabilistic logic

The technique used here is to assign each expression an infinite sequence of true/false values rather than just one true/false possibility. This is trivially different from the normal propositional calculus and all theorems hold for example (x or not x) = (*true, true, . . .*) and (x and not x) = (*false, false, . . .*). In order to make this useful a new family of logical constants are introduced. Each constant is some infinite sequence of true/false values with a fixed probability that it is true. The constants are written in the form $\tau(p)$ where p is the probability that an item in the sequence is true. So, it is always true that

$$\tau(1) = (\text{true, true, . . .}) \text{ and } \tau(0) = (\text{false, false, . . .}).$$

When evaluating a logical expression with these constants some actual value has to be assigned to them. There are a number of ways of doing this and it is convenient to assume that different constants are uncorrelated. That is the probability of a true in the sequence ($\tau_1(p)$ and ($\tau_2(q)$)) is always $p \times q$. In practice, these infinite uncorrelated sequences are likely to be approximated by finite sequences of pseudo-randomly generated bits.

To express an uncertain rule the constants can be used as follows:

$$\text{will-rain} \leftarrow \text{dark-cloud and falling-pressure and } \tau(0{\cdot}6).$$

This says that if there is dark cloud and falling barometric pressure then in 0·6 of the cases there will be rain. The second rule can be expressed as:

$$\text{will-rain} \leftarrow \text{lightning and } \tau(0{\cdot}4).$$

This has still not solved the problem that the two rules are highly correlated but they can be reformulated as:

$$\begin{pmatrix} \text{will-rain} \leftarrow \text{dark-cloud and falling-pressure and } \tau_1(0{\cdot}75) \\ \text{and} \\ \text{will-rain} \leftarrow \text{lightning and } \tau_2(0{\cdot}5) \end{pmatrix} \leftarrow \tau_3(0{\cdot}8).$$

This reformulation says that the two rules have a common case which says that they are true 80% of the time. As a result of this the probability of will-rain will only be weakly augmented when both rules fire, solving the original problem of expressing the fact the two rules are not independent of each other. By rewriting these rules in the equivalent form below it can be seen that the original $\tau(0{\cdot}6)$ has been replaced by ($\tau_1(0{\cdot}75)$ and $\tau_3(0{\cdot}8)$) and $\tau(0.4)$ by ($\tau_2(0{\cdot}5)$ and $\tau_3(0{\cdot}8)$):

$$\text{will-rain} \leftarrow \text{dark-cloud and falling-pressure and } \tau_1(0{\cdot}75) \text{ and } \tau_3(0{\cdot}8),$$

$$\text{will-rain} \leftarrow \text{lightning and } \tau_2(0{\cdot}5) \text{ and } \tau_3(0{\cdot}8).$$

In this way sets of rules with arbitrary correlations between them can be expressed. However, it is not clear that all possible ways of using the logical constants are useful. The form used above where sets of rules are enabled seems to be a reasonably intuitive way of expressing such relationships and, as will be seen later, it has some advantages when the probabilites of the constants are acquired from experience. However, the acid test for such formalisms is whether computer naive experts can express their intuitions in this form. No tests of this have been done nor has there been exploration of possible ways of "sugaring" the syntax to make it more palatable.

Probabilities can be extracted from our logical sequences by counting the number of true values in the sequence and taking the limit. The result of all this is a probability logic (Gaines, 1984), (Rescher, 1963). The logic obeys all the usual logical axioms including the tautology (x or not x).

RULE SETS

As is usual in rule based systems it is necessary in practice to restrict attention to the Horn clause subset of the logic. In the next section the logic and the types of allowable rules will be extended to cater for negation and the possibility of evidence both for and against a proposition.

The normal situation for an expert system is a fixed body of rules (Horn clauses) which encode the invariant knowledge about the problem at hand and a set of facts which describe the current situation. The existence of a probabilistic logic allows some useful extensions to this view. For example, a fact can be stated as an additional "rule" of the form:

$$\text{dark-clouds} \leftarrow \tau(1)$$

and additionally uncertain evidence (I think there is a 50% chance that those are dark clouds) can be accommodated by a rule of the form:

$$\text{dark-clouds} \leftarrow \tau(0{\cdot}5).$$

There is also nothing to stop the head of such a "factual" rule from being an intermediate deduction which is also computed elsewhere by a set of rules. For example the user might say "it is going to rain with probability 0·5 although there is nothing in the rules to support this" and enter this as a "factual" rule:

$$\text{rain} \leftarrow \tau(0{\cdot}5).$$

In order to compute effectively the consequences that result from such rule sets it seem to be necessary to restrict them to those with finite derivation trees (van Emden, 1986). That is, no recursive rules are permitted. This accords well with the normal situation in rule-based expert systems. The use of the α–β heuristic in such evaluations has been discussed for truth-functional logics of uncertainty (ibid), and can also be used here.

BIT SEQUENCES

Infinite sequences of bits can be made computationally tractable by approximating them with finite pseudo-random sequences of bits. The probabilities are in turn approximated by counting over these finite sequences. The major question this approximation raises is the accuracy with which the finite sequences represent the probabilities which would be generated by infinite length sequences. The accuracy is dependent on both the number of bits used and on the ability of the system to generate a large number of uncorrelated pseudo-random sequences. It is easily shown that the standard deviation of the estimated probability is given by $\sqrt{p(1-p)/N}$ where N is the number of bits and p is the probability of a bit being true. So for $N = 32$ and $p = \frac{1}{2}$ the results will be accurate to ±10%, for $N = 1024$ this is reduced to ±1·5%. Even the ±10% figure is well within the errors acceptable in existing systems (Shortliffe, 1976, p. 183). Conversely 1024 bits is only 32 32-bit words so the logical operations required should not be too burdensome amongst all the other activities of an expert system. It is necessary to generate a number of uncorrelated random sequences for the different τ sequences. As noted in (Gaines, 1969, Sec. 4.16) it is easy to generate a large number of uncorrelated sequences with $p = \frac{1}{2}$ using suitably long shift registers. For example, with $N = 1024$ a single 33 bit shift register can deliver 2^{22} independent sequences. These are readily combined to deliver sequences with appropriate probabilities. The stochastic computing systems described in (Gaines, 1969) are well suited to performing the types of computations needed here.

Negation

Negation poses problems for Horn clause logics in general. The essential problem is that the statement of facts about particular situations must include information that some things are true, that others are false and that some are just not known. The usual way of handling this is to make the closed world assumption that *if something cannot be proven then it is false.* This is far too draconian for the current purposes as it makes it impossible to express the fact that a particular fact is just unknown. It is ridiculous to conclude that because I do not know whether it is sunny therefore it must definitely be cloudy. A resolution of this is to extend the truth values to include *undefined* as well as true and false and to introduce two distinct forms of negation. As above the truth values are infinite sequences (this will be ignored whenever convenient). The interpretation of true is "provably true" and of false is "provably false". The two forms of negation are denoted by ~ and ¬. Their truth tables are:

	t	u	f
~	f	u	t

	t	u	f
¬	f	t	t

~ should be interpreted as "provably not" and ¬ as "not provable". So ¬ corresponds to the normal closed world notion of negation.

The truth tables for the various connectives are:

∨	t	u	f
t	t	u	f
u	u	u	f
f	f	f	f

∨	t	u	f
t	t	t	t
u	t	u	u
f	t	u	f

←	t	u	f
t	t	t	t
u	u	t	t
f	f	t	t

In the introduction the major problems of uncertain reasoning were encapsulated in the problem of ensuring that the expression (x or not x) was always true and that (x and not x) was always false. These need to be restated carefully in the new logic but there do indeed exist statements with the correct properties. For example, $x \vee \neg x$ is always true (x is provably true or not provably false) and $\sim x \vee x$ is never false. Similarly, $\neg(\neg x \wedge x)$, $\sim(\neg x \wedge x)$ and $\neg(\sim x \wedge x)$ are always true.

Note that as a result of these definitions $a \leftarrow b$ is equivalent to $a \vee \neg b$.

EXTENDED HORN CLAUSES

To accommodate these new notions the form of clauses allowed in the rule set can be extended as follows. The general form of rules is:

$$a \leftarrow b_1 \wedge b_2 \wedge b_3 \wedge \ldots \wedge b_n$$

where a, the head of the clause, can be a term of the form x or $\sim x$ and the b_i can be of the form $\sim x$, $\neg x$ or x where x is some atomic formula. Also, the b_i can be constants of the form $\tau(p)$ or $\sim\tau(p)$. $\neg x$ is not permitted in the head of a clause. Because the τ constants can only occur on the right-hand side of rules which will not "fire" if any of their terms are undefined there is no need to allow for undefined values in the infinite constants. There is no logical reason to exclude them, they just serve no useful purpose in this Horn clause logic.

The operational interpretation of these rules is that whenever x appears in the head of a rule then this forces the truth value of x to true if the body evaluates to true, and similarly if $\sim x$ appears in the head of a rule then this forces the truth value of x to false. A weaker form of the closed world assumption is needed to completely define this procedure, that is, "if a value cannot be proven true or proven false then it is undefined". This seems much more palatable than the original form. These procedures ensure that none of the rules is provably false.

CONTRADICTION

This opens the possibility that two rules will attempt to force the same conclusion to be both true and false. In systems with a single truth value such a contradiction is catastrophic, the rule set has to be rejected (or debugged). If an infinite sequence of values is available the situation is not as bad. Any position along the sequence which generates a contradiction on any value at the head of a rule will cause all conclusions at that position to be ignored. In a rule set where this happens a lot, the precision of the inferred probabilities will drop as a smaller number of positions are available to count from. Although any high probability of contradiction will be untenable, the system is at least graceful enough to allow some degree of contradiction and not fail catastrophicly.

Acquisition

DERIVING CONSTANTS FROM OBSERVATIONS

Putting logic aside for a moment we can consider the normal task of statistics which is exemplified by the question "predict whether it will rain given the current observations". Statistics tells us that what should be done is to count the number of times it has rained when the values of all the observations are the same as the current situation. The fraction of times that it has rained in these circumstances is the probability that it will rain again. However, typically there are many observations that can be made and it is unlikely that exactly the current situation has ever occurred in the past. This is now a form of the zero frequency problem, "what is the probability that something which has never occurred in the past will occur now". Statistics *per se* can give us no answer to this question it must be sought from other knowledge about the problem.

One way to do this is to select some subpart of the current situation which is judged to be particularly relevant. If this subpart is sufficiently small then the current values will have occurred before and statistics can now be brought into play. What results is effectively a rule. For example, if the features judged to be relevant to rain are dark clouds and falling barometric pressure then the statistics tell us how often the rule:

$$\text{will-rain} \leftarrow \text{dark-cloud and falling-pressure}$$

is true. More can be gained by reformulating this into a rule in probabilistic logic:

$$\text{will-rain} \leftarrow \text{dark-cloud} \wedge \text{falling-pressure} \wedge \tau_1(p).$$

p can be estimated by counting all the observations on which the statistics are based. To be more precise the value of τ_1 can be computed using:

$$\sim\tau_1 \leftarrow \text{dark-cloud} \wedge \text{falling-pressure} \wedge \sim\text{will-rain}$$

and

$$\tau_1 \leftarrow \text{dark-cloud} \wedge \text{falling-pressure} \wedge \text{will-rain}.$$

The first of these is needed to avoid contradictions. The second is adopted to maximize the utility of the rule and to avoid the trivial situation where τ_1 is always false. The number of instances where τ_1 is (provably) true or false can be used to estimate p (the cases where it is undefined need not be counted).

These two rules for computing τ_1 are particularly useful because they are in the Horn clause form introduced earlier and so can be readily computed. Unfortunately, not all uses of logical constants can be so readily inverted. I will first consider some useful cases where the inversion is possible and then discuss some of the difficulties which arise in other cases.

The general case is that which arises when using nested rules of the form

$$\begin{pmatrix} \text{will-rain} \leftarrow \text{dark-cloud} \wedge \text{falling-pressure} \wedge \tau_1 \\ \text{will-rain} \leftarrow \text{lightning} \wedge \tau_2 \end{pmatrix} \leftarrow \tau_0.$$

Following the same reasoning as above, the requirement that the rules fire as often as possible gives:

$$\tau_1 \wedge \tau_0 \leftarrow \text{dark-cloud} \wedge \text{falling-pressure} \wedge \text{will-rain},$$

$$\tau_2 \wedge \tau_0 \leftarrow \text{lightning} \wedge \text{will-rain}.$$

These two implications readily expand to the four Horn clauses:

$$\tau_0 \leftarrow \text{dark-cloud} \wedge \text{falling-pressure} \wedge \text{will-rain},$$
$$\tau_1 \leftarrow \text{dark-cloud} \wedge \text{falling-pressure} \wedge \text{will-rain},$$
$$\tau_0 \leftarrow \text{lightning} \wedge \text{will-rain},$$
$$\tau_2 \leftarrow \text{lightning} \wedge \text{will-rain}.$$

The requirement that the rules not be contradictory gives the two implications:

$$\sim(\tau_1 \wedge \tau_0) \leftarrow \text{dark-cloud} \wedge \text{falling-pressure} \wedge \sim\text{will-rain},$$
$$\sim(\tau_2 \wedge \tau_0) \leftarrow \text{lightning} \wedge \sim\text{will-rain}.$$

These cannot be so readily handled as there is ambiguity as to which of the constants should be set false when a rule fails. This ambiguity can be resolved by again saying that the rules should fire as often as possible and so only setting τ_0 false when both rules fail that is:

$$\sim\tau_1 \leftarrow \text{dark-cloud} \wedge \text{falling-pressure} \wedge \sim\text{will-rain},$$
$$\sim\tau_2 \leftarrow \text{lightning} \wedge \sim\text{will-rain},$$
$$\sim\tau_0 \leftarrow \sim\tau_1 \wedge \sim\tau_2.$$

Thus the original two Horn clauses give seven inverted Horn clauses for computing the probabilities of the constants.

This procedure is readily extended to any set of such nested rules where each constant appears only once. The rules sets may also be nested more than one deep as in:

$$\begin{pmatrix} \begin{pmatrix} a \leftarrow b \wedge \tau_2 \\ c \leftarrow d \wedge \tau_3 \end{pmatrix} \leftarrow \tau_1 \\ e \leftarrow f \wedge \tau_4 \end{pmatrix} \leftarrow \tau_0.$$

The relationships allowed by such rules are tractable in that their inverses give Horn clauses. They also seem to be able to encompass a useful set of inter-relationships between rules. It may be that a larger tractable class of rules is available but some simple examples will illustrate the problems involved. Consider the following three rules:

$$a \leftarrow b \wedge \tau_1,$$
$$a \leftarrow c \wedge \tau_2,$$
$$d \leftarrow e \wedge \tau_1.$$

The naive inverse rules for this case are:

(i) $\tau_1 \leftarrow b \wedge a,$
(ii) $\sim\tau_1 \leftarrow b \wedge \sim a,$
(iii) $\tau_2 \leftarrow c \wedge a,$
(iv) $\sim\tau_2 \leftarrow c \wedge \sim a,$
(v) $\tau_1 \leftarrow e \wedge d,$
(vi) $\sim\tau_1 \leftarrow e \wedge \sim d.$

Because τ_1 occurs twice in the head of a rule it may be forced into a contradiction.

For example, if a, b and e are true and d is false then τ_1 should be set to both true and false.

Presumably this is a signal that the rules should be debugged. However, the situation is slightly more complex than this. Because a occurs twice at the head of two of the original rules, (i) and (iii) together are too strong. If c is also true then the second rule will fire and it is not necessary to force τ_1 to be true. This situation can be resolved by the new weaker inverse rule:

(i) $\tau_1 \leftarrow b \wedge a \wedge \neg\tau_2$

Because of the very complex inter-relationships that can exist between the constants it is not clear in general how to resolve such situations and in particular how to guarantee that there are not circular relationships in the inverse rules. The problem of computing the τ_l is a satisfiability problem for the values of the τ_l with the added restriction that as many of the τ_l as possible are to be true. Satisfiability is known to be *NP*-complete, so the general problem is not computationally tractable.

EXAMPLES

A particularly interesting way of constructing the τ constants is to use example cases provided by the expert. These are presumably carefully chosen synthetic cases which are important in practice. They can be handled in the same way as normal observations with the extra possibility that they can supply the values for a number of positions in the sequences and so be more heavily weighted than cases encountered in day to day experience.

Observations where some of the facts are uncertain can also be handled in a similar way. For example it may be known only that it is cloudy with probability 0·4 and sunny with probability 0·3 (leaving an uncertainty of 0.3). When cloudy is set true, one set of values for the constants will be computed; if it is set false (or undefined) then another set will be computed. So if a number of different pseudo-observations are made where cloudy is true 40% of the time and false 30% of the time then the counts for the various constants can be incremented by a suitably small value each time. That is the same procedure can be used for evaluating the inverse rules as is used for evaluating the original rules.

Summary

The major contribution of this paper is to provide a logical framework within which uncertain rules can be expressed and their uncertainties assessed from experience. The logic on which the rules are based does not suffer from many of the problems which other more *ad hoc* schemes are heir to. Most importantly the logic allows an automatic way of updating uncertainties including the correlations and redundancies between different rules.

There are a number of unresolved questions. It is not clear how easily such rules and their relationships can be expressed and understood by experts. Similarly, the subset of rule types which can be easily inverted needs to be extended where possible and evaluated in practice.

This work was supported by the Natural Sciences and Engineering Research Council of Canada.

References

GAINES, B. R. (1969). Stochastic computing systems. In TOU, J. Ed. *Advances in Information Science,* volume II. Plenum. pp. 37–173.

GAINES, B. R. (1984). Fundamentals of decision *Studies in the Management Sciences,* **20,** 47–65.

QUINLAN, J. R. (1983) Inferno: a cautious approach to uncertain inference *The Computer Journal,* **26**(3), 255–269.

RESCHER, N. (1969). *Many-Valued Logic.* New York: McGraw-Hill.

SHAPIRO, E. Y. (1983) Logic programs with uncertainties. In BUNDY, A. Ed. *Proceeding of the 8th International Joint Conference on Artificial Intelligence.* William Kaufmann. pp. 529–532.

SHORTLIFFE, E. H. (1976). *Computer-Based Medical Consultations.* New York: Elsevier.

VAN EMDEN, M. H. (1986). Quantitive deduction and its fixpoint theory. *Journal Logic Programming,* **3**(1), 37–53.

ZADEH, L. A. (1965) Fuzzy sets *Information and Control,* **8,** 338–353.

Analysis of the performance of a genetic algorithm-based system for message classification in noisy environments

ELAINE J. PETTIT

Merit Technology, Inc., 5068 W. Plano Parkway, Plano, TX 75075-5009, U.S.A.

MICHAEL J. PETTIT

Software Consulting Services 2733 Riviera Drive Garland, TX 75040

The process of knowledge acquisition must occur continually in those knowledge-based systems which must operate in noisy, contextually rich environments. One very important application with this requirement involves the inferring of the occurrence of events which cannot be exhaustively predefined from variably noisy sensor messages. Our paper describes on-going basic research for construction of an adaptive system which can perform high-level, rapid classification of sensor messages, possibly very noisy, concerning objects in its environment. The paper concentrates on experiments to determine optimal parameters for this bi-level, genetic algorithm-based system in low, medium, and high noise environments.

1. Introduction and motivation

One of the most stubborn areas of knowledge acquisition for knowledge-based systems has been the construction of a system which can acquire knowledge during its field operation. This is particularly true of many military applications in environments which preclude the umbilical cord to the knowledge engineer that most knowledge-based systems require. Many of these applications involve the encoding and classification of messages (from sensors, intelligence, or direct observation) which a human analyst must integrate and then attempt from this fused information to infer the tactics of his adversary. Variations of this process are known as situation assessment, threat assessment, information fusion, and tactics analysis.

There are at least two roles which machine learning may play in such knowledge-based systems: (1) knowledge acquisition in the building stages of the knowledge-based system (*ab initio* knowledge acquisition), and (2) knowledge acquisition in the operational system (*in situ* knowledge acquisition). The first has received considerable attention since the popularization of expert systems has underscored the problems, both in time and quality, of getting information from experts into the system. In many ways, however, the solution of this problem is of considerably less benefit over the long term than successfully implementing *in situ* knowledge acquisition. Although the verification of the completeness and consistency of the initial knowledge base is vital to knowledge-based systems, there is no escaping the fact that everything a system will eventually encounter cannot be foreseen at the time of its creation. Without the capacity to deal with unknown situations, and even more importantly, to sieve meaningful information in the presence of noise, a real-world knowledge-based system will show sharp degradation

KNOWLEDGE-BASED SYSTEMS Vol. 1
ISBN 0-12-273251-0

under these conditions. This phenomenon of system brittleness is well-recognized in expert system technology and has been described as "falling off the knowledge cliff" (Feigenbaum as cited in Michalaski, 1986). Although the use of pre-defined generic scenarios can be useful here, without the ability to extend and refine those scenarios beyond simple instantiation the system will not be able to offer assistance when the need is most critical: when unanticipated obstacles, inoperative equipment, contradictory information, and loss of active personnel combine to present scenarios that are outside the scope of those originally envisioned. Even for routine military or civilian missions, the more closely the system knowledge resembles the real world, the more timely and reliable its advice will be.

There exist some heuristics learning systems which could possibly, with modification, exhibit on-site adaptability. However, for situation assessment problems, satisfaction of the requirement for adaptability alone is insufficient: noise immunity is critical. The process of inferring the occurrence of events from messages produced by sensors and other sources is subject to noise from two sources: (1) noise in the attributes of the patterns reported by a given sensor (a characterization problem); and (2) noise, due to the asynchronous nature of message traffic, in the number of patterns reported by sensors in a given time step (a fusion problem). Thus, an automated system for classifying these messages into meaningful events must remain immune to noise at two levels of a classification hierarchy: the message level and the event level. It must first be able to characterize accurately an event as it is occurring; second, it must be able to distinguish between noise-induced variations of an ongoing event versus transition to a new event state. The problem is one of both knowledge representation and knowledge utilization, as described in Section 2, "Related work".

To address the above issues, we are currently researching a hybrid methodology for knowledge representation and operation built upon genetic algorithms (Holland, 1975) and conceptual clustering (Michalski & Stepp, 1983). Conceptual clustering provides a means for representing and operating upon knowledge so that synergistic properties are expressible and a spectrum of deviations are tolerated. Genetic algorithms provide a strategy for highly effective pseudo-parallel search of feature hyperspace with noncommensurate, ill-behaved optimization criteria. As an initial step, we have implemented a genetic algorithm-based message classification system to explore how well it resolves the following issues:

(1) How sensitive is it to managing noise in a stable environment versus recognizing when transition to a new environment has occurred? At what point do we alter not only the low-level message input classifiers but also the higher level "conceptual classifiers" that aggregate these messages into meaningful categories? (To what degree and in what function should incremental learning occur and when should it be postponed?)

(2) Assuming a bi-level system architecture driven by genetic algorithms and operating with limited resources, should the rate of change in event recognition classifiers lag considerably behind that of message classifiers or vice versa? Increasing the rate of adaptation of the message classifiers would allow recognition over noise at the cost of nonconvergent event classification at first: increasing the rate of adaptation of the event classifiers would allow consideration of all possible interpretations quickly but with a loss of accuracy. At what point is there a tradeoff?

Concommitant with the resolution of these application issues was the need to extend, with sound theoretical underpinnings, the basis of the genetic algorithms themselves to meet certain implementation requirements. Primary among these was the need to define composite, variable length classifiers (to represent the event classifiers) and genetic operators and performance measures appropriate to these kinds of classifiers. There was a secondary need to devise a means of handling several different ongoing patterns (each of the noisy messages) in the message classifier population.

Because the research base of using genetic algorithms in stochastic environments is very sparse, we have concentrated in this paper on the definition and development of a scientific framework for empirically testing with appropriate statistical measures the behavior of various models postulated to meet the above requirements. We have begun our investigation with a "worst case" scenario: noise induction is random, adaptation cycles are below empirically derived "good" limits, and we use "offline performance" (described in Section 3) which is the least discriminating (between control and experimental performance) but probably the most realistic measure of algorithm performance in this task domain. We feel those strengths that manifest themselves under these conditions will provide the most fruitful areas of future research. We present our framework, including details of our data generation and analysis, and our research findings to date concerning the efficacy of genetic algorithms in noisy message classification.

2. Related work

There are currently two primary approaches to dealing with the problems of noise and adaptability in the development of knowledge-based systems. One is to define a set of generic objects and the processes governing their behavior, instantiate those objects with values derived from the situation, and operate on those objects in accordance with some predetermined inference strategy for resolving conflicts and focusing ambiguity. This approach is basically differential in nature, treating the "instantaneous" differences between data items or between data items and a predetermined template as the modeling basis. Several artificial intelligence techniques have been proposed to address the problem of handling preconditions established by incomplete, ambiguous, or, in the presence of an intelligent adversary, deceptive data. Most readily applicable are those which deal formally with nonmonotonic reasoning and uncertainty. These include Bayesian inference, fuzzy logic, non-monotonic logic, belief revision, endorsement theory, Dempster-Shafer theory of evidence, and cautious inference. The end result of each of these methods is that snapshots of parts of the environment are pieced together by predefined evidential reasoning algorithms to create the most probable overview at that time.

These methods quantify the contribution of the value of a data item AFTER a match has been established. What these methods cannot capture is the epistatic interaction AMONG data items in (1) establishing relevance of data at a conceptual level to the problem (as opposed to matching individual data items); and (2) reinforcing or negating the individual contributions of data items taken in all possible combinations. The problem is formally isomorphic to the optimal allocation of trials in a search of feature hyperspace for which the probability distributions of "success" and "failure" are unknown (and because of distribution overlap, empirically indecidable) for individual dimensions [Holland (1975) and DeJong (1975) provide an excellent discussion of this "k-armed" bandit problem].

Several of the foremost areas in machine learning address these problems of incorporating adaptability and noise immunity into knowledge representation and utilization paradigms. Michalski's INDUCE and CLUSTER paradigms (Michalski, 1983) (Michalski & Stepp, 1983) have demonstrated noise tolerance in training examples. However, these algorithms employ a search tapering technique based on best-first search which, from our analysis, may prove inefficient, and at worst, non-robust, in the complexity of the situation assessment domain.

Quinlan's ID3 (Quinlan, 1986) and its progeny ID4 (Schlimmer & Fisher, 1986) use a statistic based on object occurrence probability with a chi-squared distribution to determine which attributes should be rejected in constructing classification decision trees. However, both these systems are nonincremental, i.e. they require all the objects that are to be classified to be available at one time for iterative processing. They are therefore not applicable to tracking asynchronous message traffic over time. Furthermore, although Quinlan's empirical evidence indicates some utility for this statistic, one must question its use for dependent attributes since the statistic assumes independence among the variates.

Schlimmer and Granger (1986) have recently presented research which, of the alternate systems examined, is most directly applicable to the message classification problem. Their STAGGER system is designed to track "concept drift", which, as in our system, is the ability to distinguish between local noise in attributes and a global change of events. It uses Bayesian propagation of likelihood ratios to incrementally refine weights associated with individual characterizations and the structure of the characterizations themselves. The primary theoretical difference between the STAGGER method and the use of genetic operators lies in the reliance of STAGGER on the robustness of the incrementally derived probabilities to characterize what we state is an indecidable probability distribution.

Most prior research on the use of genetic algorithms themselves has concentrated on their use in a non-stochastic environment, i.e. where noise in the patterns to be tracked and learned is minimal and the patterns themselves occur consistently throughout the tracking period. Booker (1985) has studied the optimization of parameter settings for genetic algorithm-based message classifier systems in non-noisy environments: his research has served as a strong basis for the parameter settings in our system, as well as forming a basis for our application of multiple pattern tracking in the message classifier population. Smith's genetic algorithm-based poker playing system (Smith, 1980) was highly successful against Waterman's system (Waterman, 1975): he and (Grefenstette *et al.*, 1985) have supplied a basis for hierarchical composition of genetic operators. There have also been numerous successes in applying these theories in areas as diverse as dynamic system control (Goldberg, 1985), function optimization (Bethke, 1980), and object movement in a complex environment (Holland, 1986). Previous research by (Pettit & Swigger, 1983) has indicated that even in noisy environments, a genetic algorithm-based system can track the pattern of an environment in realistic, correlated flux more successfully than a computationally prohibitive model which maintains detailed statistical information regarding the status of each component.

3. Experimental methodology

3.1. BACKGROUND ON GENETIC ALGORITHMS

In one of the first mathematical analyses of the use of biologically based mechanisms in promoting adaptation in artificial systems, Holland (1975) reviewed the prolonged success of lifeforms in adapting to an environment through evolution. The biological organism is faced with testing a large set of possible genetic expressions in its offspring by means of environmental interaction with a relatively small subset of realized structures (its own genotype). Nonlinearity and epistatic interactions among gene sets complicate the problem of achieving a successful, if not optimal, genetic complement in offspring.

Holland has mathematically hypothesized that genetic operators (e.g. crossing-over, mutation, inversion) exploit the optimization of reproductive fitness (number of offspring) by a means he terms intrinsic parallelism. Intrinsic parallelism is the testing of a large pool of schemata—i.e. the set of all partitions and combinations thereof of a prototypical structure—by means of a much smaller subset of realized structures. More simply, consider the structure A consisting of a string of six binary digits, (1 0 1 1 0 1). Each binary digit may be considered to be a classifier of a binary feature vector (such as the premises in a rule-based system). Structure A is a member of a set of structures ALPHA which includes all possible strings of six binary digits. There exists a superset EPSILON which is the set of strings of length six composed of concatenations of $\{1, 0, \#\}$, where # represents a "don't care" position, i.e., its value as 0 or 1 is irrelevant. For example, let E be an element in EPSILON of the form (1 0 # # 0 1). This "E" is termed a schema, and all possible schemata compose EPSILON, the "pool of schemata". The occurrence of structure A, then, is a sample not only of A itself but of the schemata (1 0 # # 0 1), (# # # # 0 1), (1 0 1 1 0 #), etc. Thus, parallel sampling of applicable schemata can be implemented with relatively few templates.

Now consider a structure ENVIRONMENT (0 0 1 0 0 1) which represents the "state" of an environment. One—and by no means the only—measure of the fitness of Structure A in characterizing ENVIRONMENT is a computation of a metric of difference (e.g., Hamming distance) between A and ENVIRONMENT. Schemata thus represent the contribution to fitness of single detectors as well as of combinations of detectors. A subset of structures from ALPHA constitutes a population. It is, by definition, the goal of adaptation to modify these structures in order to optimize the fitness of the population. Holland and others have shown that genetic operators such as crossing-over are highly successful in (1) testing a large number of possible schemata through modifications on a much smaller number of realized structures, and (2) exploiting local optima on the way to achieving the global optimum without becoming entrapped (as often happens in hill-climbing optimization techniques).

A general algorithm for genetic adaptation is given below. Algorithms for the genetic operators themselves are presented under Section 3.3, "Experimental design".

General algorithm for genetic adaptation

Initialize population (e.g., message classifiers) and environment. Find the performance for each structure in the population and call it MU(i). Define the random variable RAND on $\{1, \ldots, M\}$ by assigning probability $(\mathrm{Mu}_i/\sum_{j=1}^{M}\mathrm{Mu}_j)$ to the ith structure in the population, where M is the number of structures in the population. Make M trials of RAND, each time storing the structure at position RAND in the population at successive positions in auxiliary list TEMP-POP. For each structure in TEMP-POP, apply mutation and other unary genetic operators with some predefined probability, and, if it is a structure in an even-numbered position, perform crossover between it and its immediate predecessor. Set the original population to TEMP-POP, and repeat all steps except initialization for the desired number of generations.

As stated in the preceding section, it has been demonstrated that such a model, with the addition of suitable selection criteria and control structures, can track the pattern of an environment in flux more successfully than a computationally prohibitive model which maintains detailed statistical information regarding the status of each component (Pettit & Swigger, 1983). At the same time the genetic model does not discard instances of new schemata when an optimum is obtained, allowing for recovery over another set of absorbing transitions when the pattern is altered in a realistic, correlated fashion. In contrast, it is possible for a component-sampling model to lock into a present optimum that was maintained over sufficient transitions: the capability for change would eventually become miniscule.

3.2. SYSTEM OVERVIEW

Figure 1 illustrates our general system design. The system resides in an environment in which objects of four types come and go. Events are the power set of the set of four object types exclusive of the null set and inclusive of the improper subset (i.e. no more than one object of one type can be present, although an extension of the system has been designed to handle this case in future research). The objects post

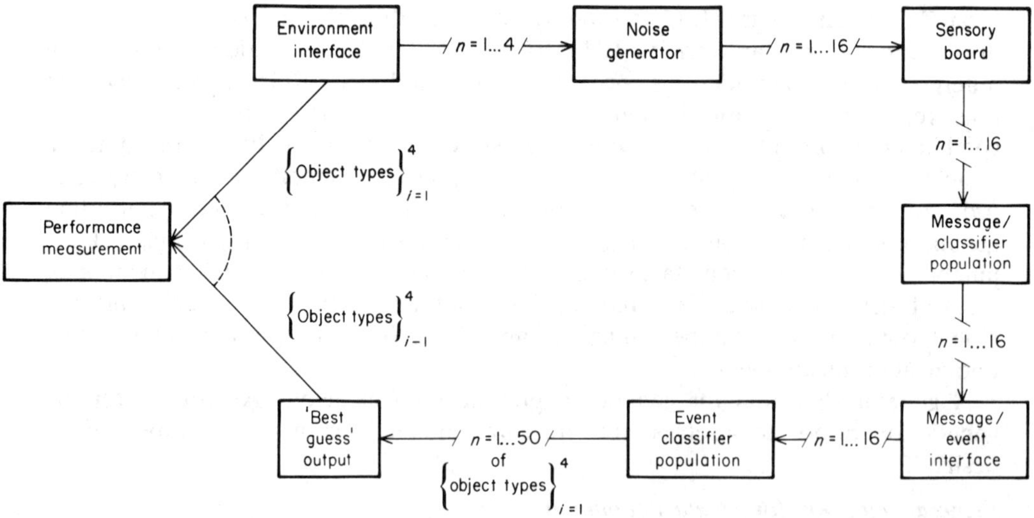

FIG. 1. System overview.

messages concerning their type and attributes to the environmental interface board. Noise via Markov chain transition matrices are induced in the values of these messages at an adjustable level. The noisy messages, ranging from 1 to 16 per object, are posted to an intermediate sensory board.

The message classifier system then uses genetic algorithms to track the noisy messages as they are posted. The results of this tracking are posted to the message-to-event-classifiers interface board. The event classifier system then uses genetic algorithms to track event components from this board, i.e. the number of objects in the environment and their types. The event classifier system posts its results to the output board. These results are compared to the actual environment for the performance measurement.

3.3. EXPERIMENTAL DESIGN

Implementation specification: All software is coded in ZetaLisp on Merit's corporate Symbolics 3670. The unoptimized compiled program consisting of five test cases and one control running under each of three noise levels with a global time period of 10 and a generation limit of between 200 and 400 iterations takes approximately 30 hours to run (since the data sets are of stochastically variable length, runtime varies). It should be noted that Holland's original design was intended for a parallel implementation. Since the paradigm involves primarily independent iterations, runtime for a parallel implementation should decrease in at least a roughly linear proportion to the number of processors available.

Functional description: The following algorithm describes the general functional operation of the system. The algorithm was adjusted as necessary to fulfill the conditions of each experiment and the control.

```
function main:
  generate an initial population (size 50) of message classifiers;
  generate an initial population (size 50) of event classifiers;
  for time-step from 1 to global_time_limit do
  begin
    generate new set of environmental objects [1 . . 4 objects of type 1, 2, 3, 4];
    set variability_modulus to desired level;
    for message_genetic_timestep from 1 to generation_limit
    do
  begin
    induce_message_noise and post noisy messages to
      sensory_board (1 to 16 per object);
          /*process through message classifiers */
    calculate the performance (P1) of each message classifier against each message
      on the sensory_board to form subpopulations (one subpopulation of 50
      classifiers per entry on sensory_board);
    apply genetic algorithms to subpopulations;
    select from each subpopulation on the basis of performance (P1) against that
      subpopulation's basis message an equal proportion (1/no. of sensory mes-
      sages) of individuals to become the 50 message classifiers of the next
```

```
        generation;
      if (message_genetic_timestep mod variability_modulus = 0) or (variability_
        modulus < 1) then;
      begin      /* process through event classifiers */
        select best individual from each message group and post to message-to-event-
          classifiers-blackboard;
        calculate the performance (P2) of each event classifier against each message
          on the message-to-event-classifiers blackboard;
        apply genetic algorithms to the event classifier population;
      end      /* process through event classifiers*/
    end      /* genetic time loop*/
    calculate performance (P3) of each event classifier to actual environment to get
      performance measure at current time_step;
    end      /* global time loop */.
  end function main.
```

Data item description: Objects in the environment consist of length 21 binary strings over the alphabet $\{0, 1\}$ with reserved bits $i \bmod 10 = 0$ or 1 indicating object type.

A random number, between 1 and 16 inclusive, of "noisy messages" are produced for each object by inverting the value of each bit with a probability of *noise-level*. The *noise-level*s used for this set of experiments were 0·25, 0·125, and 0·0625, signifying high, medium, and low levels, respectively. Noise induction by this method creates a "worst-case" model of random noise, as opposed to a correlated, realistic flux (the "systematic" noise of Schlimmer & Granger, 1986). These messages are posted to the *sensory-board*.

The first level of the system consists of a population of size 50 of message-classifying strings of length 21 over the alphabet $\{0, 1\}$. A population of size 50 has been shown by (Booker, 1985) and (DeJong, 1975) to demonstrate acceptable convergence in a non-stochastic environment.

These message-classifying strings are compared to the noisy messages posted to the *sensory-board* and are recombined in accordance with the general genetic adaptation algorithm. Performance for reproduction is computed using the M3 method of (Booker, 1985): this method, termed P1 for this model, is explained under the subtopic "Performance measures".

The second level of the system consists of a population of size 50 of event-classifying composite strings in which the string elements are themselves strings of length 21 over the alphabet $\{0, 1\}$. The composite strings consist of from 1 to 4 of these elemental strings, with only strings denoting different object types allowed in the composite.

These event-classifying strings are compared to the results posted on the message/event interface board by the message classifying strings and are recombined in accordance with the general genetic adaptation algorithm. Because of the composite nature of these strings, new genetic operators have been devised, based on the open and closed variability genetic regions postulated by evolutionary ecology (Carson, 1975) (Chapman *et al.*, 1979). Initial analysis indicates that these operators exhibit properties comparable to the intrinsic parallelism of (Holland, 1975). They are described further under the subtopic "Genetic operators". Performance for reproduction is computed using a modified M3 method from (Booker, 1985), with modifications which retain the composite nature of the strings (see subtopic "Performance measures").

After a preset number of generations, the best match score (see description of "P3" under "Performance measures") for the current event-classifier population is output, and a new set of objects is generated. Each new set of objects constitutes a time-step, and *global-time-limit* number of these time-steps are performed.

Genetic operators:

Noncomposite operators (used in the message classifiers):

(a) Mutation: Make an equiprobable trial of a random variable X on $\{1, \ldots, L\}$, where L is the number of positions in the structure undergoing mutation. Make an equiprobable trial of a random variable CHANGE on the integers 0, 1. Replace the value of the element at position x with CHANGE. The rate of application of mutation for these experiments was 0·05, slightly higher than the 0·02 (1/population_size) suggested by (DeJong, 1975), due to the operation of the system in a stochastic environment (a higher mutation rate has been shown to slow allele loss).

(b) Crossover: Make an equiprobable trial of the random variable X on $\{1, \ldots, L\}$, where L is the number of positions in each of the two structures undergoing crossover. If $X = 0$ or L, then no crossing-over takes place. Otherwise, take positions 1 through X of the first structure and append positions $X + 1$ through L of the second structure. Likewise, take positions 1 through X of the second structure and append positions $X + 1$ through L of the first structure. The rate of application of crossover for these experiments was 1·00, our rationale being that tracking in a stochastic environment would be improved with frequent exchange of pattern parts.

No other unary genetic operators were used in the noncomposite case

Composite operators (used in the event classifiers):

(a) Event-mutation: Randomly select an element from the classifier and replace it with a new, randomly-generated instance of that object type. This is a highly disruptive operation, guaranteeing no preservation of patterns that may have been converging over several generations, but one which is critical to the avoidance of local optima. Mutation rate was maintained at 0·05.

(b) Event-crossover: Three types were defined to attempt to replicate the effects of noncomposite crossover on the composite population. Some precedence for these definitions may be found in Smith (1980). These types are:

External crossover: Randomly select an element of the first structure. Randomly generate a set of object-types of which the object type of this element is a member. Exchange all elements of these object-types for which there is a corresponding element of the same object-type in the second structure. This is a conservatively mixing type of operation.

Internal crossover: Randomly select an element of the first structure. Find an element of the same type in the second structure. If there is not one, then exit with no change. Otherwise, perform an noncomposite crossover between these two elements and replace the original elements with the new ones. This provides a less conservative mixing, but a more conservative method than mutation of avoiding local optima.

Delete–insert: Randomly select an element of the first structure. If there is an element of the same type in the second structure, remove the element from the first structure and replace the same type element from the second structure with the first structure's element. Otherwise, delete the element from the first structure and add it to the second structure. If the donor structure becomes empty because of this removal, generate a new one-element classifier of the same object-type that was removed to keep the population at the same level. This operation is designed to generate composite classifiers with a variable number of composition elements.

For each pair of structures, one of these three composite crossing-over operations is selected each generation.

Performance measures: "Performance" is simply how well the structure being evaluated fulfills the objective(s) of the task domain. Three types of performance were used in this model:

P1: P1 is based upon the M3 match algorithm of (Booker, 1985). Adapted to a binary alphabet, P1 is equal to the length (L) of the structure for a perfect match, and ($m/L \wedge 2$) for a nonperfect match, where m is the number of matching positions.

P2: P2 is an adaptation of M3 for composite structures. The P1 of each element in the composite structure is calculated against each entry on the interface board; the maxima for each element are summed; and this value is returned as the performance of that classifier. P2, unlike P3, does not take into account matching the number of elements against the number of environment pattern components: these are unknown to the system. It was assumed that a classifier with extraneous elements would receive little or no competitive advantage over one with highly matching, nonextraneous elements since only one element of each object-type is allowed in the composite classifier. As we elaborate under "Results and discussion", this assumption may be invalid.

P3: P3 provides an estimation of "offline" performance and was the metric for comparison of performance among the models. "Offline performance" is a term coined by (DeJong, 1975) to indicate the maximum performance score present in a population at a given time-step. In contrast, "online performance" (DeJong, 1975) refers to the average performance score of an entire population at a given time-step. We chose to use a metric of offline performance to indicate the best information that could be expected at a given time-step with each of the models. P3 is calculated as follows:

$$\text{P3} = \frac{m}{n(1 + \text{abs}(k - n))},$$

where m = number of matching bits across all elements, n = number of objects in the environment, k = number of elements in the classifier. The factor "m" reflects point-by-point accuracy. Division by $(1 + \text{abs}(k - n))$ penalizes classifiers whose number of elements differs from the actual number of objects in the environment. Division by n normalizes all results across variable environments. P3 thus represents a proportional accuracy. The maximum value of P3 is 21, implying a complete point-by-point match and an equivalent number of elements in the event classifier as there are objects in the environment. No attempt was made in this set of experiments to determine if object types matched: it was assumed that good attribute matches implied easy object identification.

Experimental groups:

C1: Performance (P3) with no genetic algorithms invoked. This group represented expected results on the basis of chance and served as a control.

C2: Performance (P3) with genetic algorithms invoked in the message classifiers but not the event classifiers. Even classifiers were constructed from messages posted to the message to event classifiers interface board in the following manner: A random selection of object types was made to indicate the number and kinds of elements in the event classifier. If there were one or more messages of this type on the interface board, one of these was selected at random to be an element in a classifier. If not, a message was selected at random and transformed into this type. Fifty event classifiers were constructed in this manner at the end of the genetic generation cycle, and the maximum P3 value was calculated. This group was used as a secondary control to determine the efficacy of the composite genetic operators and the P2 measure. The *generation-limit* was set at 200.

C3: Performance (P3) with genetic algorithms invoked in the event classifiers and not the message classifiers. Each cycle the message classifiers with the highest P1 values were posted to the interface board. Again, this group served as a secondary control to determine how the composite genetic operators and the P2 measure affected performance. The *generation-limit* was set at 200.

T1: Performance (P3) with genetic algorithms invoked at the event classifier level at half the rate of the message classifier level. The event classifiers underwent 200 generations interleaved into 400 generations of the message classifiers.

T2: Performance (P3) with genetic algorithms invoked at the message classifier level at half the rate of the event classifiers. The message classifiers underwent 200 generations interleaved into 400 generations of the event classifiers.

T3: Performance (P3) with genetic algorithms invoked at an equal rate at the message classifier and event classifier levels. The *generation-limit* was set at 200.

Ten runs for each group were conducted under noise levels of 0·25 (high), 0·125 (medium), and 0·0625 (low).

Statistical methods: Analysis of variance (ANOVA) was used to test for differences between group means at each noise level ($A = 0{\cdot}05$). The null hypothesis tested was: C1 = T1 = C2 = T2 = C3 = T3 (each value represents a group P3 mean). Duncan's multiple range test was used to discriminate between groups when significant differences were found. All analyses were performed on the Statistical Analysis System (Barr *et al.*, 1979).

4. Results and discussion

RESULTS

The experimental results are illustrated in Figs 2–4, for the low, medium, and high noise tests, respectively. The following observations are significant.

Result 1. In all noise levels, C2, the group in which genetic algorithms were invoked at the message classifier level but not the event classifier level, performed

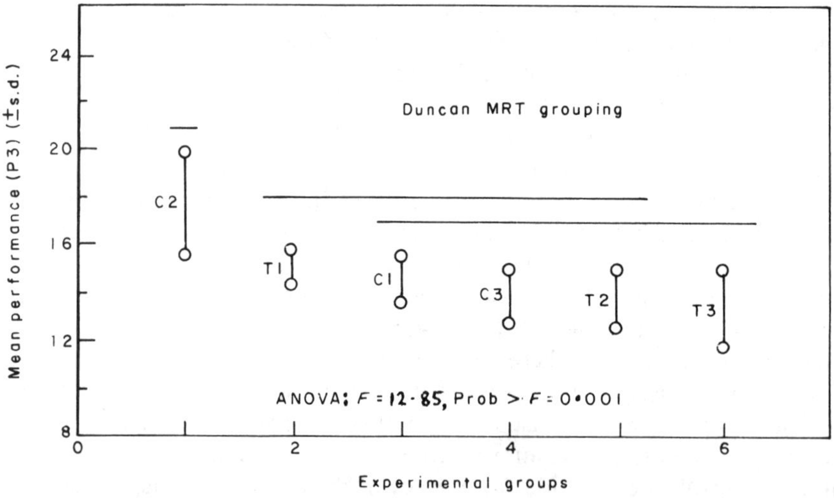

FIG. 2. Results for low noise test.

significantly better than any of the other models. In both the low and medium noise environments, C2 obtained perfect scores of 21 on two trials.

Result 2. An ANOVA of C2 performance means across all noise levels indicated no significant difference in the performance of C2 in low, medium, and high noise environments. This result is consistent with previous findings (Part 1: Test 2 in Pettit and Swigger, 1983).

Result 3. In the low and medium noise tests, T1, in which the event classifiers underwent adaptation at half the rate of the message classifiers, performed significantly better than T3, in which genetic algorithms were invoked at equal rates and for the same number of iterations for each level.

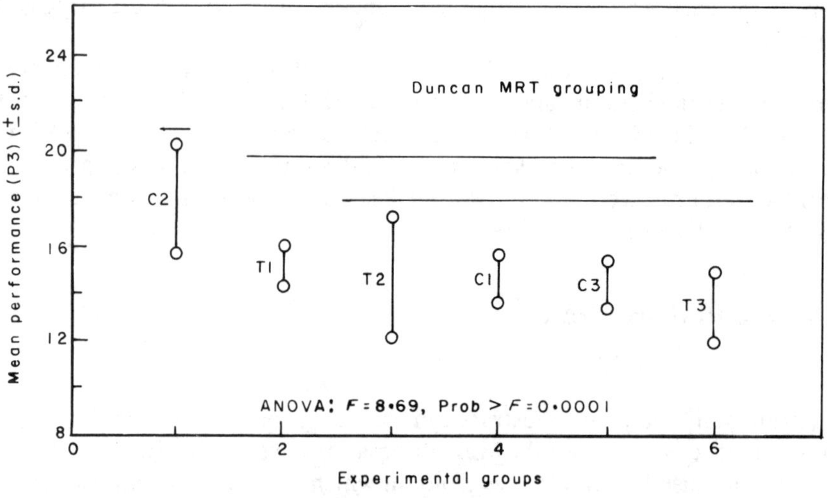

FIG. 3. Results for medium noise test.

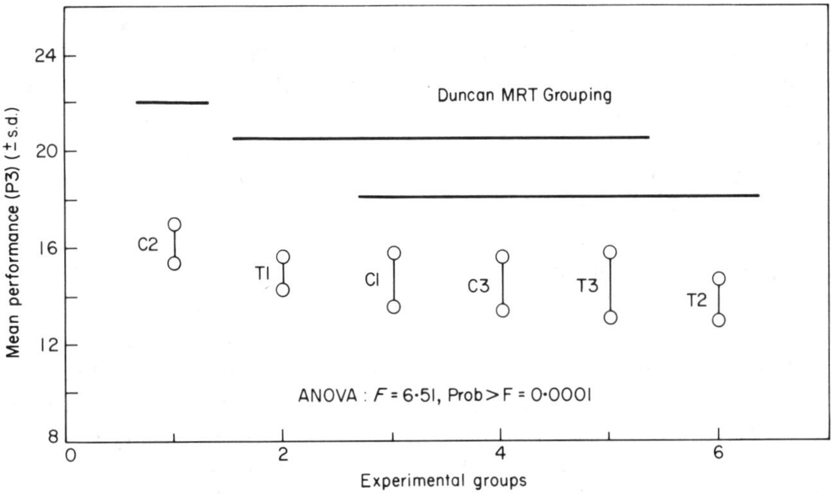

FIG. 4. Results for high noise test.

Result 4. In the low and medium noise tests, T1 (event classifiers at one-half rate, message classifiers at 400 generations) and T2 (event classifiers at twice rate, message classifiers at 200 generations) occupied the same equivalence class.
Result 5. T1 scored consistently above all the other models except C2. However, at a sample size of 10, the overlap between classes made this relative ranking statistically insignificant.

DISCUSSION

The above observations suggest the following conclusions:

The observations that C2 performed significantly better than any of the other models (Result 1) and that the noise level had no significant effect on its performance (Result 2) suggest that genetic algorithms as employed in the noncomposite case are highly robust in promoting noise filtering of message content. Further experimentation should consider determining the minimum number of iterations necessary for adequate convergence to reduce tracking time.

On the other hand, the models utilizing the composite genetic operators and performance measure P2 did not perform significantly better than the control C1. Results 3 and 4, combined with a trend suggested by Result 5, suggest that some factor(s) in the composite models is(are) actually disrupting performance. In the following analysis we seek to isolate that factor.

As explored in Holland (1975), DeJong (1975), Booker (1985) and others, the following factors present in our models affect the performance of genetic models operating without composition in nonstochastic environments: population size, mutation rate, crossover rate, selection metrics (P1 and P2 in our models), and number of iterations (*generation-limit* in our models). The first three factors were held constant across all models and can therefore be disregarded.

Increasing the number of iterations (*generation-limit*) usually improves the performance of genetic models due to their underlying convergent mechanism.

However, the observation (Result 4) that T1 and T2, with a 200% difference in *generation-limit*s, were not significantly different in even the low noise environment suggests that the problem factor(s) damp the effect of increased iterations. The two most likely sources of such an effect are the selection metric (P2) and the composite operators themselves. A highly inefficient or directionally inaccurate selection pressure would slow or impair completely the convergence mechanism. The use of composite operators inappropriate for efficiently searching the feature hyperspace would produce similar difficulties.

Result 5, although resting on tenuous statistical ground, at least weakly suggests that the higher quality of information presented by increased iterations of the message classifiers serves to offset somewhat the negative influence of the problem factor(s). This would indicate that the richness of information provided by the message classifiers (Results 1 and 2) is not being severely disrupted in the composite model. We combine this weak indicator with the statistically strong observations that: (1) in low and medium noise environments, T1 (best information quality and 200 iterations) and T2 (less information quality and 400 iterations) were not different; (2) T2 (400 iterations) and T3 (200 iterations and same information quality as T2) were not different; and (3) T1 and T3 with the same number of iterations but different information quality were different. One would expect that if the composite operators were inappropriate, information would be reshuffled to such a degree that the integrity of the search process would be lost, and T1 would never be statistically better than T3 (in contrast to Result 3). However, if the selection metric were directionally appropriate but seriously inefficient, the richness of the information would not be randomly disrupted, but the contribution of "good" relative to "meaningless" information in succeeding generations would be severely reduced. Under these circumstances, the richer the initial information the better the performance over the same number of iterations.

From this analysis, we suggest that the use of an attribute-by-attribute matching metric (as P2 and Booker's M3 are) is insufficient for composite genetic models. There is a need for a metric which can capture the underlying structure of the "events" dynamically, especially when prior information about this structure is unavailable. We are currently investigating the use of clustering techniques, including conceptual clustering, to devise this metric.

Theoretically, the composite genetic operators do not violate the assumptions of the genetic algorithms. However, since the performance of the genetic operators is primarily affected by the quality of "reproductive information" provided in the selection process we will not make conclusions concerning their empirical validity from these data.

5. Conclusion

This experiment examined the use of composite genetic operators and a composite adaptation of a noncomposite selection metric in a bi-level, genetic algorithm-based system for message classification into events in low, medium and high noise environments. Our underlying motivation is the eventual construction of a system for interpreting asynchronous, noisy message traffic in situation assessment problems where these interpretations (events) cannot be predefined but must be acquired during the operation of the system. To this end, in this experiment we have determined that (1) genetic algorithms are highly robust in filtering even random noise from message traffic; and (2) the attribute matching selection metric used extensively in genetic algorithm-based systems is insufficient for a composite model. Both of these are critical issues in the construction of such a system.

Our plans for future research in this area include: (1) devising of a suitable composite selection metric based on clustering theory; (2) examination of the effects of the number of iterations on convergence optimization; (3) incorporation of multi-objective functions for optimization, such as object identification and location; (4) introduction of a structured representation similar to (Forrest, 1985)'s use of semantic networks; (5) examination of the online performance of our model in tracking objects which gradually come and go in the environment; and (6) recognition by our model of multiple instances of the same kind of object.

This research was supported in part under Merit Technology's internal research and development program in machine learning and in part by the graduate research program in computer science at North Texas State University. We would like to thank Dr Kathleen Swigger of NTSU for her generous support, and also Gregg Jernigan and his artificial intelligence staff at Merit for their encouragement and suggestions.

References

BARR, A. J., GOODNIGHT, J. H., SALL, J. P., BLAIR, W. H. & CHILKO, D. M. (1979). *SAS User's Guide*. Raleigh, NC: SAS Institute.

BETHKE, A. D. (1980). Genetic algorithms as function optimizers. PhD dissertation. Ann Arbor: University of Michigan.

BOOKER, L. B. (1985). Improving the performance of genetic algorighms in classifier systems. In: GREFENSTETTE, J. J. Ed. *Proceedings of an International Conference on Genetic Algorithms and their Applications*. Pittsburgh: Carnegie–Mellon University. pp. 80–92.

BRINDLE, A. (1981). Genetic algorithms for function optimization. Ph.D. dissertation, University of Alberta.

CARSON, H. L. (1975). The genetics of speciation at the diploid level. *The American Naturalist,* **109**(965), 83–92.

CHAPMAN, R. W., AVISE, J. C. & ASMUSSEN, M. A. (1979). Character space restrictions and boundary conditions in the evolution of multistate characters. *Journal of Theoretical Biology,* **80,** 51–64.

DEJONG, K. A. (1975). An analysis of the behavior of a class of genetic adaptive systems. PhD dissertation. Ann Arbor: University of Michigan.

FORREST, S. (1985). Implementing semantic network structures using the classifier system. In: GREFENSTETTE, J. J. Ed. *Proceedings of An International Conference on Genetic Algorithms and their Applications*. Pittsburgh: Carnegie–Mellon University. pp. 24–44.

GOLDBERG, D. E. (1985). Genetic algorithms and rule learning in dynamic system control. In: GREFENSTETTE J. J. Ed. *Proceedings of An International Conference on Genetic Algorithms and their Applications*. Pittsburgh: Carnegie–Mellon University. pp. 8–15.

GREFENSTETTE, J. J., GOPAL, R., ROSMAITA, B. & VAN GUCHT, D. (1985). Genetic algorithms for the traveling salesman problem. In: GREFENSTETTE J. J., Ed. *Proceedings of An International Conference on Genetic Algorithms and their Applications*. Pittsburgh: Carnegie–Mellon University. pp. 160–168.

HOLLAND, J. H. (1975). *Adaptation in Natural and Artificial Systems*. Ann Arbor: University of Michigan Press.

HOLLAND, J. H. (1986). Escaping brittleness: the possibilities of general-purpose learning algorithms applied to parallel rule-based systems. In: MICHALSKI, R. S., CARBONELL, J.

& MITCHELL, T., Eds. *Machine Learning: An Artificial Intelligence Approach, Volume II.* Los Altos: Morgan Kaufman Publishers. pp. 625–646.

MICHALSKI, R. S. (1983). A theory and methodology of inductive learning. In: MICHALSKI, R. S., CARBONELL, J. & MITCHELL, T. Eds. *Machine Learning: An Artificial Intelligence Approach, Volume I.* Palo Alto: Tioga Publishing Co. (now Morgan Kaufmann). pp. 83–134.

MICHALSKI, R. S. (1986). Understanding the nature of learning. In: MICHALSKI, R. S., CARBONELL, J. & MITCHELL, T. Eds. *Machine Learning: An Artificial Intelligence Approach Volume II.* Los Altos: Morgan Kaufman Publishers. pp. 3–26.

PETTIT, E. J. & SWIGGER, K. (1983). An analysis of genetic-based pattern tracking and cognitive-based component tracking models of adaptation. In: *Proceedings of the National Conference on Artificial Intelligence,* (Washington, D.C.). Los Altos: William Kaufmann. pp. 327–332.

QUINLAN, J. R. (1986). The effect of noise on concept learning. In: MICHALSKI, R. S., CARBONELL, J. & MITCHELL, T. Eds. *Machine Learning: An Artificial Intelligence Approach Volume II.* Los Altos: Morgan Kaufman Publishers. pp. 149–166.

QUIRIN, W. L. (1978). *Probability and Statistics.* New York: Harper & Row.

SCHLIMMER, J. C. & FISHER, D. (1986). A case study of incremental concept induction. In: *Proceedings of the Fifth National Conference on Artificial Intelligence, Volume 1.* Los Altos: Morgan Kaufmann. pp. 496–501.

SCHLIMMER, J. C. & GRANGER, R. H., Jr. (1986). Beyond incremental processing: tracking concept drift. In: *Proceedings of the Fifth National Conference on Artificial Intelligence,* Volume 1. Los Altos: Morgan Kaufman. pp. 502–507.

SMITH, S. F. (1980). A learning system based on genetic adaptive algorithms. PhD dissertation. Ann Arbor: University of Michigan.

WATERMAN, D. (1975). Adaptive production systems. In: *Proceedings 4th International Joint Conference on Artificial Intelligence.*

Index